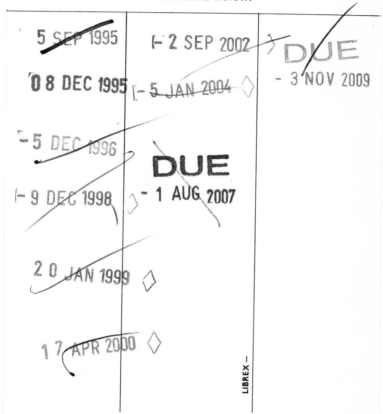

The Technology of Vitamins in Food

Edited by

P. BERRY OTTAWAY
Berry Ottaway and Associates Ltd
Hereford

Published by
**Blackie Academic & Professional, an imprint of Chapman & Hall,
Wester Cleddens Road, Bishopbriggs, Glasgow, G64 2NZ**

Chapman & Hall, 2–6 Boundary Row, London SE1 8HN, UK

Blackie Academic & Professional, Wester Cleddens Road, Bishopbriggs, Glasgow G64 2NZ, UK

Chapman & Hall Inc., 29 West 35th Street, New York NY10001, USA

Chapman & Hall Japan, Thomson Publishing Japan, Hirakawacho Nemoto Building, 6F, 1-7-11 Hirakawa-cho, Chiyoda-ku, Tokyo 102, Japan

DA Book (Aust.) Pty Ltd., 648 Whitehorse Road, Mitcham 3132, Victoria, Australia

Chapman & Hall India, R. Seshadri, 32 Second Main Road, CIT East, Madras 600 035, India

First edition 1993

© 1993 Chapman & Hall

Typeset in 10/12 pt Times New Roman by ROM Data Corporation, Cornwall, England.

Printed in Great Britain by Hartnolls Ltd, Bodmin, Cornwall

ISBN 0 7514 0092 0

Apart from any fair dealing for the purposes of research or private study, or criticism or review, as permitted under the UK Copyright Designs and Patents Act, 1988, this publication may not be reproduced, stored, or transmitted, in any form or by any means, without the prior permission in writing of the publishers, or in the case of reprographic reproduction only in accordance with the terms of the licences issued by the Copyright Licensing Agency in the UK, or in accordance with the terms of licences issued by the appropriate Reproduction Rights Organization outside the UK. Enquiries concerning reproduction outside the terms stated here should be sent to the publishers at the Glasgow address printed on this page.
  The publisher makes no representation, express or implied, with regard to the accuracy of the information contained in this book and cannot accept any legal responsibility or liability for any errors or omissions that may be made.

A catalogue record for this book is available from the British Library

Preface

The role of vitamins in human nutrition is an important aspect of nutrition and food science, and our knowledge in this area is still advancing. Over the last few years there has been a considerable consumer awareness of nutrition and healthy eating which has been enhanced by numerous articles in the media and by nutritional labelling on products. As a consequence, the food industry has had to become more concerned with the nutritional value of products and the maintenance of guaranteed micronutrient levels.

While the food industry has the responsibility of producing foods that provide a realistic supply of nutrients, including vitamins, it is now also required to offer products with a high degree of convenience and long shelf lives.

Vitamins are a group of chemical compounds which are relatively unstable. They are affected by a number of factors such as heat, light and other food components and also by the processes needed to preserve the food or to convert it into consumer products (e.g. pasteurisation, sterilisation, extrusion, irradiation, etc.). The result of these interactions may be a partial or total loss of the vitamins.

Food technology is concerned with both the maintenance of vitamin levels in foods and the restoration of the vitamin content to foods where losses have occurred. In addition, foods designed for special nutritional purposes such as baby and infant foods and slimming foods need to be enriched or fortified with vitamins and other micronutrients.

In addition to being essential nutrients, some vitamins have an important role as technological additives in foods and can be found acting as colours, antioxidants and texture improvers.

This book is aimed at all who are involved in the product and process development, quality control and analysis of foods, particularly those foods where the levels of vitamins have to be maintained or claimed. It provides a comprehensive survey of all facets of vitamins in foods but places a special emphasis on the practical technological aspects of handling vitamins in a wide range of foods. It also critically reviews the methods of analysis used to determine the levels of vitamins in the various groups of foods.

The aim of the book is not only to provide background knowledge of vitamins in foods but also to provide a practical manual for people who work with vitamins as either nutrients or technological aids.

PBO

Contents

1 Biological functions of vitamins — 1
J. MARKS

1.1	Introduction	1
1.2	Retinol (vitamin A) and the provitamin carotenoids	1
1.3	Vitamin D	3
1.4	Vitamin E (tocopherol)	3
1.5	Vitamin K	4
1.6	Thiamin	5
1.7	Riboflavin	5
1.8	Pyridoxin	7
1.9	Niacin	8
1.10	Vitamin B_{12}	9
1.11	Folates	10
1.12	Pantothenic acid	11
1.13	Biotin	12
1.14	Taurine and choline	12
1.15	Carnitine	13
1.16	Ascorbic acid	13
1.17	Other organic trace constituents of food	14
1.18	Vitamins and 'free radical' damage	15
	Further reading	17

2 Natural occurrence of vitamins in food — 19
H. CRAWLEY

2.1	Fat soluble vitamins and carotenoids		19
	2.1.1	Vitamin A and the carotenoids	20
	2.1.2	Determination of vitamin A in foods	21
	2.1.3	Natural sources of retinoids and carotenoids	21
	2.1.4	Vitamin D	23
	2.1.5	Vitamin E	25
	2.1.6	Vitamin K	27
2.2	Water soluble vitamins		28
	2.2.1	Thiamin	28
	2.2.2	Riboflavin	31
	2.2.3	Niacin	31
	2.2.4	Vitamin B_6	34
	2.2.5	Vitamin B_{12}	35
	2.2.6	Pantothenic acid	35
	2.2.7	Biotin	35
	2.2.8	Folic acid	35
	2.2.9	Vitamin C	39
	References		40

3 Nutritional aspects of vitamins   42
D. H. SHRIMPTON

   3.1 Vitamin deficiency diseases   42
       3.1.1 Introduction   42
       3.1.2 Fat soluble vitamins   43
       3.1.3 Water soluble vitamins: the vitamin B-complex   45
       3.1.4 Water soluble vitamins: vitamin C   48
   3.2 Recommended daily allowances   49
       3.2.1 Introduction   49
       3.2.2 International concepts of the function of RDAs / RDIs   50
   3.3 Safety   53
       3.3.1 Introduction   53
       3.3.2 Issues   54
       3.3.3 Attitudes   55
       3.3.4 Need   55
       3.3.5 Adventitious acquisition   57
       3.3.6 Safety and RDAs (DRVs)   57
       3.3.7 Need and consumption   57
       3.3.8 Possible guidelines for safety   58
   References   59

4 Industrial production   63
M. J. O'LEARY

   4.1 Introduction   63
       4.1.1 History   63
       4.1.2 Current situation   64
       4.1.3 Future production   66
   4.2 Vitamin production   66
       4.2.1 Vitamin A   66
       4.2.2 Provitamin A: β-carotene   68
       4.2.3 Vitamin B_1: thiamine   70
       4.2.4 Vitamin B_2: riboflavin   72
       4.2.5 Niacin   73
       4.2.6 Pantothenic acid   74
       4.2.7 Vitamin B_6: pyridoxine   75
       4.2.8 Folic acid   78
       4.2.9 Vitamin B_{12} 79
       4.2.10 Vitamin C   79
       4.2.11 Vitamin D   83
       4.2.12 Vitamin E: α-tocopherols   83
       4.2.13 Vitamin F group   85
       4.2.14 Biotin (vitamin H)   85
       4.2.15 Vitamin K   88
   References   88

5 Stability of vitamins in food   90
P. BERRY OTTAWAY

   5.1 Introduction   90
   5.2 Fat soluble vitamins   91
       5.2.1 Vitamin A   91
       5.2.2 Vitamin E   92
       5.2.3 Vitamin D   93
       5.2.4 Vitamin K   93

	5.2.5	β-carotene (provitamin A)	94
5.3	Water soluble vitamins		94
	5.3.1	Thiamin (vitamin B_1)	94
	5.3.2	Riboflavin (vitamin B_2)	96
	5.3.3	Niacin	96
	5.3.4	Pantothenic acid	97
	5.3.5	Folic acid	97
	5.3.6	Pyridoxine (vitamin B_6)	98
	5.3.7	Vitamin B_{12}	98
	5.3.8	Biotin	99
	5.3.9	Vitamin C	99
5.4	Vitamin–vitamin interactions		101
	5.4.1	Ascorbic acid–folic acid	101
	5.4.2	Ascorbic acid–vitamin B_{12}	101
	5.4.3	Thiamin–folic acid	102
	5.4.4	Thiamin–vitamin B_{12}	102
	5.4.5	Riboflavin–thiamin	102
	5.4.6	Riboflavin–folic acid	102
	5.4.7	Riboflavin–ascorbic acid	102
	5.4.8	Other interactions	102
5.5	Processing losses		103
	5.5.1	Vegetables and fruits	103
	5.5.2	Meat	104
	5.5.3	Milk	104
5.6	Irradiation		105
5.7	Food product shelf life		106
5.8	Protection of vitamins		110
References			112

6 Vitamin fortification of foods (specific applications)   114
A. O'BRIEN and D. ROBERTON

6.1	Addition of vitamins to foods		114
	6.1.1	Introduction	114
6.2	Beverages		115
	6.2.1	Vitiminisation of instant beverages	116
	6.2.2	Vitiminisation of concentrates, nectars and juice drinks	116
	6.2.3	Vitamin stability	116
	6.2.4	Vitamin incorporation	117
6.3	Cereal products		118
	6.3.1	Breakfast cereals	118
	6.3.2	Bread	124
	6.3.3	Pasta	127
6.4	Dairy products		128
	6.4.1	Milk	128
	6.4.2	Yoghurt	132
	6.4.3	Ice cream	135
	6.4.4	Margarine	136
6.5	Confectionery		138
	6.5.1	Hard boiled candies	138
	6.5.2	Chocolate	139
	6.5.3	Fondant	139
	6.5.4	Marshmallows	140
	6.5.5	Pectin jellies	140
	6.5.6	Starch jellies	140
	6.5.7	Chewing gum	141
References			142

7 Vitamins as food additives   143
J. N. COUNSELL

- 7.1 Ascorbic acid (vitamin C) — 143
  - 7.1.1 Properties — 144
  - 7.1.2 Fruit, vegetables and fruit juices — 144
  - 7.1.3 Soft drinks — 147
  - 7.1.4 Beer — 148
  - 7.1.5 Wine — 148
  - 7.1.6 Flour and bread — 148
  - 7.1.7 Pasta — 150
  - 7.1.8 Meat processing — 150
- 7.2 Carotenoids (provitamins A) — 154
  - 7.2.1 Properties — 155
  - 7.2.2 Fat based foods — 156
  - 7.2.3 Water based foods — 159
- 7.3 Riboflavin (vitamin B_2) — 168
- 7.4 Niacin — 169
- 7.5 dl-α-tocopherol (vitamin E) — 169
  - 7.5.1 Oils and fats — 170
- References — 171

8 Vitamin analysis in foods   172
I.D. LUMLEY

- 8.1 Introduction — 172
  - 8.1.1 Laboratory environment — 172
- 8.2 Oil soluble vitamins — 173
  - 8.2.1 Vitamin D — 173
  - 8.2.2 Vitamin A — 179
  - 8.2.3 Provitamin A carotenoids — 183
  - 8.2.4 Vitamin E — 186
- 8.3 The B-group vitamins — 190
  - 8.3.1 Microbiological assays — 191
  - 8.3.2 Thiamin–vitamin B_1 — 193
  - 8.3.3 Riboflavin–vitamin B_2 — 200
  - 8.3.4 Niacin — 206
  - 8.3.5 Vitamin B_6 — 212
  - 8.3.6 Folates — 218
  - 8.3.7 Vitamin B_{12} — 223
  - 8.3.8 Pantothenic acid — 224
  - 8.3.9 Biotin — 224
- 8.4 Vitamin C — 224
  - 8.4.1 Introduction — 224
  - 8.4.2 Extraction of vitamin C — 225
  - 8.4.3 Determination of vitamin C — 225
  - 8.4.4 Summary — 227
- References — 228

9 Food fortification   233
D.P. RICHARDSON

- 9.1 General policies for nutrient additions — 234
- 9.2 Legislation concerning addition of nutrients to foods — 234
  - 9.2.1 Food for special dietary uses — 235
  - 9.2.2 Foods having lost nutrients during manufacture — 235

	9.2.3 Food resembling a common food	236
	9.2.4 Staple foods	236
9.3	Claims for nutrients and labelling of fortified foods	237
9.4	Restrictive regulations and policies on health claims	238
9.5	The stability of vitamins	239
9.6	Additions of iron sources to foods and drinks	240
9.7	Communicating nutrition	243
9.8	Conclusion	243
	References	244

Appendix 1: Chemical and physical characteristics of vitamins   **246**

Appendix 2: Recommended nutrient reference values for food labelling purposes   **261**

Index   **265**

Contributors

Mr Peter Berry Ottaway	Berry Ottaway and Associates Ltd, 1A Fields Yard, Plough Lane, Hereford HR4 0EL
Mr John N. Counsell	22 Haglis Drive, Wendover, Buckinghamshire HP22 6LY
Ms Helen Crawley	112 Queens Road, Wimbledon, London SW19 8LS
Mr Ian Lumley	Laboratory of the Government Chemist, Queens Road, Teddington TW11 0LY
Dr John Marks	Girton College, Cambridge CB3 0JG
Mrs Amanda O'Brien	Roche Products Ltd, Vitamin and Chemical Division, PO Box 8, Welwyn Garden City, Hertfordshire AL7 3AY
Dr Michael J. O'Leary	Roche Products Ltd, Dalry, Ayrshire KA24 5JJ
Dr David P. Richardson	The Nestle Company Ltd, St George's House, Croyden, Surrey CR9 1NR
Mr David Roberton	High Brooms, Letchworth Lane, Letchworth, Herts SG6 3NF
Dr Derek H. Shrimpton	Crangon, Little Eversden, Cambridge CB3 7HL

1 Biological functions of vitamins
J. MARKS

1.1 Introduction

There is no simple, overall statement which embraces each and every biological function of the vitamins. This is scarcely surprising because vitamins have themselves remarkably little in common, save for their organic nature and the chance recognition of their biological importance at a particular time in history. In general terms, vitamins act as either coenzymes or prohormones but, like all simplifications, this omits certain other vital functions which they subserve.

Originally they were regarded as being only available in the diet (we now know that some can be synthesised in the body); to be ingested in small amounts (but many other essential organic factors in the diet which are present in small amounts are not classified as vitamins) and to be necessary as coenzymes (one is now best regarded as a prohormone). In view of this great variety of activities the only reasonable way to detail their biological functions is unavoidably to take each vitamin serially and separately. For convenience they will be taken within their two separate chemical categories, first the four oil soluble members of the group and then the water soluble ones. The latter group embraces several compounds which are often grouped together under the term vitamin B-complex (or vitamin B-group), the other member of the water soluble series being vitamin C. The vitamin B-complex vitamins have perhaps the greatest level of biological similarity, for they play a vital role in enzyme reactions which are necessary for carbohydrate, fat and protein metabolism.

1.2 Retinol (vitamin A) and the provitamin carotenoids

Vitamin A can be obtained from two sources: preformed retinol, usually in the form of retinyl esters which can be found in and therefore absorbed only from animal tissue and as carotenoids, mainly from plant tissues. These carotenoids can be cleaved in the body to yield retinol. The main source of carotenoid derived retinol is β-carotene, though other carotenoids can also be converted to retinol.

Retinol is essential for growth and for the normal differentiation and development of tissues. In many peripheral tissues retinol has a general metabolic effect, particularly in epithelial cells where it is irreversibly converted into retinoic acid.

The biochemical effect of retinol (or more correctly retinoic acid) in the peripheral tissues by means of which the integrity of the epithelial cells is preserved is far from clear. It appears to be related to distortions in nitrogen metabolism and to amino-acid balances within the tissues. This effect on nitrogen metabolism is probably effected through RNA synthesis, perhaps by retinol serving as a carrier of sugars in the formation of specific glycoproteins. This effect is seen particularly in various mucus-secreting tissues which become keratinised.

Human vitamin A deficiency, particularly in infants and children can give rise to increased mortality during intercurrent infections. One of the most obvious of the peripheral epithelial effects of a deficiency gives rise to the dryness of the conjunctiva and the cornea (xerophthalmia) and subsequent ulceration which can lead on to permanent eye damage.

The visual effect within the retina is specific to retinal, the aldehyde derived from retinol. The rods contain a pigment (rhodopsin) consisting of a specific protein (photopsin) bound to 11-*cis* retinal. When exposed to light the 11-*cis* retinal is converted to the all-*trans* form of retinol which no longer forms a stable complex with photopsin (Figure 1.1). The energy generated has the effect of closing sodium channels via a reduction in the cyclic GMP. The generation of energy in the optic nerves which is interpreted as scotopic (dim-light) vision is derived from the change in the conformation of the specific protein, probably via changes in intracellular cyclic GMP, and hence in calcium ion permeability although sodium ion uptake has also been implicated.

The possible relationship of retinol and β-carotene to protection against free radical damage is considered later and may be an important function not only of retinol but also of β-carotene in its own right.

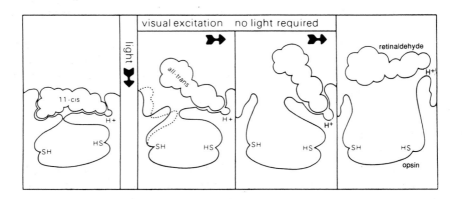

Figure 1.1 Schematic representation of the role of retinal in scoptic vision (based on an original diagram by Wald).

1.3 Vitamin D

The naturally occurring form is vitamin D_3 (cholecalciferol) which is formed in the skin by the action of ultraviolet (UV) irradiation on 7-dehydrocholesterol. However, ergocalciferol (vitamin D_2), which is derived from UV irradiation of ergosterol in plants, is a food additive, is consumed from the food and can be used by the body as an alternative to vitamin D_3. When there is adequate exposure of the skin to sunlight, oral intake of ergosterol is unnecessary, indeed it may lead to toxic manifestations. It comes into its own when sunlight is scarce or the skin is heavily shrouded, as for example in certain ethnic communities. Vitamin D is converted to 25-hydroxycalciferol in the liver and subsequently to 1,25-dihydroxycalciferol (1,25-DHCC) in the kidney under the control of parathyroid hormone. This 1,25-DHCC (calcitriol) is the active form of vitamin D, as a hormone in calcium homeostasis. Calcitriol is transported to distal target tissues. Here it enters the cell, binds to a specific receptor and is taken up by the nucleus. Within the nucleus it acts like any steroid hormone stimulating the production of protein via transcription of messenger RNA. It is a calcium-binding protein which is formed by this process which probably facilitates the transport of calcium ions across membranes. However, a different mechanism may be responsible. Whatever the mechanism, the hormone increases calcium absorption (and indirectly phosphate absorption) from the intestine. The effect within bone is complicated. By increasing circulating calcium ions it assists calcification but the direct effect is to increase the release of calcium and phosphate from bone, probably by activating osteoclasts.

There is evidence that calcitriol functions in the kidney too, increasing renal absorption of calcium. It may also feed back on the parathyroid glands, regulating the secretion of parathyroid hormone. It is not clear whether the muscle weakness which is a feature of low vitamin D levels is the result of a low circulating level of calcium ions or whether the muscle membrane movement of calcium ions is also calcitriol dependent.

In infants and children vitamin D deficiency gives rise to rickets, with skeletal deformity, bone pain and tenderness and muscle weakness. Once skeletal formation is complete, vitamin D deficiency in adult life leads to osteomalacia with bone demineralisation, bone tenderness and muscle weakness.

1.4 Vitamin E (tocopherol)

It is only relatively recently that the essential nature of vitamin E has been accepted. Vitamin E activity is manifested by two series of compounds, the tocopherols and the tocotrienols but the physiologically most important form is α-tocopherol.

Vitamin E has a powerful antioxidant effect within the animal body, particularly within lipids and this antioxidant property may be its most important or perhaps even sole function. Derived from this, the important physiological action may be stabilisation of various lipid membranes including those of mitochondria and this links it with the prevention of free radical damage (see below). Although there is now a substantial body of evidence which links the biological effects of tocopherol with lipid membrane stabilisation and prevention of free radical damage, other biological functions may be involved. For example, it may also have a direct effect on various enzyme systems rather than an indirect effect via organelle protection, since other antioxidants cannot substitute completely for tocopherol. The effect may be mediated via an alteration in enzyme synthesis. Tocopherol also influences the activity of the immune system.

Deficiency of vitamin E is rare in any group other than infants, except where there is a pronounced lipid absorption defect. In infancy (particularly premature infants) encephalomalacia, cerebellar ataxia, pigmentary retinopathy and anaemia may be manifestations of tocopherol inadequacy. Current experiments raise the possibility that some degenerative disorders, particularly during the process of ageing, may also be the result of inadequate levels of vitamin E in lipid membranes. There are divergent views about the role of vitamin E in the prevention of the cardiovascular changes of atherosclerosis.

1.5 Vitamin K

Vitamin K activity is demonstrated both by absorbed plant phylloquinone and also by the related menaquinones which are synthesised by intestinal bacteria and then absorbed. Each provides about half the daily requirement in man.

Vitamin K is required for the post-translational carboxylation of certain glutamate residues in proteins to γ-carboxyglutamate. At one time it was considered that this activity took place in the liver in relation only to four factors necessary for blood coagulation, namely factors II (prothrombin), VII, IX and X, together with proteins C, S, Z and M. Each of these factors contains metabolically important γ-carboxyglutamate residues, the formation of which depends on both NADH and vitamin K. Many of these proteins can be looked at as zymogens which can be activated by specific proteases, the modified proteins in turn activating still other zymogens. This concept leads to the cascade theory of coagulation. It is now known, however, that in addition to these factors, vitamin K is also necessary for at least two coagulation inhibitors and for several proteins which are present in various tissues. The function of most of these latter proteins is unknown.

The main effect of deficiency of vitamin K is a haemorrhagic diathesis. This is seen frequently in infants in the early days of life, unless they are supplemented with vitamin K (or the mother is supplemented just before the infant is born). In older children or adults a deficiency occurs due to either a poor absorption or poor utilisation of vitamin K, which results from the administration of a competitive anticoagulant and not as a result of a deficiency in the food. The only natural vitamin K deficiency in adults is that associated with obstructive jaundice leading to poor absorption.

1.6 Thiamin

Otherwise known as vitamin B_1 or previously aneurin, thiamin is required mainly for metabolism of carbohydrates, although it also plays a role in the metabolism of fat and alcohol. Thiamin diphosphate (previously termed thiamin pyrophosphate) and thiamin triphosphate are the active forms of the vitamin which are formed by the phosphorylation of thiamin by ATP in a reaction which is catalysed by thiamin pyrophosphokinase. Thiamin phosphates are required as cofactors in a substantial variety of reactions (they are cofactors for at least 20 different enzymatic reactions) but especially those involving decarboxylation to form activated aldehyde species which may or may not undergo further oxidation as they are transferred to an acceptor. It is involved, for example, in those that are catalysed by transketolase and pyruvate decarboxylase. Hence it is important in transketolation of glucose in the pentose shunt, in the conversion of pyruvate to acetyl CoA (coenzyme A) and within the citric acid cycle itself. Each of these reactions is of central importance to the adequate generation of ATP. Hence a deficiency of the active form of thiamin will mainly affect those tissues that have a high energy requirement, particularly that derived from carbohydrate sources, such as the nerves, heart and kidney. This effect is seen particularly in the peripheral nerves where stimulation results in a decrease in the level of thiamin diphosphate and thiamin triphosphate and a release of thiamin monophosphate and free thiamin into the extracellular spaces. It is currently thought that thiamin cofactors are involved in the sodium and potassium ions gating mechanism via an ATPase.

With this biochemical background it is scarcely surprising that the main physiological function of thiamine is seen in the nerves and heart and that the main pathology also occurs in these parts of the body during a deficiency.

Beri-beri, the deficiency disease associated with a low thiamin intake is, not surprisingly, particularly liable to occur in those areas of the world where carbohydrate forms a significant proportion of the energy intake. The clinical manifestations include nerve degeneration associated with both sensory and motor nerves giving rise to paraesthesia and muscle weakness; inadequate cardiac function leading to signs of heart failure and digestive disturbances. In industrially developed countries deficiency (Wernicke–Korsakov syndrome) is usually associated with alcoholism, not only due to reduced food intake but to the higher requirement for thiamin phosphates for the metabolism of alcohol as the main substrate and poor absorption and utilisation of thiamin due to alcoholic gastroenteritis and liver damage.

1.7 Riboflavin

Otherwise known as vitamin B_2, riboflavin plays an essential role in all mammalian oxido-reduction processes and in those of many other organisms.

Riboflavin is phosphorylated in the intestinal mucosa as part of the absorption process. Small quantities are stored in the liver, spleen, kidney and cardiac muscle but as with most water soluble vitamins there are no major body stores.

Riboflavin is converted into two tissue coenzymes, flavin mononucleotide (FMN) and more commonly flavin adenine dinucleotide (FAD).

The flavin nucleotides are prosthetic groups and remain bound, either covalently or non-covalently, depending on the enzyme, to the protein during the oxido–reduction reactions. Thus unlike the coenzyme derived from nicotinamide (NADH, see section 1.9) flavins do not transport hydrogen atoms from one part of the cell to another. The flavins act as oxidising agents within the cell by virtue of their ability to accept a pair of hydrogen atoms through two successive one-electron transfers with the formation of intermediate semiquinones. This means that the flavin prosthetic group may exist in any of the redox states: the oxidised flavin; the one-electron reduced state (semiquinone) and the two-electron reduced state (hydroquinone).

The flavoproteins show a wide range of redox potentials and hence can play a wide variety of roles in oxidative metabolism (Table 1.1). Flavoproteins within the electron transport chain accept electrons direct from the NADH generated in reactions of glycolysis and the tricarboxylic acid cycle, and pass them on down the chain. Some flavoproteins function outside the electron transfer chain as intermediate electron carriers (e.g. the enzyme dihydrolipoamide reductase uses a flavoprotein cofactor in the oxidative decarboxylation of pyruvate and 2-oxoglutarate dehydrogenase complexes). When used as cofactors in flavoprotein oxidases (e.g. xanthine oxidase) the electrons are transferred to oxygen with the formation of hydrogen peroxide. Still other flavoprotein enzymes (e.g. succinate dehydrogenase and acyl-CoA dehydrogenase) react with the reduced substrate and pass the electrons directly into the electron transfer chain with which they are chemically directly associated. Thus by these mechanisms the flavin nucleotides are central to the oxidative metabolism of both carbohydrates and fats.

More specifically, flavin nucleotides can act as oxidising agents in reactions which form oxoacids and generate ammonia from amino acids. Amino acid oxidases use flavine coenzymes for this purpose but glutamate dehydrogenase uses a nicotinamide coenzyme for a similar purpose. They also act in the oxidation of aldehydes, the oxidation of aliphatic side chains and the transfer of hydrogen from pyridine nucleotides (NADH, NADPH) to cytochrome c. Riboflavin is also involved, with pyridoxin, in the formation of niacin from tryptophan.

Table 1.1 Some enzymes which are riboflavin coenzyme-dependent (some of these are only found in plant tissues)

Aerobic	Warburg's yellow oxidase
	Glucose oxidase
	D-amino oxidase and L-amino oxidase
	Xanthine oxidase.
Anaerobic	Cytochrome reductase
	Succinic dehydrogenase
	Acyl-coenzyme A dehydrogenase
	Fatty acid synthetases
	Mitochondrial α-glycerophosphate dehydrogenase
	Erythrocyte glutathione reductase.

With such a wide diversity of biochemical reactions which are central to metabolism of carbohydrate, fat and protein, it is surprising that there are so few obvious clinical manifestations of riboflavin deficiency. Those that are seen are associated with the mouth and eye. They include cheilosis (dry cracked lips), angular stomatitis (raw fissures at the angles of the lips) and tongue surface denudation, together with photophobia, lachrymation and a sensation of 'grittiness' under the eyelids. With a severe deficiency, seborrhoeic dermatitis may develop around the nasolabial folds; the eye may show corneal vascularisation and in addition vulval or scrotal dermatitis may occur.

1.8 Pyridoxin

The older term, vitamin B_6, covers a group of metabolically interchangeable and related compounds, pyridoxol (the alcohol), pyridoxal (the aldehyde) and the amine (pyridoxamine). Vitamin B_6 in any form is rapidly converted in the body into the active coenzyme pyridoxal-5-phosphate.

Pyridoxal-5-phosphate forms the prosthetic group of enzyme–coenzyme systems in general metabolic processes, but particularly those for protein metabolism. Altogether the number of enzymes which are known to require pyridoxal phosphate totals more than sixty.

Within the field of protein metabolism the reactions with which these are involved include transamination, amino acid oxidative decarboxylation, racemisation, cleavage, synthesis, desulphhydration, trans-sulphurations and dehydration (Table 1.2). The aromatic L-amino acid decarboxylase and glutamic acid decarboxylase also require pyridoxal phosphate. Within the field of carbohydrate metabolism, pyridoxal phosphate is the coenzyme for glycogen phosphorylase. It is also involved in lipid metabolism but the mechanism is not yet established, although it may involve the pyridoxine effects on hormone levels. Pyridoxal phosphate is also required for phosphorylation and in the synthesis of haem.

As a result of these various activities, pyridoxin has an important role to play in brain metabolism. In particular it is required for the formation of the whole group of amines that act as synaptic transmitters in various regions of the brain (noradren-

Table 1.2 Some pyridoxin-dependent enzymes in the synthesis, degradation and interconvertion of amino acids

Amino acid decarboxylases—various specific ones
   some, e.g. glycine decarboxylase, also need a flavin-linked
   dehydrogenase and tetrahydrofolate
Amino transferases
Amino acid racemases
5-aminolevulate synthetase (ALA synthetase)
β-carbon replacement reactions (e.g. tryptophan synthetase)
β-carbon elimination–deamination reactions (e.g. tryptophanase)
γ-carbon replacement and elimination–deamination reactions

aline, tyramine, dopamine and 5-hydroxytryptamine) and for the inhibitory transmitter substance γ-amino butyric acid.

Pyridoxin, with riboflavin, also plays a role in the niacin economy of the body because it is involved in the conversion of tryptophan to niacin.

Pyridoxin is so widely present within the diet that a natural deficiency is rare save in infancy. In adults a deficiency is usually iatrogenic, precipitated, for example, by contraceptive steroids, by a tuberculostatic (isoniazid), an antibiotic (cycloserine), an antihypertensive (hydralazine) and by penicillamine (used in the management of the rare Wilson's disease). The only clinical manifestations of deficiency are convulsions in infants (probably due to low levels of γ-amino butyric acid (GABA), peripheral neuritis, anaemia, some skin reactions similar to those seen with riboflavin / nicotinic acid deficiency and malaise/depression. Except in a marked deficiency the manifestations are very non-specific.

1.9 Niacin

The term niacin covers both nicotinic acid and the related nicotinamide. Part of the niacin requirements come preformed from the food but niacin can also be formed in the body from dietary tryptophan. It is broadly accepted that 60 mg tryptophan is equivalent to 1 mg niacin. There is considerable variation in the proportion which is taken into the body preformed. Additionally the conversion of tryptophan is reduced by oestrogens, while the enzymes kynureninase and kynurenine hydroxylase, which are involved in the conversion have a requirement for pyridoxin and riboflavin, further complicating the vitamin status.

Within the body niacin is converted to two coenzymes: nicotinamide adenine dinucleotide (NAD – previously known as coenzyme I or diphosphopyridine nucleotide (DPN)) and nicotinamide adenine dinucleotide phosphate (NADP – previously known as coenzyme II or triphosphopyridine nucleotide (TPN)). Both are cofactors of redox reactions in the energy-generating metabolism of cells.

Nicotinamide adenine dinucleotides are involved in a very large number of oxido–reduction reactions both in the cytosol and in mitochondria, being converted reversibly to NADH and NADPH. Although usually regarded as coenzymes they are not tightly bound to the enzymes and are probably better regarded as true cell substrates. The major role of NADH is to transfer electrons from metabolic intermediates into the electron transfer chain, whilst NADPH acts as a reducing agent in a large number of biosynthetic processes. It is believed that the hydrogen acceptor function of these two coenzymes resides in the *para* position of the nicotinamide component. Thus, in the electron transfer chain one of the links is NADH and the current theory is that within the mitochondrial membrane there is a functional organisation of a linear sequence of carriers in which the hydrogen atoms from NADH can reduce oxygen.

In general the $NAD^+/NADH$ system is mainly involved in the oxidative processes of catabolism (e.g. tricarboxylic acid cycle, electron transfer, glycolysis)

Table 1.3 Some important dehydrogenases which use niacin coenzymes

---
Alcohol dehydrogenase
Lactate dehydrogenase
Malate dehydrogenase
Steroid reductase
Dihydrofolate reductase
Aldehyde dehydrogenase
Glutamate dehydrogenase

---

while the $NADP^+$/NADPH system is involved in reductive processes of anabolism. A list of the most important dehydrogenases using the nicotinamide coenzymes is given in Table 1.3. In addition to these, NAD is the coenzyme by which galactose 1-phosphate is converted to glucose-1-phosphate and hence to glycogen. It is currently thought that one important role of NAD is in the synthesis and repair of DNA.

Niacin deficiency leads to the florid manifestations of pellagra, often summarised in the mnemonic 'diarrhoea, dermatitis and dementia'. The gastrointestinal symptoms are often the first to appear and include not only mouth changes of glossitis and stomatitis with a swollen beefy red tongue but also anorexia, abdominal discomfort and diarrhoea. The skin abnormality consists of an initially dry, discoloured and scaly symmetrical dermatitis in the areas that are exposed to sunlight and these later desquamate. The early mental symptoms include lassitude, depression and loss of memory but it may develop into a confusional state very similar to that seen with Wernicke encephalopathy. Occasionally the mental disorder may occur as a single feature (pellagra sine pellagra) and this may be very difficult to recognise.

1.10 Vitamin B_{12}

There are several related chemical structures consisting of a corrin nucleus surrounding a cobalt atom which show vitamin B_{12} activity. Collectively they can be described as the cobalamins. They are not found in plants and the sole natural source are microorganisms. The active compounds in mammals are hydroxycobalamin, methylcobalamin and adenosylcobalamin. Most of the normal human intake is derived from animal sources and strict vegans need to be supplemented with cyanocobalamin.

The absorption of the colabamins from the terminal ileum requires binding to a salivary haptocarrin and subsequently to a glycoprotein of molecular weight about 50 000 secreted by gastric parietal cells (the so-called intrinsic factor). Deficiency states in the human are more commonly due to poor absorption as a result of an absence of the gastric intrinsic factor than to dietary inadequacy.

The conversion of cyanocobalamin to its coenzyme form is catalysed by an enzyme system, cyanocobalamin coenzyme synthetase. The process involves a reaction with a deoxyadenosyl moiety derived from ATP and also requires diol or dithiol, a reduced flavin or reduced ferredoxin as the biological alkylating agent.

Vitamin B_{12} provides the prosthetic group for two classes of enzymes. The first (utilising adenosyl cobalamin coenzymes) is involved in various intramolecular group transfers (isomerisations) such as the breakdown of valine via methylmalonyl CoA and succinyl CoA. The second (using methyl cobalamin) consists of methyltransferase reactions involved in the recycling of folate coenzymes (see below) and in the synthesis of methionine from homocysteine.

Human deficiency leads to a megaloblastic anaemia (large irregular red blood cells) and irreversible neurone damage (subacute combined degeneration of the cord). The red cell abnormality is probably caused by an inability to produce sufficient S-adenosylmethionine in the marrow for methyl group transfer in the formation of neucleotides for DNA synthesis. The neuropathy is probably due to a lack of methionine for methyl transfer to form the choline for the phospholipids and sphingomyelin which are required for the formation of myelin.

Because lassitude is a feature of the anaemia and the administration of vitamin B_{12} leads to an intense feeling of well-being in the very early days of the treatment of pernicious anaemia, it has been suggested that this vitamin could itself act as a general 'tonic' but there is no conclusive evidence for this effect.

1.11 Folates

Pteroyl glutamic acid (folic acid) is the parent compound for a series of related chemical substances which show similar biochemical properties and which are known collectively as the folates. Any of these dietary folates are converted during absorption from the intestinal tract such that only tetrahydrofolic acid enters the portal blood and is subsequently transported to the tissues. Tetrahydrofolate is transported by an energy-dependent process into bone marrow cells and other cells where it exerts its specific function.

Tetrahydrofolic acid forms the prosthetic group of many enzymes that are involved in the transfer of 'one carbon' moiety. These may be in the metabolically interconvertible forms of formyl (–CHO), active carboxyl (–COOH) or hydroxymethyl (–CH$_2$OH). These 'one carbon' moieties arise mainly from the beta carbon of serine, the alpha carbon of glycine, from carbon-2 of the imidazole ring of histidine and also from formate itself.

The reactions involving this transfer of one-carbon units carried by tetrahydrofolic acid occur in amino acid and nucleotide metabolism and include:

- conversion of homocysteine to methionine which also uses vitamin B_{12} during the reaction (see below),
- methylation of transfer RNA,
- *de novo* purine nucleotide synthesis by the provision of carbons 2 and 8 of the purine ring,
- pyrimidine nucleotide synthesis.

The regulatory features of one-carbon metabolism in mammalian tissues are quite complex. The overall pathway is confined to the cytoplasm and involves the interconversion and metabolism of a number of relatively unstable folate coenzymes. These regulatory controls are poorly understood. Folate polyglutamates of different chain length may be preferred as substrates by different folate-requiring metabolic pathways, the reason for which is unknown.

The relationship of the pathophysiology of cobalamins and folates probably depends upon the fact that the conversion of N^5-methyltetrahydrofolate to tetrahydrofolate requires methylcobalamin, during which reaction homocysteine is methylated to methionine.

Folic acid deficiency resembles that of vitamin B_{12} deficiency save that folic acid deficiency does not lead to subacute combined degeneration of the cord. It does, however, produce a megaloblastic bone marrow with a macrocytic anaemia which is indistinguishable from pernicious anaemia. Folic acid deficiency can also result in a psychosis with personality changes such as hostility and paranoid behaviour. In young children, growth may be restricted.

1.12 Pantothenic acid

Pantothenic acid is extremely widely distributed in animal and plant tissues and in consequence a natural dietary deficiency is virtually unknown. Panthothenic acid is a constituent of coenzyme A which is present in all tissues and is very important for all intermediary metabolism (Figure 1.2). It links carbohydrate, fat and amino acid metabolism and forms the common normal pathway for these substances to enter the citric acid cycle. It is essential for all acetylation processes and for the formation of the steroid hormones via cholesterol. Coenzyme A can be considered

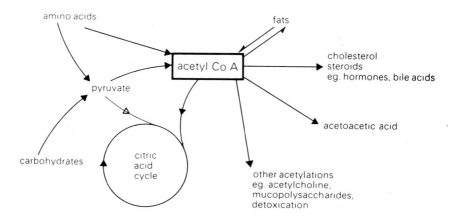

Figure 1.2 Central role of coenzyme A in metabolism.

to be a derivative of adenosine diphoshate (ADP) in which there is an additional phosphate at the 3-position of the ribose component and esterification of the 5-diphosphate group with pantothenic acid linked to 2-amino ethanthiol. It is this terminal –SH group which confers on coenzyme A its coenzyme property, since the terminal thiol group forms acyl esters with carboxylic acids. These acyl CoA compounds then act as acylating agents in a variety of intracellular reactions essential for both anabolism and catabolism.

Despite the central role in metabolism, experimental depletion studies have revealed remarkably few clinical manifestations save for vague malaise, muscle weakness and gastrointestinal disturbances, a further example that clinical effects do not correlate well with the physiological functions.

1.13 Biotin

Biotin is one of the essential organic constituents of the diet which although not originally classified as a vitamin is now usually placed in that general class.

In animals and humans biotin was first recognised as a protective agent against 'egg white disease' which occurs when large amounts of raw egg albumin are ingested. A glycoprotein avidin which is present in egg white binds biotin with high affinity leading to the manifestations of deficiency.

Biochemically biotin is a coenzyme for a substantial number of enzymes involved in carboxylation. Carboxylation of biotin to form N-1'-carboxybiotin yields an 'active form' of carbon dioxide which can carboxylate subsequent substates. The main reactions in which biotin is involved include carboxylation of pyruvic acid via methylmalonyl CoA and direct to oxalo-oxalate; of propionyl CoA to methylmalonyl CoA; transcarboxylation in the catabolism of certain amino acids; conversion of acetyl CoA to malonyl CoA in the formation of long-chain fatty acids and a role in pyridine and pyrimidine synthesis.

With the exception of patients on inadequate total parenteral nutrition for long periods or on haemodialysis, the only other groups of people who experience low biotin intake are those who consume substantial numbers of uncooked eggs and also possibly infants. There are two related skin disorders in infants, seborrhoeic dermatitis and desquamative erythroderma (Leiner's disease) that appear to be related to biotin deficiency. There is also a rare biotin dependency disorder in which an abnormality in the carboxylation of propionic acid leads to a ketotic hyperglycinaemia. Experimental biotin depletion produces a dry scaly dermatitis, glossitis, anorexia, hallucinations, malaise, depression and hypersomnolence.

1.14 Taurine and choline

Taurine, which can be synthesised in adults, may rarely be deficient in premature infants and can perhaps be regarded as a vitamin in this particular group. In fact this group, with immature enzyme activity, poor reserves, limited food intake and high requirements for growth, is that in which many of these accessory food factors have been first recognised.

In the absence of taurine supplementation, bile salt production is poor and fat absorption is limited. Taurine may also play a role as a scavenger of hypochlorous acid (see section 1.18 on free radical disorders)

The features which have been ascribed to taurine apply with almost equal force to choline. It can be formed in the body by the methylation of ethanolamine and since the widely distributed lecithin, as an important precursor is rarely, if ever, deficient, natural low levels are rare. It is a factor in bile-salt production, but the main functions of choline are as a source of labile methyl groups for the biosynthesis of other methylated compounds and in the formation of acetyl choline, the neurotransmitter.

1.15 Carnitine

Carnitine is another substance which is normally synthesised in adequate amounts but for which it is now recognised that there are isolated circumstances in which there is inadequate synthesis which may be coupled with a low dietary intake.

Carnitine is required for the transport of fatty acids, particularly into mitochondria for metabolism. This reaction is catalysed by carnitine acyltransferase.

Apart from poor transport in infants, several cases have now been reported of carnitine deficiency in children and adults. They can be divided into two distinct groups. The first show a deficiency of muscle carnitine (myopathic carnitine deficiency) and these present with a history of muscle cramps, exercise fatigue and myoglobinuria. The second group (systemic carnitine deficiency) shows a widespread cellular deficiency, probably the result of inadequate liver synthesis. The liver deficiency is the presenting disorder with an enlarged fatty liver, hypoglycaemia, hyperammonaemia and sometimes after fasting and/or severe infections, coma. These secondary deficiency states can be managed by carnitine supplementation.

1.16 Ascorbic acid

There is no doubt that vitamin C (ascorbic acid) is an essential dietary ingredient in humans although the vast majority of animal species can synthesise the substance and do so to a tissue-saturated level. What is still in dispute is the full range of roles that ascorbic acid plays in the human physiological economy and associated with this, the assessment of the ideal dietary intake.

The enediol groups at the second and third carbon atoms are sensitive to oxidation and can easily convert into a diketo group forming dehydroascorbic acid. Within the body this reaction may perform an important function as a redox system probably in association with glutathione and cysteine.

Table 1.4 Some important biochemical roles of ascorbic acid in the body

---

Hydrogen / electron transport—as redox system
Hydroxylation of
   proline and lysine in collagen synthesis
   in steroid, lipid and drug metabolism
Carnitine biosynthesis
Metabolism of L-tyrosine and catecholamine formation
Immunological and antibacterial activity
Iron metabolism particularly iron absorption
Modification of prostaglandin synthesis
Prevention of histamine accumulation
Protection against free radical damage

Knowledge is still incomplete about all the biochemical and physiological roles which ascorbic acid plays in the body. Some of the reactions which are believed to be relevant are shown in Table 1.4.

Among the most important reactions is the post-translational hydroxylation of proline in the formation of collagen with its importance in growth and wound healing. Specifically collagen hydroxylases are ascorbate dependent. Ascorbate also protects the enzyme p-hydroxyphenylpyruvate oxidase which is responsible for tyrosine catabolism. Another important reaction is that catalysed by dopamine β-mono-oxygenase in the formation of catecholamines within the nervous system and this is also ascorbate dependent.

Other properties which are worthy of attention are the improvement in iron absorption, an increase in physical working capacity of young adolescents as determined by their maximum oxygen utilisation and a beneficial effect on the blood cholesterol level. The important current aspect of all these properties is that a beneficial effect is apparent at levels of intake up to a level which leads to tissue saturation but there is no evidence of pharmacological effects from higher dosage. The role of ascorbate in the protection of the body from free radical damage is considered later.

An insufficient intake of vitamin C results in scurvy, a disease which is characterised by multiple haemorrhages. In adults the earliest symptoms are lassitude, weakness and muscle pains but these are followed by bleeding gums, gingivitis and loosening teeth. Hyperkeratotic follicular papules and spiral and unerupted hairs may often be detected if they are sought.

These early signs are followed by minute haemorrhages under the skin, particularly in sites of local trauma or pressure. In more severe cases extensive haemorrhages occur, either into the tissues or through the membranes of, for example, the intestinal tract.

1.17 Other organic trace constituents of food

At various times other substances which may be important in the tissue metabolic economy have been proposed to be classed as vitamins on the basis that they are organic in nature, have an essential role in growth and the maintenance of health at relatively low concentration and have a definite dietary requirement which cannot be met by body synthesis. While this last property is, of course, not now necessarily true for all the substances which have previously been included in the vitamin category, it would appear to be illogical to include fresh substances that do not fit all these requirements. In consequence these additional substances, which include ornithine, ubiquinones and the bioflavinoids, are not currently classed as vitamins and have not been considered herein. Whether taurine, choline and carnitine should be regarded as 'vitamins' in a broad sense is still a matter of dispute but they have been included in this chapter.

1.18 Vitamins and 'free radical' damage

During the past decade information has accumulated which implicates 'free radical' damage as a potent source of disease in human beings. If this concept is supported by further studies then it is extremely likely that several of the vitamins will have a valuable place in the prevention of these diseases.

Free radicals are highly reactive oxygen-containing molecules which develop in the body as a result of both normal and abnormal metabolic processes. The term free radical, which is the preferred one for the biological sciences, implies the existence of one or more unpaired electrons, but other reactive forms of oxygen (e.g. singlet oxygen, hydrogen peroxide) together with various 'oxygenating' enzymes can also exert similar effects, so that the alternative term 'oxygen species' is chemically more accurate. The main oxygen species are shown in Table 1.5.

Free radicals are formed within the tissues as a result of the normal metabolic activity of the body. Indeed those who currently do not accept the concept of diseases being produced by free radicals point out, quite correctly, that these substances are a normal and vital feature of many tissue metabolic processes. Many of the biochemical reactions involving the vitamins which have been described in this chapter involve oxygen species as a vital component (e.g. electron transport system, autoxidation of flavins and thiols).

The important feature appears to be that not only can free radicals be generated by these metabolic processes but they can react with other cellular constituents.

Table 1.5 Main oxygen species of possible pathological importance

$O_2^{\cdot}$	Superoxide anion
H_2O_2	Hydrogen peroxide
$HO^{\cdot}$	Hydroxyl radical
$HO_2^{\cdot}$	Perhydroxy radical
$RO^{\cdot}$	Alkoxy radical
$ROO^{\cdot}$	Peroxy radical
O_2^*	Singlet oxygen

Table 1.6 Some exogenous factors which encourage free radical accumulation

Drug oxidations
  e.g. paracetamol, carbon tetrachloride

Redox-cycling substances
  e.g. paraquat, alloxan, nitro-aromatics

Radiations
  sunlight / heat
  ionising, e.g. X-rays

Pollutants
  Cigarette smoke, tar products

These in turn can release other reactive substances. Thus the intracellular generation of free radicals is self perpetuating unless adequate amounts of free radical quenchers are present.

Perhaps more importantly from the pathological point of view, it is clear that free radical production is encouraged by several exogenous factors at least some of which are a prominent feature of modern life and known to be disease producing (Table 1.6).

As highly reactive chemicals existing within the tissues, extraneously high concentrations of free radicals have the ability to bond to most normal cellular components and such bonding can be shown to produce tissue damage (Table 1.7). These changes have been implicated in a large number of diseases and for several of these recent experimental and epidemiological evidence appears to be sufficiently good for the clinical relevance to require attention (Table 1.7). This particularly applies to some of the chronic diseases that have become more prevalent in recent years.

The effect of oxygen species within the body is antagonised by constituents of the cells and extracellular spaces but also by antioxidants absorbed with the food. Within the cells, the antioxidant defences depend largely on two specific enzymes, glutathione peroxidase and superoxide dismutase. Glutathione peroxidase requires

Table 1.7 Cellular effects attributed to free radical damage and their disease implications

Cell component	Damage	Possible clinical relevance
Nucleic acids	Mutagenicity	Some forms of cancer
	Carcinogenicity	
Proteins	Enzyme damage	Radiation injury
Lipids / lipoproteins	Cell wall damage	Atherosclerosis
	Organelle damage	Ageing
Carbohydrates	Allergenicity	Auto-immune reactions
	Receptor damage	Nervous system degenerations
Macromolecules	Connective tissue damage	Inflammatory-immune disorders
		Cataract formation

Table 1.8 Main nutrients which are regarded as important in protection against free radical damage. The vitamins are shown in **bold type**.

Substances	Action
Glucose	Hydroxyl radical scavenger in cytosol.
Vitamin E	Reacts directly with peroxyl, hydroxyl, superoxide radicals and singlet oxygen. Mutually regenerates with vitamin C and mutually spares with selenium. An important lipid membrane antioxidant.
Vitamin C	Water-soluble extracellular antioxidant which reacts directly with hydroxyl, superoxide radicals and singlet oxygen. Mutually regenerates with vitamin E.
Vitamin A	Fat-soluble weak free radical scavenger.
Beta-carotene	Lipid-soluble antioxidant and very powerful singlet oxygen quencher.
Selenium	Present in cytosol glutathione peroxidase. Interacts with vitamin E.
Copper	Present in cytosol superoxide dismutase and extracellular ceruloplasmin.
Zinc	Present in cytosol superoxide dismutase.
Manganese	Present in mitochondrial superoxide dismutase.
Taurine	Hypochlorous acid scavenger.

NADPH and, as already explained, reactions of this type which link in to the electron transport chain need a variety of the water-soluble vitamins.

In the extracellular space the specific enzyme activity is either low or absent and various chemical antioxidants become important. These include a variety of nutrients, particularly antioxidant vitamins (Table 1.8). However, many of these vitamin antioxidants only influence the situation adequately in the presence of various non-dietary defences and mineral constituents of the diet, the range of which is outside the scope of this chapter.

The idea that free radical damage is a cause of various diseases and particularly of those in the second half of life must currently remain a hypothesis. The necessary further evidence, particularly that relating to the protective effect of antioxidants will only become available over the next few years. However, even the current evidence appears to be sufficiently convincing to suggest that one of the most important general functions of many of the vitamins lies in the field of antioxidant protection.

Further reading

Most of the material which forms the basis of this review is now available within textbooks and rather than clutter the text with numerous references it was considered more appropriate to define relatively recent textbooks and monographs (i.e. within about the last ten years) within which the topics can be examined in greater detail and which provide extensive bibliographies, *viz*:

Barker, B.M. and Bender, D.A. (1980) *Vitamins in Medicine*, 2 Vols, 4th edn. Heinemann Medical, London.

Counsell, J.N. and Hornig, D.H. (1981) (eds) *Vitamin C: Ascorbic Acid*. Applied Science, Englewood.

Halliwell, B. and Gutteridge, J.M.C. (1985) *Free Radicals in Biology and Medicine*. Clarendon Press, Oxford.
Hanck, A. (1983) (ed.) *Vitamins in Medicine: Recent Therapeutic Aspects*. Huber, Bern.
Machlin, L.J. (1984) (ed.) *Handbook of Vitamins: Nutritional, Biochemical and Clinical Aspects*. Marcel Dekker, New York.
Marks, J. (1985). *The Vitamins: Their Role in Medical Practice*. MTP Press, Lancaster.
Martin, B.R. (1988). *Molecular Regulation: A Molecular Approach*. Blackwell, London.
Neuberger, A. and Jukes, T.H. (1982) (eds) *Human Nutrition. Current Issues and Controversies*. MTP Press, Lancaster.
Newsholme, E.H. and Leech, A.R. (1983) *Biochemistry for the Medical Sciences*. John Wiley, Chichester.
Seib, P.A. and Tolbert, B.M. (1982) (eds) *Ascorbic Acid: Chemistry, Metabolism and Uses*. Advances in chemistry series 200, American Chemical Society, Washington.

In addition there have been some recent review articles on individual vitamins which have provided an extensive bibliography, *viz*:

Beisel, W.R. (1990) Vitamins and the immune system. *Ann. New York Acad. Sci.* **587**, 5–8.
Bendich, A. (1990) Antioxidant vitamins and their functions in immune responses. *Adv. Exp. Med. Biol.* **262**, 35–55.
Bremer, J. (1990) The role of carnitine in intracellular metabolism. *J. Clin. Chem. Clin. Biochem.* **28**, 297–301.
Chanarin, I., Deacon, R., Lumb, M. and Perry, J. (1989) Cobalamin-folate interrelations. *Blood Rev.*, **3**, 211–215.
Chytil, F. and Sherman, D.R. (1987) How do retinoids work? *Dermatologia* **175** (suppl 1), 8–12.
Dabek, J. (1990) An emerging view of vitamin D. *Scand. J. Clin. Lab. Invest.* (suppl) **201**, 127–133.
Dakshinamurti, K. and Chauhan, J. (1989) *Biotin. Vitam. Horm.* **45**, 337–384.
England, S. and Seifter, S. (1986) The biochemical functions of ascorbic acid. *Ann. Rev. Nutr.* **6**, 365–406.
Marks, J. (1989) Clinical implications of free radicals. *Naringsforskning* **33**, 130–137.
Padh, H. (1991) Vitamin C: newer insights into its biochemical functions. *Nutr. Rev.* **49**, 65–70.
Sies, H. and Murphy, M.E. (1991) Role of tocopherols in the protection of biological system against oxidative damage. *J. Photochem. Photobiol.* **8**, 1166–1173.
Sinclair, A.J. Barnett, A.H. and Lunec, J. (1990) Free radicals and antioxidant systems in health and disease. *Brit. J. Hosp. Med.* **43**, 334–344
Sklan, D. (1987) Vitamin A in human nutrition. *Prog. Food. Nutr. Sci.* **11**, 39–55.
Votila, L. (1990) The metabolic functions and mechanism of action of vitamin K. *Scand. J. Clin. Lab. Invest.* (suppl) **201**, 109–117.

2 Natural occurrence of vitamins in food
H. CRAWLEY

All the vitamins required by man are available from the food supply, and to ensure an adequate intake of vitamins from food alone, it is advantageous to eat a variety of foods from both animal and vegetable sources. Some vitamins are concentrated in a small number of foods, others are widely distributed in nature, but may only occur in small quantities, and most vitamins are considered to be micronutrients. The presence of vitamins in foodstuffs is variable depending on a diverse range of factors, for example, the variety of plants, the season of the year, the growing conditions, the breed, maturity and feeding of livestock or the storage or treatment of the food. In addition, the values commonly presented for the vitamin contents of foodstuffs in food composition tables will vary depending on the analytical method used to assess the vitamin content, as well as the sample of foods chosen for the analysis. It is important to appreciate that samples of the same or similar foods may vary in composition, and that food table values generally provide an average for use when analysing the diets of groups of people, rather than individuals. One of the major differences between samples of similar foods will be variations in the moisture content, which will strongly influence the proportion of all nutrients present. Where vitamins are associated with a particular macronutrient in foods, such as fat or protein, variations in the content of these fractions will also affect the vitamin content.

This chapter aims to provide an overview of the nature and concentrations of vitamins naturally occurring in foodstuffs commonly consumed by man. The tables showing vitamin contents of common foodstuffs have been compiled from a number of food tables and published articles from different geographical regions, and aim to represent 'typical' values for the foods specified. Those references most commonly consulted for all tables were: FAO (1990); Holland *et al.* (1988, 1989, 1991); Paul and Southgate (1978); Pennington (1989); Posati and Orr (1976); Scherz and Kloos (1981); West *et al.* (1988); Wills (1987). Additional material used for specific vitamins is given for each table as appropriate.

2.1 Fat soluble vitamins and carotenoids

The vitamins generally classified as the fat soluble vitamins are vitamins A, D, E and K and the carotenoids, a group of organic substances widely distributed in fruits and vegetables, a number of which act as precursors for vitamin A. The vitamin activity of all the vitamins in this group comes from more than one biologically active component, and values given in food composition tables generally reflect the overall vitamin activity of all these biologically active components. Since the presence of fat soluble vitamins is associated with the fat component of foods, the macronutrient composition of foods will also influence the relative proportions of fat soluble vitamins. Where the fat content of animals is seasonally variable this will also exert a seasonal effect on the fat soluble vitamin content.

As with other vitamins, recent analytical advances have provided a more accurate assessment of the biological activity of some of the fat soluble vitamins in foods. The development of high performance liquid chromatographic (HPLC) techniques for identifying biologically active components has enhanced our knowledge of the food contents of these vitamins in particular. HPLC methods are generally more sensitive, and have increased specificity for the different biologically active vitamers, allowing a more detailed understanding of the true 'vitamin' content of foodstuffs.

2.1.1 Vitamin A and the carotenoids

Vitamin A can be characterised as a long chain alcohol, which exists in a number of isometric forms in foods. The vitamin A activity of foods is due to a number of related compounds which differ in biological activity and which fall into two basic groups:

(i) preformed vitamin A or retinoids, which include retinol isomers and esters, retinaldehyde and dehydroretinol.
(ii) those carotenoids that can be converted by the body into retinol, the most important of which is β-carotene.

Most of the naturally occurring retinol in foods is in the form of esters, which are hydrolysed in the gut to the corresponding alcohol. *Cis* and *trans* retinol esters are both found as naturally occurring retinoids, with all-*trans* retinol as the predominant form in most foods, and 13-*cis* retinol as the most common of the isomers. 13-*cis* retinol has about 75% of the activity of all-*trans* retinol. *Cis* isomers occur in foods in varying amounts, with fish liver oils and eggs containing as much as 35 and 20%, respectively, of their total retinol in this form. Retinaldehyde or retinal is found in fish roes and hen's eggs, which contain about 90 and 10% of their vitamin A activity in this form. Dietary retinaldehyde has about 90% of the activity of all-*trans* retinol. Dehydroretinol is the predominant form of vitamin A in the livers and flesh of freshwater fish, and has about 40% of the biological activity of all-*trans* retinol.

β-carotene is the main, and sometimes only, source of vitamin A activity in most vegetables, although there are more than 500 known carotenoids present in foods, 32 of which are thought to have vitamin A activity (Bauernfeind, 1972). Other fractions with known vitamin A activity are α-carotene, δ-carotene, β-cryptoxanthin and β-carotene-5, 6-epoxide, all of which have approximately half

Table 2.1 Biological activity of vitamin A-active compounds found in foods

Compound	Biological activity (%) of all-*trans*, retinol activity
All-*trans* retinol	100
13-*cis* retinol	75
Retinaldehyde	90
Dehydroretinol	40
β-carotene	16.7
α-carotene	
γ-carotene	
β-cryptoxanthin	8.35
β-carotene-5,6-epoxide	

the biological activity of β-carotene. Some active carotenoids are only found in a specific group of foods, 4-oxo-β-carotene for example, which is thought to have 44–50% of the activity of β-carotene, is found only in some algae, sea-urchins, red sponges, shrimps and crustaceans. Other carotenoids found commonly in foods, but which are inactive as vitamin A precursors, include lutein which is found commonly in dark green vegetables, sweetcorn, potatoes, many fruit, poultry and eggs, lycopene which is found particularly in tomatoes and some other foods such as pink grapefruit, watermelon, apricots and green peppers, and capsanthin which is found in red peppers and paprikas.

The presence of inactive carotenoids has meant that food composition data published in the past has often overestimated the contribution to total vitamin A activity made by some vegetable foods, by failing to separate active and inactive components, or conversely, the total biological activity may have been underestimated by only considering β-carotene content.

Table 2.1 outlines the most important vitamin A-active compounds found in foods, and their biological activity.

2.1.2 Determination of vitamin A in foods

Many of the literature values for the vitamin A activity of foods published before the 1980s, were based on the colorimetric Carr–Price reaction following adsorption chromatography (Carr and Price, 1926), or a partition chromatographic technique (Bell, 1971), followed by ultraviolet spectrophotometry. More recently, the development of high performance liquid chromatography (HPLC) techniques have allowed a more selective determination of the various retinoids (Reynolds and Judd, 1984), and carotenoids (Simpson, 1983) in foods, which suggest that older values often overestimated the total vitamin A activity.

2.1.3 Natural sources of retinoids and carotenoids

Retinoids only occur in animal products, and the best sources are liver, oily fish

Table 2.2 Retinoid content of common foods[a]

Food	Retinol activity (µg / 100 g)
Milk, whole, pasteurised	38 – 55
Milk, semi-skimmed, pasteurised	21 – 25
Milk, skimmed, pasteurised	tr
Cream, double	455 – 600
Butter	674 – 1062
Cheese, cheddar	260 – 440
Cheese, cottage	45
Cheese, Brie, Camembert	240 – 350
Cheese, Roquefort	295 – 310
Eggs, whole	159 – 210
Beef, lean	10
Kidney, pig	63 – 265
Liver, pig[b]	560 – 14 200
calf	8300 – 31 700
Herring	39 – 46
Mackerel	37 – 55
Sardines, canned in oil, fish only	5 – 11
Salmon, Atlantic	7 – 22
Cod roe	35 – 121
Cod liver oil	18 000

[a] Additional references used Sivell *et al.* (1982, 1984).
[b] Liver from animals fed with vitamin A supplemented feed will contain considerably more retinol.

and fish liver oils, dairy products, kidney, offal products and eggs. The vitamin A content, expressed as total retinol equivalents, of some common foods are given in Table 2.2.

Carotenoids are found in plants, particularly in orange, green and yellow vegetables, and some fruits. Intact carotenoids can also be deposited in the organs of animals, such as in the yolk of eggs or in liver, and β-carotene is also found in milk, cream, cheese, butter and margarine, with a high concentration in unrefined palm oil. The carotenoid content of vegetables will vary depending particularly on the variety and the maturity of the vegetable. The dark green outer leaves of a cabbage will have a higher carotenoid content than the pale green heart, for example, and the number of outer leaves discarded will influence the carotenoid content of the cooked cabbage portion. Foods such as sweet potato can vary in colour from white to deep orange depending on the variety, and the total carotenoid content can range from less than 100 µg, to more than 16 000 µg per 100 g (Holland *et al.*, 1991). Similarly, lettuce varieties with darker green leaves will contain substantially more carotenoids than paler varieties and white cassava will contain no carotenoids while yellow varieties will contain approximately 760 µg / 100 g (Holland *et al.*, 1991). In ripening fruit, the decrease in chlorophylls is often accompanied by an increase in carotenoid content making mature fruit better sources. A study of carrot varieties in the US, showed variations in the total carotene content from 6300–58 000 µg / 100 g depending on the depth of colour develop-

Table 2.3 Contents of carotenoids in common foods (µg / 100 g[a])

Food	α-carotene	β-carotene	α-crypto-xanthin	β-carotene equivalent	Retinol equivalent
Milk, whole	12	-	-	12	2
Cream, double	238	-	-	238	40
Cheese, cheddar	126	-	-	126	21
Oil, palm	-	500	-	500	83
Broad beans	9	165	0	170	28
Green beans	39	310	0	315	53
Peas	19	290	0	300	50
Asparagus	10	310	0	315	53
Beetroot	20	10	0	20	3
Broccoli	0	675	15	685	114
Carrot, old	2425	6905	0	8115	1355
Carrot, new	1765	4355	190	5330	890
Curly kale	0	3130	32	3145	525
Pepper, green	9	260	0	265	44
red	135	3165	1220	3840	640
yellow	55	135	0	185	31
Pumpkin	29	940	0	955	160
Spinach	0	3515	35	3535	589
Sweetcorn	Tr	19	155	97	16
Tomato	0	620	35	640	105
Apple	0	12	0	12	2
Apricot	0	560	0	560	93
Pear	0	17	0	17	3
Orange	19	38	0	48	8
Grapefruit	0	2	3	4	Tr
Peach	0	86	51	115	19
Banana	12	14	0	20	3
Kiwi	0	43	4	45	8
Grapes	0	33	0	33	6
Plum	0	430	0	430	72
Strawberry	0	9	0	9	2

[a]Additional references used Bauernfeind (1972); Wills *et al.* (1987); Heinonen *et al.* (1989); Sivell *et al.* (1984).

ment (Simon and Wolff, 1987). Studies in the UK suggest a range of 4300–11 000 µg/100 g amongst carrots typically grown there (Holland *et al.*, 1991). The carotene content of butters both produced and imported to the UK show variations depending on the season, with β-carotene contents varying from 174 to 1000 µg/100 g, and generally lower values in summer than at other times of the year (Buss *et al.*, 1984).

Table 2.3 shows the carotenoid content of some common foodstuffs, expressed as both µg of carotenoids per 100 g, and as µg of retinol equivalents.

2.1.4 Vitamin D

Vitamin D is present in nature in several forms, occurring predominantly in animal products. Vitamin D_2 or ergocalciferol is derived by ultraviolet (UV) irradiation of the ergosterol which is widely distributed in plants and fungi. Vitamin D_3 or cholecalciferol is derived similarly from UV irradiation of 7-dehydrocholesterol found in the skin of animals, including man. Although cholecalciferol is widely distributed among animals, there are few useful dietary sources commonly consumed. The highest concentration occurs in fish oils, and fatty fish and eggs can be useful sources, however, fortified foods often provide the majority of vitamin D in western diets. Plants are extremely poor sources, fruit and nuts contain no vitamin D, and grains and vegetable oils contain only negligible amounts of ergocalciferol. If cholecalciferol is present in plants it is at levels which make detection difficult, but it has been reported in green leafy vegetables (Lawson, 1985).

The low concentrations have in the past made determination of the vitamin D contents of food unreliable, but recent developments in HPLC techniques have allowed a greater precision (Sivell *et al.*, 1982).

Total vitamin D activity in eggs is the sum of cholecalciferol and any 25-hydroxycholecalciferol or other metabolites which may be present. Dietary supplements given to hens can increase the vitamin D content of eggs considerably. Levels of vitamin D in battery eggs in Britain have been shown to vary from 0.5–2.1 µg / 100 g depending on the season, with the lowest values in June, and the highest values in July and August. Since the breed and age of the bird, rate of laying, size of egg and relative proportion of yolk will affect the vitamin D content, it is likely that some combination of these factors may have been involved in influencing the differences in the summer months (Sivell *et al.*, 1982). Exposure of free range hens to natural light can also affect the vitamin D content of their eggs (Devaney *et al.*, 1933).

Table 2.4 outlines the vitamin D content of some common foods.

Table 2.4 Common sources of vitamin D[a]

Food	Vitamin D (µg / 100 g)
Milk, whole, pasteurised	0.03
Milk, semi-skimmed, pasteurised	0.01
Milk, skimmed, pasteurised	0
Cream, double	0.27
Butter	0.8
Cheese, cheddar	0.3
Cheese, cottage	0.03
Cheese, Brie, Camembert	0.2
Cheese, Roquefort	0
Eggs[b]	1.5 (0.5–2.1)
Herring	7.5–42.5
Mackerel	2.5–25
Salmon, canned	5–20
Sardines, canned in oil, fish only	6
Cod liver oil	210
Herring liver oil	3180

[a]Additional references used Sivell *et al.* (1982); Lawson (1985).
[b]Eggs from hens fed a vitamin D supplement will have a considerably higher vitamin D content.

2.1.5 Vitamin E

The total vitamin E activity of foods is primarily due to the presence of α-tocopherol, although other biologically active tocopherols have been defined. α-Tocopherol has the greatest biologically activity of all the tocopherols, with β-tocopherol having an activity of 40%, and α-tocotrienol having an activity of about 30% of α-tocopherol. γ-Tocopherol has about 10% of the activity of the α form, and the other forms have activities of 5% or less. Food tables often give values only for the α-tocopherol content of foods, highlighting the presence of other biologically active compounds only if they make a significant contribution to total vitamin E, or where data is available. Vegetable oils are generally the richest source of vitamin E, and wheatgerm oil has the highest content of α-tocopherol, containing 0.85–1.25 mg / g. Some oils such as palm oil contain significant amounts of α- and γ-tocotrienol as well as α-tocopherol, although some vegetable oils are poor sources of vitamin E, coconut and linseed oils for example.

Table 2.5 outlines the tocopherol contents of a number of different vegetable oils.

α-Tocopherol is the primary source of vitamin E in dairy products, although marked seasonal variations occur. During the periods of maximal fresh grass consumption, the α-tocopherol content of cow's milk can rise to 0.12 mg / 100 g milk, compared to an average content of 0.07–0.09 mg / 100 g (Holland *et al.*, 1989; McClaughlin and Weihrauch, 1979). Pasteurisation results in a reduction in the tocopherol content of milk, as does skimming. The α-tocopherol content of margarine will be highly variable depending on the oils used in manufacture, and typically ranges from 10 to 75 mg / 100 g (McClaughlin and Weihrauch, 1979). Cereal grains and nuts are good sources of tocopherol, and vegetables and fruit are frequently good sources of tocopherols although there appears to be a wide variability in both the type and amount of tocopherols and tocotrienols present (Bauernfeind, 1980).

Table 2.5 Tocopherol content of refined vegetable oils, (mg / 100 g oil[a])

Oil	α-T	β-T	γ-T	δ-T	α-T-3	β-T-3	γ-T-3	Vit. E activity[b]
Coconut	0.35	0	0.17	0.35	1.29	0.10	1.32	0.78
Corn	14.26	0.38	64.9	2.75	0.58	0	0	21.1
Cotton seed	35.26	0	29.98	0	0	0	0	38.26
Olive	11.92	0	0.72	0	0	0	0	11.99
Palm	18.32	0	0	0	11.46	0	5.75	21.82
Peanut	11.62	0	12.98	0.33	0	0	0	12.92
Rapeseed	17.65	0	27.04	0.04	0	0	0	20.35
Safflower seed	34.05	0	3.5	0.49	0	0	0	34.4
Soyabean	10.99	0	62.4	20.4	0	0	0	17.43
Sunflower seed	59.5	0	3.54	0	0	0	0	59.85
Wheatgerm	149.44	81.19	0	0	0	0	0	183.0

α-T = α-tocopherol; β-T = β-tocopherol; γ-T = γ-tocopherol; δ-T = δ-tocopherol; α-T-3 = α-tocotrienol; β-T-3 = β-tocotrienol; γ-T-3 = γ-tocotrienol.
[a] Values taken from McClaughlin and Weihrauch (1979).
[b] Total vitamin E activity = α-tocopherol + 0.4 (β-tocopherol) + 0.1 (γ-tocopherol) + 0.01 (δ-tocopherol) + 0.3 (α-tocotrienol) + 0.05 (β-tocotrienol) + 0.01 (γ-tocotrienol).

Non-α-tocopherols are situated mainly in non-green tissues such as nuts, fungi and cereal grains, and tocotrienols have been isolated in carrots, kale, broccoli and mushrooms (Piironen *et al.*, 1986). Tocopherols are more concentrated in the leaves of plants than the roots, and more prevalent in dark green mature leaves than pale or immature leaves, with the lowest amount in the roots and in colourless fruits. The vitamin E content of plants will also vary depending on the growing conditions, such as the intensity of sunlight and soil state. The vitamin E content of meats and offals is low, although the α-tocopherol content does exhibit seasonal variation in all meats except pork, due to seasonal feeding differences.

Table 2.6 outlines the α-tocopherol content of some common foodstuffs.

Table 2.6 α-Tocopherol content of some common foods (mg / 100 g[a])

Food	α-tocopherol
Milk, whole, pasteurised	0.09
Milk, semi-skimmed, pasteurised	0.03
Milk, skimmed, pasteurised	0
Double cream	1.1
Butter	0.5–5.0
Cheese, cheddar	0.6
Eggs	1.1
Beef, average	0.2–0.5
Pork, average	0.1–0.5
Chicken	0.3
Cod	0.2
Herring	0.2–1.1
Prawn	0.9–2.9
Broad bean	0.05–0.5
Green bean	0.02–0.2
Peas	0.13–0.21
Asparagus	1.5
Brussels sprouts	0.9
Carrot	0.5
Lettuce	0.5
Parsley	1.7–3.6
Potato	0.06
Spinach	1.75
Watercress	1.3
Apple	0.4
Banana	0.25
Mango	1.1
Orange	0.3
Wheat, wholegrain	0.58–5.2
Oats, wholegrain	1.1
Rice, polished	0.11
Almonds	23–96
Brazil nuts	6.5
Peanuts	8.3
Sunflower seeds	50

[a] Additional references used are Bauernfeind (1980); McClaughlin and Weihrauch (1979); Piironen *et al.* (1985, 1986).

2.1.6 Vitamin K

Vitamin K is the term used to describe a group of quinone compounds, all of which are related to 2-methyl-1,4-napthoquinone, and these compounds are widely distributed in nature. The form of the vitamin isolated from plants is generally called phylloquinone, and those forms synthesised by bacteria and in animal products are members of the menaquinone series. Relatively few values for the vitamin content of foods are available since dietary deficiency is not observed, and values for the vitamin K content of foods do not generally appear in all food tables. The determination of the vitamin K content of foods has classically been determined by a chick bioassay, but the presence in foods of interfering substances, and the very small concentrations of the vitamin present make sensitive analyses difficult. For this reason many of the values quoted for the vitamin K content of foods should be viewed with caution, and at best it is possible only to suggest ranges of contents in foods with confidence. HPLC assays of phylloquinone have mostly been limited to vegetables (Parrish, 1980) and milks (Haroon *et al.*, 1982). Menaquinones

Table 2.7 Vitamin K content of some common foods (μg / 100 g[a])

Food	Vitamin K
Milk, whole, pasteurised	4
Milk, skimmed, pasteurised	4
Butter	10–50
Cheese, average	10–50
Egg, whole	50
Egg, yolk	147
Corn oil	60
Soybean oil	540
Meat, average	4
Liver, average	80–104
Wheat bran	83
Flour, wholemeal	30
Wheatgerm	39
Oats	63
Chick peas	264
Green beans	28
Lentils	223
Mung beans	170
Peas	81
Soya beans	190
Potato	16
Broccoli	132
Cabbage, green	149
Carrot	13
Seaweed, dried	17 000
Spinach	266
Tomato	23
Watercress	57
Tea, green, dry	712

[a] References used Parrish (1980); Haroon *et al.* (1982); USDA (1986).

produced by the intestinal bacteria may also be important in animals, since recent HPLC analyses have suggested that menaquinones are stored in the liver by man (Department of Health, 1991)

In general, seaweeds, green and leafy vegetables are the best sources of the vitamin, although there are significant changes in the vitamin K content of plants during growth and maturity. Spinach, kale, broccoli and cabbage appear to be particularly good sources, and vitamin K appears to be associated with chlorophyll in plants. Soya beans and soybean oil are better sources than other grains and oils, and beef liver has been found to contain substantially more vitamin K than the livers of other animals, and other organs (Parrish, 1980).

Table 2.7 suggests ranges of values for the vitamin K content of common foods.

2.2 Water soluble vitamins

The vitamins commonly considered in this group are thiamin, riboflavin, niacin, vitamin B_6, vitamin B_{12}, folates, pantothenic acid, biotin and vitamin C. Again, many of these vitamins occur naturally in more than one biologically active form, and since these vitamins are water soluble, they are generally more labile than the fat soluble vitamins. Most vitamins in this group, with the exception of vitamin B_{12}, are widely distributed in both animal and vegetable foods, although the amounts present are often very small. Again, the development of more sensitive and specific HPLC analysis techniques has improved the determination of many of the vitamins in this group. Values from more conventional analytical methods, which generally appear in food tables, are often in good agreement with values determined by HPLC, although HPLC provides a more efficient analysis in most cases.

2.2.1 Thiamin

Thiamin is present in practically all plant and animal tissues, however the content in many foods is small, and food preparation may result in considerable losses. In most animal products, thiamin occurs in a phosphorylated form (thiamin mono-, di- and triphosphates), with 80–85% as the diphosphate, the active coenzyme form. In plant products, thiamin occurs predominantly in the non-phosphorylated form. Assay methods require a suitable extraction procedure to release the thiamin from the coenzyme in animal products. Foods rich in thiamin include dried brewer's or baker's yeast (the former being the highest known source), pork, particularly pig's liver, cereal germs, whole grains and wholegrain products, nuts and dried legumes. Thiamin is not found in oils and fats, and is present at only low levels in green vegetables, fruit and seafood. In cereal grains the thiamin is unevenly distributed, being low in the endosperm and high in the germ and scutellum (the thin layer between the germ and endosperm), thus leading to substantial losses from grain products on milling. The rinds of cheese such as Brie and Camembert are good sources of thiamin, generally containing about 0.4 mg / 100 g. Most methods of analysis are based on the alkaline oxidation of thiamin by ferricyanide which produces thiochrome, measured by its intense blue fluorescence. Microbiological assays are, however, simple and inexpensive, and values from both methods are well accepted. More recently HPLC determination of thiamin has been developed, and recent analyses of vegetables in the UK by both microbiological and HPLC methods, gave markedly higher values when HPLC was used (Kwiatkowska et al., 1989), although other reports suggest that the increased sensitivity of HPLC techniques give generally lower values for other foods (Skurray, 1981).

Table 2.8 shows the thiamin contents of common foods.

Table 2.8 Thiamin content of common foods (mg / 100 g[a])

Food	Thiamin
Milk, whole, pasteurised	0.04
Milk, semi-skimmed, pasteurised	0.04
Milk, skimmed, pasteurised	0.04
Sterilised whole milk	0.03
Cheese, cheddar	0.03
Cheese, Brie, Camembert	0.04[b]
Yoghurt, low fat	0.05
Eggs, whole	0.09
Beef, average	0.04–0.09
Lamb, average	0.09–0.15
Pork, average	0.58–0.89
Chicken	0.1
Liver, pigs	0.31
Kidney, pigs	0.32
Cod	0.08
Herring	0.13
Flour, wholemeal	0.47
Flour, brown[c]	0.3
Flour white[c]	0.1
Wheatgerm	2.01
Rice, brown	0.59
Rice, polished	0.08
Oatmeal	0.5
Buckwheat	0.28
Bulgur wheat	0.48
Broad beans	0.04
Butter beans, haricot beans	0.45
Lentils, red split	0.5
Red kidney beans	0.65
Peas	0.74
Potato	0.21
Lettuce	0.12
Onion	0.13
Spinach	0.07
Sweetcorn	0.2
Yeast, baker's, dried	2.3
Yeast, brewer's, dried	15.6

[a] Additional references used Skurray (1981); Ang and Moseley (1980); Fellman et al. (1982); Finglas and Faulks (1984); Scott and Bishop (1986, 1988).
[b] The rind of Brie and Camembert contain 0.4 mg thiamin/100 g. [c] These values are for unfortified flours.

Table 2.9 Riboflavin content of some common foods (mg / 100 g[a])

Food	Riboflavin
Milk, whole, pasteurised	0.17
Milk, semi-skimmed, pasteurised	0.18
Milk, skimmed, pasteurised	0.18
Milk, sterilised	0.15
Double cream	0.16
Cheese, cheddar	0.38–0.52
Cheese, cottage	0.17–0.26
Cheese, Brie, Camembert	0.34–0.67
Cheese, Roquefort	0.59–0.65
Yoghurt, low fat	0.21–0.25
Egg, whole	0.47
Beef, average	0.22
Lamb, average	0.25
Pork, average	0.23
Chicken	0.2
Kidney, average	2.0
Liver, average	3.1
Cod	0.02–0.16
Herring	0.09–0.33
Salmon	0.06–0.22
Sardines, canned in oil, fish only	0.36
Shrimps	0.03
Cod roe	1.0
Wheat bran	0.36
Wheatgerm	0.72
Oatmeal	0.1
Flour, wholemeal	0.09
Flour, white	0.03
Rice, brown	0.07
Rice, white	0.02
Broad beans	0.04
Butter beans, haricot beans	0.13
Chick peas	0.24
Lentils, red split	0.2
Red kidney beans	0.19
Peas	0.02
Potato	0.04–0.19
Asparagus	0.06
Broccoli	0.06
Brussels sprouts	0.11
Carrots	0.01
Cassava	0.04
Mushrooms	0.09
Plantain	0.06
Seaweed, Irish moss	0.47
Spinach	0.09
Sweetcorn	0.05
Yam	0.01
Apple	0.02
Avocado	0.15
Banana	0.06
Melon	0.02
Orange	0.03
Peach	0.05
Strawberry	0.05
Raisins	0.08
Almonds	0.92
Hazelnuts	0.18
Peanuts	0.10
Yeast, baker's dried	4.0

[a] Additional references used Skurray (1981); Ang and Moseley (1980); Fellman *et al.* (1982); Scott and Bishop (1986, 1980).

2.2.2 Riboflavin

Riboflavin is one of the most widely distributed of the vitamins, and all plant and animal cells contain it, although there are few rich food sources. Riboflavin is present in foods in two bound forms, riboflavin mononucleotide and flavin adenine dinucleotide, which must be hydrolysed before analysis. Plants and many bacteria and fungi synthesise riboflavin, and it also occurs in good amounts in dairy products. The best sources of riboflavin are yeast, milk, egg white, fish roe, kidney, liver, heart and leafy vegetables. Grains and legumes contain riboflavin but are not particularly good sources, although there is an increase in riboflavin content during germination. In green vegetables the leafy portions and growing parts contain most riboflavin, but as the leaves get older and drier the riboflavin content decreases. Milk from cows fed on young grass contains more riboflavin than those fed on dried grass or root crops, so that riboflavin in milk is generally higher in summer.

The riboflavin content of foods is determined by fluorimetric or microbiological assay, following acid or enzymic hydrolysis to release riboflavin from the bound forms present in foods, and values from both assay methods are well accepted. HPLC assays of riboflavin, when compared with microbiological determinations, have shown HPLC to give somewhat lower values for similar foods (Kwiatkowska et al., 1989; Skurray, 1981).

Riboflavin is sensitive to light and milk exposed to light in summer can lose 70% of its riboflavin content in full sunshine within two hours (Scott et al., 1984). Even when exposed to low light intensity, 30% loss is likely in 24 hours, although hardly any losses occur if food is stored in the dark.

Table 2.9 outlines the riboflavin content of common foods.

2.2.3 Niacin

The generic term 'niacin' describes the content of both nicotinic acid and nicotinamide in foods, although the term nicotinic acid is still commonly used to describe the activity of both these compounds, and niacin is sometimes used to describe nicotinamide alone. Both forms of niacin are equally active as vitamins, and both are present in a variety of foodstuffs. Nicotinic acid is often present in foods, particularly in maize, in a bound form 'niacytin', bound to polysaccharides and peptides in the outer layers of cereal grains, which makes it unavailable to man (Mason et al., 1973). Studies have shown that in immature corn, the niacin appears predominantly unbound and is biologically available, but as the corn matures, bound niacin predominates (Wall et al., 1987).

Nicotinic acid and nicotinamide are present in most foods apart from fatty substances, and are particularly rich in meat, fish, wheat and wholegrain cereals. Coffee beans contain significant quantities of niacin, which increase on roasting.

Assay methods generally liberate nicotinic acid from bound forms by acid or enzymic hydrolysis, and values given are based on the determination of both nicotinic acid and nicotinamide in foods. In most cases no allowance is made for the unavailability of nicotinic acid from some cereal foods, and food table values may therefore overestimate the contribution made by cereal foods.

Table 2.10 gives the niacin content of some common foods.

Table 2.10 Niacin content of common foods (mg / 100 g[a])

Food	Niacin
Milk, whole, pasteurised	0.08
Milk, semi-skimmed, pasteurised	0.09
Milk, skimmed, pasteurised	0.09
Milk, whole, sterilised	0.09
Soya milk	0.11
Double cream	0.04
Yoghurt, low fat	0.12–0.19
Cheese, cheddar	0.04–0.11
Cheese, cottage	0.14–0.26
Cheese, Brie, Camembert	0.48–1.57
Cheese, Roquefort	0.57–0.74
Egg, whole	0.07
Beef, average	4.5
Pork, average	5.2
Lamb, average	5.0
Chicken, meat only	7.8
Kidney, average	8.0
Liver, average	14.0
Cod	1.7
Herring	2.0–6.0
Wheat bran	29.6
Bulgur wheat	4.5
Oatmeal	1.0
Flour, wholemeal	5.7
Flour, white	0.7
Broad beans	3.2
Butter beans, haricot beans	2.5
Lentils, red, split	2.0
Soya beans	2.2
Potato, average	0.4–5.0
Asparagus	1.0
Broccoli	0.9
Brussels sprouts	0.2
Carrots	0.2
Lettuce	0.3–0.6
Mushrooms	3.2
Plantain	0.7
Seaweed, Irish moss	0.6
Spinach	1.2
Sweet potato	0.5
Sweetcorn	1.9
Tomato	1.0
Yam	0.2
Yeast, baker's, dried	36
Coffee, ground, roast	10–40

[a] Additional references used Skurray (1981); Scott and Bishop (1986, 1988).

Table 2.11 Vitamin B_6 content of common foods (mg / 100 g[a])

Food	Vitamin B_6
Milk, whole, pasteurised	0.06
Milk, semi-skimmed, pasteurised	0.06
Milk, skimmed, pasteurised	0.06
Milk, sterilised	0.04
Soya milk	0.07
Double cream	0.03
Yoghurt, low fat	0.08
Cheese, cheddar	0.06–0.13
Cheese, cottage	0.08
Cheese, Brie, Camembert	0.15–0.28
Cheese, Roquefort	0.1
Eggs	0.12
Beef, average	0.25
Lamb, average	0.2
Pork, average	0.4
Chicken, meat only	0.4
Kidney, average	0.3
Liver, average	0.5
Cod	0.3
Herring	0.5
Shrimps	0.1
Cod roe	0.32
Wheat bran	1.38
Oatmeal	0.12
Flour, wholemeal	0.5
Flour, white	0.15
Rice, brown	0.55
Rice, white	0.17–0.3
Broad beans	0.06
Butter beans, haricot beans	0.53
Chick peas	0.53
Green beans	0.05
Lentils, red, split	0.6
Red kidney beans	0.4
Soya beans	0.38
Potato, average	0.4
Asparagus	0.09
Broccoli	0.14
Brussels sprouts	0.37
Carrot	0.14
Lettuce	0.03–0.08
Mushrooms	0.18
Spinach	0.17
Sweetcorn	0.15
Yam	0.16
Apple	0.03
Banana	0.51
Melon	0.04–0.07
Orange	0.06
Raisins	0.3
Yeast, baker's, dried	2.0

[a] Additional references used Scott and Bishop (1986, 1988).

Table 2.12 Vitamin B_{12} content of common foods (μg / 100 g[a])

Food	Vitamin B_{12}
Milk, whole, pasteurised	0.4
Milk, semi-skimmed, pasteurised	0.4
Milk, skimmed, pasteurised	0.4
Milk, whole, sterilised	0.1
Double cream	0.2
Yoghurt, low fat	0.2
Cheese, cheddar	0.8–1.5
Cheese, cottage	0.7
Cheese, Brie, Camembert	1.1–3.1
Cheese, Roquefort	0.5
Egg, whole	2.5
Egg, yolk	6.9
Beef, average	1–2
Lamb, average	1–2
Pork, average	1–3
Chicken	tr
Kidney, average	30–50
Liver, average	50–100
Brain, calves, lambs	9
Cod	1–2
Herring	6
Mackerel	10
Sardines, canned in oil, fish only	28
Cod roe	10
Oysters	15
Seaweed, dried	2.5–47
Miso, fermented soybean paste	0.2
Tempeh, fermented soybean curd	0.1[b]
Marmite, yeast extract	0.5

[a] Additional references used Scott and Bishop (1986, 1988).
[b] Vitamin B_{12} may be as high as 1.6 µg per 100 g with bacterial contamination.

2.2.4 Vitamin B_6

Vitamin B_6 is the name given to the biologically active derivatives of 3-hydroxy-2-methylpyridine, and the three naturally occurring forms, pyridoxal, pyridoxine and pyridoxamine, and their 5'-phosphates which are all interconvertible. Pyridoxine and pyridoxamine are the forms generally found in animal products, and pyridoxal that in plant products, and all forms have been demonstrated to be equiactive in rats. The vitamin is widely distributed in foods, and meats, whole grain cereals and nuts are good sources of vitamin B_6, as are cheeses with extensive mould growth, since this vitamin can be synthesised in relatively large amounts by bacteria. Assay of total vitamin B_6 activity requires a preliminary acid extraction procedure, and values are usually obtained by microbiological assay, although this is a lengthy procedure complicated by the existence of the related forms in complex organic matrices. HPLC methods allow the examination of different vitamin B_6 vitamers, and allows greater precision.

Table 2.11 gives the vitamin B_6 contents of common foods.

2.2.5 Vitamin B_{12}

Compounds with vitamin B_{12} activity consist of a corrinoid ring surrounding an atom of cobalt and vitamin B_{12} is often referred to as cyanocobalamin, although the naturally occurring forms are methyl-, adenosyl-, hydroxo- and aquo-cobalamins. Adenosyl- and hydroxo-cobalamin are the most common forms in food, although methylcobalamin is found in egg yolk and cheese. Almost all animal products contain vitamin B_{12} and it can also be synthesised by certain algae (e.g. seaweeds) and bacteria. The vitamin is almost entirely absent from higher plants, although compounds with vitamin B_{12} activity have been isolated from peas, potato and bean plants. The amounts found in foods are generally small, with offal, especially liver, kidney and heart, meats and fatty fish being the best sources. Fermented products such as miso and tempeh made from soya beans are also non-animal sources of this vitamin, and since bacteria can synthesise vitamin B_{12}, 'spoiled' food can also be a source of vitamin B_{12}. Microbiological assay is generally the preferred method of determination in foods.

Table 2.12 gives the content of vitamin B_{12} in foods.

2.2.6 Pantothenic acid

First isolated as a yeast growth factor, pantothenic acid is part of the coenzyme A molecule, and is widely distributed in plants and animals. In foods it appears either in the free form or bound as a larger coenzyme molecule, and enzymic hydrolysis is required prior to microbial assay. Rich sources of pantothenic acid are yeast, heart, brain, liver, avocado, meat, eggs and some green vegetables. Although pantothenic acid is present in all living matter, 'refined' foods such as sugar, fats and oils, sago, tapioca and cornstarch do not contain any.

Table 2.13 gives the pantothenic acid contents of foods.

2.2.7 Biotin

Biotin appears in nature as eight stereoisomers of a cyclic derivative of urea, but only one isomer, d-biotin, is biologically active. Yeasts or bacteria of many species either make or retain biotin, and in addition biotin is widely distributed in small concentrations in all animal and plant tissues, liver and kidney being particularly good sources. Yeast, pulses, nuts, whole grain cereals, eggs, and some vegetables are also good sources, and other meats, dairy produce and fruit are relatively poor sources. Analyses are usually by microbiological assay.

Table 2.14 gives the biotin contents of common foods.

2.2.8 Folic acid

Folic acid or pteroylglutamic acid is the parent molecule for a large number of derivatives known as folates or folacin, but the stable pteroylglutamic acid itself

Table 2.13 Pantothenic acid content of common foods (mg / 100 g^a)

Food	Pantothenic acid
Milk, whole, pasteurised	0.35
Milk, semi-skimmed, pasteurised	0.32
Milk, skimmed, pasteurised	0.32
Double cream	0.19
Low fat yoghurt	0.45
Cheese, cheddar	0.3–0.5
Cheese, cottage	0.2–0.4
Cheese, Brie, Camembert	0.36–1.4
Cheese, Roquefort	0.5–1.73
Egg, whole	1.77
Egg, yolk	4.6
Beef, lamb, average	0.6
Pork, average	0.7–1.1
Chicken	1.2
Kidney, average	3–3.85
Liver, average	8
Heart, lamb's	2.5
Brain, calve's, lamb's	2.0–2.6
Wheat bran	2.4–2.9
Flour, wholemeal	0.8
Flour, white	0.3
Rice, brown	1.1
Rice, polished	0.4
Oatmeal	1.0
Wheatgerm	1.2–1.9
Broad beans	4.9
Butter beans	1.3
Soya beans	0.8–1.7
Potato	0.4
Broccoli	1.2
Cabbage	0.2
Carrot	0.25
Mushroom	2.0
Sweetcorn	0.7
Yam	0.3
Avocado	1.1
Dates, dried	0.8
Watermelon	1.6
Almonds	0.5
Hazelnuts	1.2
Peanuts	2.7
Yeast, baker's, dried	11.0

^{a} Additional references used Walsh *et al.* (1981); Scott and Bishop (1986, 1988).

does not appear naturally. The main naturally occurring folates are tetrahydrofolate (THF), 5-methyltetrahydrofolate (5-MeTHF) and 10-formyltetrahydrofolates. Polyglutamates based mainly on 5-MeTHF predominate in fresh food, but on storage these slowly break down to monoglutamates and oxidise to less available folates. Many estimates of the folate contents of foods determined by microbial

Table 2.14 Biotin content of common foods (µg / 100 g[a])

Food	Biotin
Milk, whole, pasteurised	2
Milk, semi-skimmed, pasteurised	2
Milk, skimmed, pasteurised	2
Milk, whole, sterilised	1.8
Double cream	1.1
Yoghurt, low fat	2.9
Cheese, cheddar	1.7–3.6
Cheese, cottage	3
Cheese, Brie, Camembert	2.8–7.6
Cheese, Roquefort	2.3
Egg, whole	11–20
Beef, average	tr
Lamb, average	1–2
Pork, average	1–3
Chicken	2
Kidney, average	24–37
Liver, average	27–41
Liver, chicken	210
Cod	3
Herring	10
Wheat bran	14–45
Flour, wholemeal	7–9
Flour, white	1
Wheatgerm	25
Broad beans	3
Soya beans	65
Tempeh, fermented soya curd	53
Potato	tr
Green beans	1
Cauliflower	2
Mushrooms	12
Quorn, myco-protein	9
Spinach	tr
Tomato	2
Apple	1
Currant, black	2
Orange	1–2
Almonds	18–23
Chestnuts	1–2
Walnuts	2–37
Yeast, baker's, dried	85–200

[a] Additional references used Scott and Bishop (1986, 1988).

assay are likely to be unreliable, and the development of HPLC assessment techniques is likely to improve estimations. The availability of the different forms and conjugates of folic acid are still poorly understood, and many values reported in the literature or in food tables should be viewed with caution. Liver is the richest source of folates, in which most is present as the more available 5-MeTHF, but most other foods contain predominantly polyglutamates. Relatively speaking, most food tables show that apart from liver, yeast extract, wheat bran and wheatgerm, egg yolk, some cheeses and green leafy vegetables are rich sources of folates.

Table 2.15 gives ranges of folate content of foods.

Table 2.15 Folate content of common foods (μg / 100 g[a])

Food	Folates
Milk, whole, pasteurised	6
Milk, semi-skimmed, pasteurised	6
Milk, skimmed, pasteurised	5
Milk, whole, sterilised	0.2
Milk, soya	17
Yoghurt, low fat	17
Soybean oil	450–630
Cheese, cheddar	16–42
Cheese, cottage	12–27
Cheese, Brie, Camembert	56–100
Cheese, Roquefort	45–49
Egg, whole	50
Egg, yolk	130
Beef, average	tr
Lamb, average	1–2
Pork, average	1–3
Chicken	2–4
Kidney, average	24–37
Liver, average	27–210
Herring	10
Wheat bran	80–260
Flour, wholemeal	57
Flour, white	22
Rice, brown	49
Rice, polished	20
Oatmeal	60
Wheatgerm	330–370
Broad beans	145
Lentils, red, split	35
Miso, fermented soy bean paste	33
Potato	20–35
Green beans	14–80
Peas	10–60
Asparagus	57–175
Broccoli	65–200
Brussels sprouts	135
Carrots	10–20
Cassava	24
Lettuce	55
Spinach	130–330
Sweetcorn	41
Tomatoes	11–24
Watercress	120
Yeast, baker's, dried	4000

[a] Additional references used Scott and Bishop (1986, 1988).

2.2.9 Vitamin C

Vitamin C or L-ascorbic acid in foods is easily oxidised to dehydro-L-ascorbic acid, and both forms are likely to be present in equilibrium in foods. In fresh foods the reduced form is the major one present, but cooking, processing and storage increase the proportions of the dehydro form. Vitamin C is widely distributed in nature, and occurs in significant quantities in vegetables and fruit and in animal organs such as liver and kidney. Very small amounts are found in other meats and in milk. Plants rapidly synthesise L-ascorbic acid from carbohydrates, and variations occur in content due to the species of plants, ripeness, place of origin, storage conditions

Table 2.16 Vitamin C content of some common foods (mg / 100 g[a])

Food	Vitamin C
Milk, whole, pasteurised	0.8–1
Milk, semi-skimmed, pasteurised	0.8–1
Milk, skimmed, pasteurised	0.8–1
Milk, whole, sterilised	tr
Kidney, average	7–14
Liver, average	10–23
Brain, calf and lamb	23
Potato, new	16
Potato, old, freshly dug	21
Potato, stored 3 months	9
Potato, stored 9 months	7
Asparagus	12–15
Broad beans	32–41
Green beans	12–32
Broccoli	87–150
Cabbage	21–60
Carrots	5–10
Cassava	31–41
Curly kale	110
Pepper, green	120–200
Spring greens	180
Sweetcorn	6–12
Tomatoes	17–33
Watercress	62–101
Yam	4
Apples	5–30
Banana	10–19
Cherries	5–18
Grapefruit	35–45
Guava	200 (20–600)
Melon	5–34
Orange	40–60
Peach	7–14
Pineapple	17–40
Raspberry	14–35
Rosehips	1000
Strawberries	40–90
Tangerines	30
Parsley, fresh	114–190

[a] References also used Scott and Bishop (1986, 1988).

and handling. The vitamin C content of foods varies from species to species as well as in different samples of the same species, and its labile nature makes determination in foods very variable. Some plants accumulate L-ascorbic acid in the tissues making them particularly good sources, and this is the case for fresh tea leaves, some berries, guava and rosehips. The richest sources in the diet are fruit usually eaten raw, particularly citrus fruits, and the growing points of vegetables. Unsprouted cereals and their products contain practically no vitamin C. Parsley is a good source of vitamin C, and the dried herb also contains substantial amounts, although most other herbs are poor sources.

In the past dehydro-L-ascorbic acid has not always been included in vitamin C analyses, but 2,3-di-oxo-L-gulonic acid formed from the hydrolysis of dehydro-L-ascorbic acid, and which is inactive, can be erroneously included. Previously, vitamin C content was estimated from titration after an oxidation–reduction reaction or by a fluorimetric method, but newer HPLC techniques have improved determination and proved to be very reproducible (Bushway *et al.*, 1989). Substantial losses of vitamin C occur on the storage of vegetables, with 80% of ascorbate reported lost from freshly picked green vegetables after 24 hours market storage in Sri Lanka, for example (Kailasapathy and Koneshan, 1986), and a 66% reduction in the vitamin C content of potatoes after 9 months storage has been reported (Finglas and Faulks, 1984).

Table 2.16 outlines the vitamin C contents of some common foods.

References

Ang, C.Y.W. and Moseley, F.A. (1980) Determination of thiamin and riboflavin in meat and meat products by HPLC. *J. Agric. Food Chem.* **28**, 3, 483–489.

Bauernfeind J.C. (1972) Carotenoid vitamin A precursors and analogs in foods and feeds. *J. Agric. Food Chem.* **20**, 3, 456–473.

Bauernfeind, J.C. (1980) *Vitamin E, a Comprehensive Treatise* (ed. L.M. Machlin). Marcel Dekker, New York, Basel.

Bell, J.G. (1971) Separation of oil-soluble vitamins by partition chromatography on Sephadox LH20. *Chem. Ind. (London)* 201–202.

Bushway, R.J., King, J.M., Perkins, B. and Krishnan, M. (1981) High performance liquid chromatographic determination of ascorbic acid in fruits, vegetables and juices. *J. Liq. Chromatogr.* **11**, 16, 3415–3423.

Buss, D.H., Jackson, P.A. and Scuffam, D. (1984) Composition of butters on sale in Britain. *J. Dairy Res.* **51**, 637–641.

Carr, F.H. and Price, E.A. (1926) Colour reactions attributed to vitamin A. *Biochem. J.* **20**, 497–501.

Department of Health (1991) *Dietary Reference Values for Food Energy and Nutrients for the United Kingdom*, Report of the panel on dietary reference values of the Committee on medical aspects of food policy, HMSO, London.

Devaney, G.M., Munsell, H.E. and Titus, H.W. (1937) The effect of sources of vitamin D on storage of the antirachitic factor in egg. *Poultry Sci.* **2**, 215.

Fellman, K.J., Artz, W.E., Tassinan, P.D., Cole, C.L. and Augustin, J. (1982) Simultaneous determination of thiamin and riboflavin in selected foods by HPLC. *J. Food Sci.* **47**, 2048–2050.

FAO (1990) *Roots, Tubers, Plantains and Bananas in Human Nutrition.* Food and Agriculture Organization of The United Nations, Rome.

Finglas, P.M. and Faulks, R.M. (1984) Nutritional composition of UK retail potatoes, both raw and cooked. *J. Sci. Food Agric.* **35**, 1347–1356.

Haroon, Y., Shearer, S., Rahim, W.G., Gunn, G., McEnery, G. and Barkhan, P. (1982) The content of phylloquinone (vitamin K-1) in human milk, cow's milk and infant formula foods determined by high-performance liquid chromatography. *J. Nutr.* **112(6)**, 1105–1117.

Heinonen, M.I., Ollilainen, V., Linkola, E.K., Varo, P.T. and Koivistoinen, P.E. (1989) Carotenoids in Finnish Foods: Vegetables, fruits and berries. *J. Agric. Food Chem.* **37**, 655–659.

Holland, B., Unwin, I.D. and Buss, D.H. (1988) *Third supplement to McCance and Widdowson's The Composition of Foods: Cereal and Cereal Products.* Royal Society of Chemistry, Letchworth, HMSO, London.

Holland, B., Unwin, I.D. and Buss, D.H. (1989) *Fourth supplement to McCance and Widdowson's The Composition of Foods: Milk Products and Eggs.* Royal Society of Chemistry, Letchworth, HMSO, London.

Holland, B., Unwin, I.D. and Buss, D.H. (1991) *Fifth supplement to McCance and Widdowson's The Composition of Foods: Vegetables, Herbs and Spices.* Royal Society of Chemistry, Letchworth, HMSO, London.

Kailasapathy, K. and Koneshan, T. (1986) Effect of wilting on the ascorbate content of selected fresh green leafy vegetables consumed in Sri Lanka. *J. Agric. Food Chem.* **34**, 259–261.

Lawson, E. (1985) Vitamin D. In: *Fat Soluble Vitamins*, (ed: A.T. Diplock) Heinemann, London, pp. 76–153.

Mason, J.B., Gibson, N. and Kodicek, E. (1973) The chemical nature of the bound nicotinic acid of wheat bran: studies of nicotinic acid containing molecules. *Brit. J. Nutr.* **30**, 297–311.

McClaughlin, P.J. and Weihrauch, J.C. (1979) Vitamin E content of foods. *J. Am. Diet. Assoc.* **75**, 649–665.

Parrish, D.B. (1980) Determination of vitamin K in foods: a review. *Crit. Rev. Food Sci. Nutr.* **13**, 337–352.

Paul, A.A. and Southgate, D.A.T. (1978) *McCance and Widdowson's The Composition of Foods*, 4th edition, HMSO, London.

Pennington, J.A.T. (1989) *Bowes and Church's Food Values of Portions Commonly Consumed*, 15th edition. J. P. Lippincott, Philadelphia.

Piironen, V., Syväoja, E.-L., Varo, P., Salminen, K. and Koivistoinen, P. (1985) Tocopherols and tocotrienols in Finnish foods: meat and meat products. *J. Agric. Food Chem.* **33**, 1215–1218.

Piironen, V., Syväoja, E.-L., Varo, P., Salminen, K. and Koivistoinen, P. (1986) Tocopherol and tocotrienols in Finnish Foods: Vegetables, fruits and berries. *J. Agric. Food Chem.* **34**, 742–746.

Posati, L.P. and Orr, L.M. (1976) *Composition of foods - Dairy and Egg Products* Agricultural handbook No. 8-1, U.S Department of Agriculture, Washington D.C.

Reynolds, S.L. and Judd, H.J. (1984) Rapid procedure for the determination of vitamins A and D in fortified skimmed milk powder using high-performance liquid chromatography. *Analyst* **109**, 489–492.

Scherz, H. and Kloos, G. (1981) *Food Composition Tables 1981/82* (Eds S. W. Souci, W. Fachmann and H. Kraut) Wiss. Verlags Gmbtl. Stuttgart.

Scott, K.J. and Bishop, D.R. (1986) Nutrient content of milk and milk products: Vitamins of the B complex and vitamin C in retail market milk and milk products. *J. Soc. Dairy Technol.* **39**, 32–35.

Scott, K.J. and Bishop, D.R. (1988) Nutrient content of milk and milk products: Vitamins of the B complex and vitamin C in retail cheeses. *J. Sci. Food Agric.* **43**, 193–199.

Simon, P.W. and Wolff, X.Y. (1987) Carotenes in typical and dark orange carrots. *J. Agric. Food Chem.* **35**, 1017–1022.

Simpson, K.L. (1983) Relative values of carotenoids as precursors of vitamin A. *Proc. Nutr. Soc.* **42**, 7.

Sivell, L.M., Wenlock, R.W. and Jackson, P.A. (1982) Determination of vitamin D and retinoid activity in eggs by HPLC. *Human Nutr: Appl. Nutr.* **36A**, 430–437.

Sivell, L.M., Bull, N.L., Buss, D.H., Wiggins, R.A., Scuffam, D. and Jackson, P.A. (1984) Vitamin A acitivity in foods of animal origin. *J. Sci. Food Agric.* **35**, 931–939.

Skurray, G.R. (1981) A rapid method for selectively determining small amounts of niacin, riboflavin and thiamin in foods. *Food Chem.* **7**, 77–80.

U.S Department of Agriculture (1986) *Vitamin K Contents of Foods.* Provisional tables, June 1986, Washington D.C.

Wall, J.S., Young, M.R. and Carpenter, K.J. (1987) Transformation of niacin containing compounds in corn during grain development: relationship to niacin nutritional availability. *J. Agric. Food Chem.* **35**, 752–758.

West, C.E., Pepping, F. and Temalilwa, C.R. (1988) *The Composition of Foods Commonly Eaten in East Africa.* Wageningen Agricultural University, The Netherlands.

Wills, R.B.H. (1987) Composition of Australian fresh fruit and vegetables. *Food Technol. Aust.* **39**, 523–526.

3 Nutritional aspects of vitamins
D. H. SHRIMPTON

3.1 Vitamin deficiency diseases

3.1.1 Introduction

The concept of a disease caused by a nutrient deficiency has its origin in the studies of Lind in 1753 (see Stewart, 1953) on the cause of scurvy and in the demonstration by Cook (1776) and quoted by Carpenter (1986), of the practical value of Lind's hypothesis. By the end of the eighteenth century the impact on the navy of the control of scurvy was so great that it was asserted that it was equivalent to doubling the fighting force (Dudley, 1953). A century later the classic publications of Eijkman (1890, 1896), Funk (1912) and Hopkins (1912) established a generalised theory that a number of 'specific accessory constituents', later termed vitamins by Drummond (1920), were uniquely associated with the occurrence of specific diseases with overt clinical symptoms.

Currently there is much debate and some study of what may be an optimal condition of health and of the diet to achieve it. In such a concept it could be argued that states of health which did not achieve the optimum were states of disease and that, where they were nutritionally related to consumption of vitamins that were less than those associated with optimal health, then those too should be considered as vitamin deficiency diseases.

A concept of metabolic perfection and a high degree of immunocompetence is a seductive one, but there is insufficient hard data from which a generalised theory can be constructed and from which quantified definitions of optimal consumption can be set. For this reason this section of the chapter is confined to the occurrence of overt disease related to deficiencies of specific vitamins and to what is known of their incidence in the world. The concept of 'optimisation' will be discussed in section 3.3 on safety.

While dietary deficiencies of specific vitamins can have serious consequences for health, it must always be considered against the perspective that the dominant deficiency in the world is energy (Blaxter, 1986a), i.e. starvation, and that many overt symptoms have this as their primary cause, even though the diet in such circumstances is also likely to be deficient in many vitamins. The important corollary is that the correction of a vitamin deficiency is of little use to a starving individual unless the energy supply is corrected first. Hence, where starvation is a problem, the priorities are first to supply food and second, to correct nutritional deficiencies and imbalances.

3.1.2 Fat soluble vitamins: A, D, E, K

3.1.2.1 Vitamin A. Three compounds, retinol, retinaldehyde and retinoic acid occur naturally and have vitamin A activity. The standard for reference is retinol and all further reference will be to this compound. It is essential for vision, growth, cellular differentiation and proliferation, reproduction and the integrity of the immune system (Moore, 1957). The most obvious clinical signs of deficiency, which relate uniquely to retinol, are those concerned with vision and it is these that are referred to here.

The dominant clinical signs of deficiency relate to vision, being those of xerophthalmia and keratomalacia (Sommer, 1982). In the former there is a progression from mild night blindness through to corneal ulceration. These changes can be reversed by the administration of retinol; but still more acute deficiency results in irreversible lesions of the cornea (keratomalacia) and permanent blindness.

The occurrence is primarily in the developing nations of the world in children up to the age of six years. The statistical data is certainly incomplete and the best indication of the extent of the problem is given in a Report of the World Health Organisation (WHO, 1982).

Depressingly this study group formed the opinion that the worldwide incidence of xerophthalmia was increasing, quoting 100 000 annual incidents in the early 1970s, rising to 250 000 annual incidents by 1978. Its estimate of the total extent of keratomalacia was some 8–9 million cases at any one time. The data on a national basis at that time was:

- Africa
  Little data
- Asia

  Bangladesh: 17 000 annual cases of blindness.
  China: incidence of blindness in those examined was up to 8.76% in Guangzhau, 8.57% in Beijing, but low in Henan.
  India: at any one time there are 7.4 million cases of xerophthalmia and 0.22 million of keratomalacia. It was estimated that the annual rate of blindness was 52 000 and of partial blindness was 110 000 to 132 000.
  Indonesia: a risk by the age of five years of developing xerophthalmia of 52% and of keratomalacia of 2%.
- Middle East
  No data
- South and Central America
  Generally a low incidence.

3.1.2.2 Vitamin D. In humans the two main forms are ergocalciferol (vitamin D_2) and cholecalciferol (vitamin D_3). Both are precursors of the physiologically active forms 25-hydroxy vitamin D, dominant in the plasma, and 1,25-dihydroxy vitamin D, the hormone involved in calcium homeostasis (Fraser, 1975). Activity is calculated on the basis of cholecalciferol and this term will be used hereafter for all forms of vitamin D.

The clinical disease associated with a deficiency of vitamin D is rickets. Dominantly a disease of young children, but also recognised in adolescence and in pregnancy (Stephen, 1975). The disease is debilitating, but not fatal. Consequently it is not listed in the World Health Organization classification of diseases and hence there is no hard international data on the occurrence of rickets.

Classically rickets is a disease of young children in the cities of industrial countries (Arneil, 1975). In the United Kingdom, a relatively recent study by the government on current infant feeding practice (DHSS, 1988) reported that the introduction of welfare foods in the early 1940s resulted in the almost total abolition of the disease. Rickets reappeared in the 1960s and 1970s, mainly in the children of immigrant families; but it appears that this trend has been successfully reversed following the introduction of adequate supplementation (Dunnigan *et al.*, 1985).

In the developing countries rickets is stated to be widespread in parts of North Africa and the eastern Mediterranean and to be increasing in Mexico (WHO, 1990a).

This contrasts with the absence of rickets reported by medical missionaries at the end of the last century in Morocco, and also in China, India, Japan and Sri Lanka (Palm, 1890) and possibly reflects the advance of industrialisation and urbanisation in these regions.

3.1.2.3 Vitamin E. Many polymers of tocopherol and tocotrienol have vitamin E activity (McClaughlin and Weihrauch, 1979) In mammals, including humans, α-tocopherol is the most biologically active isomer, even though γ-tocopherol often predominates in the diet (Traber and Kayden, 1989). Vitamin E activity is expressed as units of α-tocopherol and is the isomer referred to in this chapter as vitamin E.

Only two clinical disorders are specifically associated with a deficiency of vitamin E. Neither is common and no worldwide statistics are available on the incidence of either.

The vitamin E status of the newborn is generally considered to be reduced compared with that of the adult (Muller, 1987) and, in premature infants, it can be sufficiently reduced for the clinical syndrome of haemolytic anaemia to occur (Hassan *et al.*, 1966). In adults who have suffered from a prolonged (5–10 years) inability to absorb fats a specific vitamin E deficiency syndrome can occur in which the symptoms are primarily neurological (Jeffrey *et al.*, 1987).

Both of these clinical deficiency conditions arise from an interference in the balance of the formation and destruction of reactive oxygen species in intermediary metabolism. The hydroperoxyl radical is one of the critical products of the free radical oxidative processes (Halliwell, 1987) and the tocopherols are the most abundant and efficient scavengers of these radicals in biological membranes (Di Mascio et al., 1991).

Vitamin E is not the only vitamin with antioxidant properties and its metabolic activities are also influenced by the presence of selenium. Given such complexity of possible interactions, it is not surprising that there is conflicting data relating to the proposition that: an 'excess' of free radicals is a major factor in the etiology of cancers and coronary heart disease and is the mitigating influence on the development of 'oxygen scavengers' in general and of vitamin E in particular (Diplock, 1991).

3.1.2.4 Vitamin K. Three groups of compounds have vitamin K activity: phylloquinones, which are synthesised by plants; menaquinones, which are synthesised by bacteria; and a synthetic group, the menadiones. They are termed respectively: K_1, K_2 and K_3. In humans, the main circulating form is K_1, while the storage form is K_2 (Usui et al., 1990).

This vitamin is necessary for the synthesis of prothrombin and at least five other proteins involved in controlling the clotting of blood. A deficiency results in prolonged clotting times and in severe cases of bleeding.

Clinical deficiencies have only been reported in infants. It may occur spontaneously as haemorrhagic disease of the newborn or as an idiopathic late-onset form at three to eight weeks of age. The idiopathic form is serious and usually results in death if the haemorrhage is intercranial.

There are no published international statistics of the incidence of haemorrhagic disease, but its occurrence is only occasionally reported in the industrialised nations and there is virtually no other epidemiological data.

3.1.3 Water soluble vitamins: the vitamin B-complex

3.1.3.1 Choline, inositol, p-aminobenzoic acid, citrin. At various times, each of these substances has been claimed as a vitamin of the B-group; but, as the detail of intermediary metabolism became known it was evident that these metabolites, although important, were not vitamins within the accepted definition. Since the identification of vitamin B_{12} as cyanocobalamin there have been claims for a variety of substances to be characterised as vitamins of the B-group, identified as B_{13} through to B_{17}. Although metabolic roles have been identified for some, none satisfies the criterion of a vitamin. Accordingly they will not be discussed in this chapter and this section will be concerned with the eight vitamins that are now recognised as being the components of the vitamin B complex, namely: B_1, B_2, niacin, B_6, B_{12}, biotin, folic acid and pantothenic acid.

3.1.3.2 Vitamin B_1: thiamin. Thiamin is intimately asociated with carbohydrate metabolism and disorders arising from a dietary deficiency are exacerbated by carbohydrate-rich diets, and especially by large intakes of alcohol, when severe neuropathies can develop. Clinical deficiency disease is consequently characteristic of tropical populations living on carbohydrate-rich diets, particularly those based on milled rice and on cassava; and also of alcoholics in 'western style' societies (Leevy and Baker, 1968).

The generic name of beri-beri refers to all forms of disease arising from a deficiency of thiamin. There are four well differentiated types.

Acute cardiac beri-beri. Characterised by sudden onset of epigastric pain, nausea and vomiting, followed by paralysis of the limbs and trunk and signs of cardiac failure. Slight exertion can result in sudden death. Those who survive become oedematous and chronic.

Chronic dry beri-beri. In this type the lower limbs are paralysed and there is a progressive atrophy of the muscles. Up to 1960 it was common in South East Asia among young adults and lactating women (Nicholls, 1938a)

Infantile beri-beri. Traditionally a common cause of death in those areas of the tropics where the diet was based on rice and is caused by a deficiency of thiamin in the mother's milk because of thiamin deficiency in the maternal diet (Bray, 1928). It is, in fact, the only serious disease of malnutrition associated with breast feeding and is characterised by vomiting and oedema leading to heart failure.

Mental beri-beri. This embraces Wernicke's encephalopathy and Korsakov's psychosis. In both there is serious mental degeneration resulting from haemorrhagic lesions in the central nervous system. It has recently been described as the Wernicke–Korsakov syndrome (Department of Health, 1991a) and is a chronic condition of alcoholics. There is no published international data of incidence, but Sinclair (1955) commented that it was associated with the drinking of crude and poorly refined spirits and rarely occurred amongst beer drinking populations.

3.1.3.3 Vitamin B_2: riboflavin. This vitamin is an essential component of the oxidising systems on which man depends for life; but a deficiency does not result in death. The symptoms of clinical deficiency are relatively mild and result in lesions of the mucocutaneous surfaces of the mouth, skin lesions and vascularisation of the cornea (Department of Health, 1991b).

This apparent anomaly arises from the fact that in deficiency, the consumption of oxygen results in a compensatory reduction in general metabolic activity. In such circumstances other deficiencies arise which result in more dramatic clinical conditions, such as pellagra.

Because of the complex and fundamental consequences of a deficiency of riboflavin there are few epidemiological studies in which the clinical symptoms observed can be attributed solely to the absence of riboflavin.

One such has been made in a rural farming region of the Gambia (Bates, 1988) where it was found that riboflavin deficiency was both common and severe amongst pregnant and lactating women, and thus infants. At present there is no statistical data on the occurrence of riboflavin deficiency in the world.

3.1.3.4 *Niacin.* This term includes two compounds, nicotinic acid and nicotinamide, the latter being the active component of the two coenzymes involved in oxido–reduction systems. Consequently a deficiency results in a fundamental disturbance of intermediary metabolism and can result in death.

The clinical disease resulting from a deficiency of niacin is pellagra (Harris, 1941). It is characterised by a dermatitis which occurs principally in regions exposed to the sun or to pressure, for example at joints. It is commonly, but not invariably, accompanied by diarrhoea and, in advanced cases, by dementia.

Nicotinamide can be synthesised in the body from the essential amino acid tryptophan. Hence, the traditional regions for a prevalence of pellagra are those in which the diet is deficient in both niacin and tryptophan. These are diets where maize or sorghum is the major source of food.

Until 1914 pellagra was widespread in Africa, Asia, Central and South America and in the littoral regions of the Mediterranean (Nicholls, 1938b). An exception was Mexico, where it has been suggested that, because maize is principally consumed as tortilla (where steeping in limewater is part of the preparative process), the tryptophan, which is acid labile but alkali stable, is protected and thus spares the niacin (Laguna and Carpenter, 1951).

Pellagra still occurs in parts of Africa and Asia, but there are no statistics from which a reliable estimate of incidence can be made. However, Blaxter (1986b) reported that there was still a significant incidence of pellagra in the Deccan of India because of the strong adherence by the population to the traditional diet based on jowah, a porridge prepared from a species of sorghum (*S. vulgaris*).

3.1.3.5 *Vitamin B_6: pyridoxine.* There are many closely related forms of pyridoxin that are all metabolically and clinically active. The principal ones are pyridoxal, pyridoxine, pyridoxamine and their $5'$-phosphates.

It is unusual for a clinical deficiency condition to be recognised as specifically caused by a dietary deficiency of vitamin B_6. Because of its fundamental role in energy metabolism, a deficiency of the vitamin is commonly associated with deficiencies of other B-vitamins. However, there was an unusual incidence of clinical symptoms resulting from a deficiency of vitamin B_6 in infants in the 1950s (Bessey et al., 1957).

The deficiency was characterised by abdominal distress and neurological symptoms, many convulsing, because the infants had been fed on a formula milk that had been severely heat damaged during processing (Coursin, 1964).

3.1.3.6 *Vitamin B_{12}: cyanocobalamin.* This is the only member of the cobalamin group of cobalt-containing compounds to have significant activity for the human and all reference will be to cyanocobalamin.

Deficiency symptons usually arise from an inability to absorb the vitamin rather than from a direct dietary deficiency. In these circumstances the clinical consequences are pernicious anaemia and, in severe cases of depletion, irreversible neurological disorders resulting from the demyelination of the spinal chord and brain, peripheral nerves and the optic nerves (Lindenbaum et al., 1988). Megaloblastic anaemia can also occur in deficiency because of the involvement of the vitamin in the recycling of the folate coenzymes (Department of Health, 1991b).

Dietary deficiency, even in strictly rigorous vegetarian sects such as veganism, is rare (Armstrong et al., 1974). Against this background it is surprising that the World Health Organization has recently advocated the consumption of animal foods (WHO, 1990b) to ensure an adequate supply. However, the basis for adequacy developed by WHO relates to an assessment of the overall haematological position embracing iron and folate together with vitamin B_{12} (FAO, 1988).

3.1.3.7 Folic acid: pteroylglutamic acid (PGA). It is now recognised that many compounds are metabolically active in the role of PGA and these are collectively termed folate or folacin. A deficiency of the vitamin is clinically identified by the development of megaloblastic anaemia (Rodriguez, 1978).

There are no published statistics of the incidence of clinical deficiency in developing countries and it has been suggested by Schorah and Habibzadeh (1986) that individuals in industrialised countries are closer to the threshold of overt deficiency of folic acid than for any other nutrient.

Macrocytic anaemias are found in 10% of acute geriatric admissions in the UK (Schorah, 1991). There is also growing evidence that a deficiency of the vitamin in preconception is a major factor in the occurrence of neural tube defects in the newborn (Wald and Polani, 1984, Milunsky et al., 1989).

3.1.3.8 Biotin and pantothenic acid. It is generally accepted that neither vitamin is associated with clinical symptoms of deficiency in humans except under experimental conditions.

3.1.4 Water soluble vitamins: vitamin C

L-ascorbic acid and its oxidised form, dehydroascorbic acid have comparable vitamin C activity (Sabry et al., 1958). Clinical deficiency, scurvy, is characterised by a weakening of collagenous structures which in turn leads to capillary haemorrhaging and, in severe cases, to death (Hornig, 1975).

Historically, scurvy has been primarily associated with sailors and it has been suggested that it should be regarded as one of the foremost of the world's occupational diseases (McCord, 1971). Carpenter (1986), quoting McCord, reports that two million sailors died from scurvy in the three centuries up to 1900 with the implication that the total was substantially higher because only the English, Russian, Portuguese and Spanish kept reliable records.

Currently its occurrence in the developing nations is in areas afflicted with long term drought, particularly in Africa (WHO, 1990b). In industrialised nations it is unusual, although low plasma concentrations have been reported in some groups of elderly persons (Newton *et al.*,1985), and Schorah (1991) has quoted an incidence as high as 1% in those over seventy years of age in the UK.

In infants, clinical scurvy has not been observed in those that have been fully breast-fed, even in communities where the intake of the mother has been very low (Department of Health, 1991b). However, a biochemical deficiency has been reported in the Gambia during the rainy season when the intake of vitamin C by the mother is low (Bates, 1988).

3.2 Recommended daily allowances

3.2.1 Introduction

Recommended daily allowance (RDA) is used interchangeably with the term recommended daily intake (RDI). However, the confusion that has arisen is not over the use of two closely similar names for the same purpose, but over the many different uses that have evolved over time for a quantified statement of nutrient requirements.

Previous reviews of the evolution of nutritional standards (Leitch, 1942; Trusswell, 1983) attribute the origin to a Dr E. Smith in nineteenth century England. He was concerned by the cotton famine in Lancashire at that time and set himself the task of answering the question: 'What is the least cost per head per week for which food can be bought in such quantity and in such quality as will avert starvation-disease from the unemployed population.' Dr Smith's standard of 1862, expressed in terms of basic foods, must, in practice, have had something in common with those whose concern it had been to victual ships and other isolated groups of individuals in previous centuries. For example, Samuel Pepys established victualling allowances for ships' captains in 1686 (Bryant, 1933).

Trusswell (1983) described the move from food to nutrient made in Britain by Lusk during the First World War to provide a standard for energy to feed the army and nation and its further development by the BMA in 1933 to maintain health and working capacity during the depression.

The scientific disciplines on which nutritional theory is based were developing rapidly, enabling the League of Nations (1938) to publish a comprehensive list of allowances for 12 nutrients. These were:

- energy, protein, fats
- calcium, phosphorus, iron, iodine
- vitamins A, B_1, B_2, C and D

This publication marked both the end of the period in which the motivation was that of preventing hunger and opened the way for the development of a new concept of using allowances to promote health by identifying and specifying key nutrients.

In the USA in 1941, the National Research Council transformed the concept of nutrient allowances by avowing that the aim was, as quoted by Buss (1986), to achieve buoyant health ... and the building up of our people to a level of health and vigour never before attained or dreamed of ...'

This was the concept behind the first publication of dietary allowances in the USA (National Research Council, 1943).

3.2.2 International concepts of the function of RDAs / RDIs

In the developing nations there has been little opportunity for the mass of the population to exert choice from a variety of foods. For the majority of these populations the staple food has been, and for many remains, the available diet. In these circumstances an earlier concept of the League of Nations, that of relating dietary standards to agriculture, has been retained in an attempt to develop indigenous resources towards the goal of providing sufficient food of an appropriate nutritional value for the population.

Such an attempt was formalised by the Indian Medical Research Council and revised by Patwarden in 1960. As pointed out by Davidson and Passmore (1965), they were a compromise between physiological need and agricultural feasibility and, if achieved, would have resulted in a substantial improvement in the nutrition of the population.

This approach has been developed by Gopalan (1989) to include those who are at risk of becoming victims of the diseases of affluence (see Shrimpton, 1993 in press). In such an approach it is not only the RDAs that are considered in relation to the available food supply, but also agricultural developments are considered in relation to the production of nutritionally desirable commodites. Such a linking of nutritional standards with agricultural policy has also been proposed by some health-conscious groups in Britain where, for example, a reform is advocated of the EC's Common Agricultural Policy to match production to nutritional ideals. A different approach is to promote nutritional education to a level where individuals, understanding the implications of RDAs, choose appropriately; and this would appear to be the current choice of the UK government (Department of Health, 1991a)

In 1980 the US National Academy of Sciences published the ninth edition of its RDAs and discussed the uses to which it thought they could be appropriately applied. These were:

- to population groups rather than to individuals
- as standards or guides to serve as a goal for good nutrition
- to assess and interpret dietary surveys of populations
- as guidelines for planning and procuring food supplies
- as guidelines in establishing policies for health and welfare programmes
- in setting standards for new products
- in nutrition labelling
- in regulating the nutritional quality of foods
- in nutrition education.

Table 3.1 Differences in RDA values from a published analysis

Vitamin	Range*
A	1.3
B_1	1.3
B_2	1.25
B_3	2.2
B_{12}	2.5
C	2.7
D	10.0 (excluding zero values)

* Highest as a multiple of the lowest

By 1983, there were at least forty national publications and also one from FAO / WHO of RDAs (Trusswell *et al.*, 1983). All differed from each other, as illustrated in Table 3.1 calculated from a published analysis of selected nutrients (Buss, 1986).

The most recent publication of RDAs is that from the UK (Department of Health, 1991b). It differs from both its national and international predecessors in two respects: the composition of the reviewing body and the theoretical approach to the recommendations. These differences have important implications.

3.2.2.1 Composition. Uniquely in a national committee to review RDAs, observers were invited from outside the country: from the European Commission and from Denmark and the Netherlands.

3.2.2.2 Theoretical approach. Although it has become customary to use RDA and RDI interchangeably, they have different connotations and, together with differences in the uses to which such recommendations are used, the habit has contributed to the misunderstandings and confusions which abound in this topic. RDA refers to provision, RDI to need.

The committee returned to the fundamental statistical nature of the recommendations (Buss, 1986) and recognised three levels of reference from the normal distribution curve which describes the nutrient requirements of a group of healthy individuals.

Because it is assumed that the distribution of requirement is normal, the mean and the median are coincident. Hence the approximation is valid that two standard deviations either side of the mean will correspond to the 2.5 and 97.5 percentiles.

The reviewing committee introduced a new nomenclature to identify these three points in nutritional terms as follows:

- The mean : EAR (estimated average requirement)
- 2.5 percentile : LRNI (lowest reference nutrient intake)
- 97.5 percentile : RNI (reference nutrient intake)

This approach, which is not new, being the principle on which all previous estimates of the requirements of populations have been made, is novel in its application.

Furthermore, it points up both the strengths and the limitations of the approach to nutrient requirements through the study of populations. The strength is the high degree of probability that, by the criteria of adequacy chosen, the majority of the population will have a satisfactory intake of nutrients if the guidelines are followed. A weakness is that 2.5% of the population may receive too little and that 2.5% may receive significantly more than is considered necessary by the criteria of adequacy chosen for the purpose.

A different use of the same data is to consider that the normal distribution represents requirement. On this basis 2.5% of the population will have a requirement that will not be satisfied by the criteria chosen and 2.5% will have requirements that are so low that they will almost certainly consume more than is considered appropriate by the chosen criteria of need. The issue of need and the corollary of safety will be discussed in section 3.3 of this chapter.

The issue of a minority being at risk through underconsumption has been met by the authors of the report in an unusual manner. They suggest that the individuals making up the extremes of the distribution curve are continually changing, while the distributions of intake and requirement remain constant. Consequently they recommend that the RNI should be used as the basis for the recommendation of nutrient intakes for populations. Thus the RNI corresponds numerically with previous publications of RDA and RDI.

The authors also recommend that the lowest value, the LRNI, should be regarded as that level which a population should receive to avoid clinical signs of deficiency. The mean or median value is recommended as the most appropriate value to use for labelling purposes. It is implicit in the statistical approach that there is a correlation between the distribution of the intakes of nutrients by a group with the requirements of that group for those nutrients.

Lastly, concerning terminology, this most recent publication on RDAs suggests that, because all three of the terms introduced have a function in describing desirable intakes of nutrients for populations, there should be a new collective term to embrace all three and this they suggest should be dietary reference value (DRV). Thus:

DRV embraces LRNI: EAR: RNI

Although a new terminology in an area where there is already so much misunderstanding may, at first sight, contribute to yet more confusion, the recognition that the current terminology was inadequate and the provision of a rational system for professional use is a major innovation with considerable potential. In particular it makes clear the difference between the approach to assessing the requirements of a population and of an individual. Further it provides a rational way of accomplishing the difficult task of nutritional labelling. However, for these ambitions to be realised it will be necessary for a considerable programme of nutritional education to be undertaken which must be characterised by clarity, consistency and a total absence of ambiguity.

3.2.2.3 Labelling. The convention has become established that, for the purpose of labelling, the RDA of the adult male will be used as the marker. Products may then be labelled to indicate the proportion of the RDAs of particular nutrients which are contributed by unit weight of the product, most commonly 100 g or a single helping of specified weight.

Amongst the shortcomings of this approach are:

- the RDA is a population recommendation, not an individual one
- the RDA is not a statement of requirement (except for energy), but a mean requirement of a population plus two standard deviations to provide a 'safety factor'
- the use of one class (the adult male) can be misleading as a basis for labelling particular foods, for example, baby foods.

Using the DRV approach, it is suggested (Department of Health, 1991b) that the mean value of the requirement of a population is more appropriate, that is the EAR. It is also suggested that the EAR should be used that is appropriate to the market for which the product is intended.

There is an appeal in this approach, but it is unlikely to be meaningful to the public unless it is adopted throughout the EC after 1992 and only after a campaign of education that has been demonstrably successful.

3.3 Safety

3.3.1 Introduction

Westernised communities have become increasingly sensitive to the possibility that discomforts and, in extreme cases, dangers that were once accepted as 'part of life' can now be attributed to a shortcoming in the performance of a process, of an individual or of an organisation. In part as a consequence of this attitude, there has developed a public desire for safety to be guaranteed by legislation and regulation.

In the field of nutrition, and most especially in the usage of vitamins, this presents substantial problems. The knowledge of function and requirement is incomplete in both the general case and the specific one of any particular individual. Yet, coincident with incomplete knowledge, there is a very rapid rate of acquiring new information, some of which appears to upset hypotheses that had appeared to be satisfactory.

As a consequence there is a continuing debate amongst professionals that is right and proper and is the chosen method of developing understanding in the scientific community, but when popularised and publicised the debate spreads confusion and increases the level of misunderstanding amongst the public. At worst, this can be misinformation, a longstanding problem in the communication of nutritional concepts to the public (McKenzie, 1968).

In such circumstances the legislator has a problem. The basis of his decision must be capable of being understood by 'the intelligent layman'. His decision must not favour any one sector of industry without an apparently unchallengeable reason. Above all, his decision must not be seen to expose the consumer to any risk. In this situation there is an understandable temptation to seek solutions that can be presented as general cases based on logic and expressed in simple terms. Unfortunately this can give rise to mistakes and to limitations in the exploitation of scientific advance for the public benefit.

3.3.2 Issues

The topic is confused by many claims and counterclaims relating to the perceived benefits of specified intakes of particular vitamins in the absence of unequivocal and objective evidence.

The greater part of the debate centres around the issues of health and is dominantly a problem of 'westernised' societies. Amongst the questions debated are:

- how is adequacy determined?
- are metabolic markers appropriate?
- if so, which ones?
- is 'optimal' health a realistic target?
- if so, how is it defined?
- do vitamins have a part to play in the control or prevention of currently intractable diseases?
- if so, is this a nutritional or a medical issue?
- is there an ideal balance between vitamins and between any one vitamin and other nutrients?
- if there are such balances, over what period should the balance be maintained: for example within every eating occasion, within a day, within a week or longer?

At present there are no answers to any of these questions, either on a population group basis or for individuals. Hence they all converge on two issues:

- is there a risk to health if an individual consumes one or more vitamins in amounts which differ significantly from any published RDA?
- conversely, may an individual lose the opportunity to improve personal health through 'optimisation' if no vitamin is consumed in any amount which significantly exceeds a published RDA?

Therefore attention must be focussed on a specific and quantified answer to the first question so that the second can be pursued to the point that case law through experience will provide guidelines until such time as a rigorous theoretical exposition becomes available.

3.3.3 Attitudes

The most prevalent attitude is to relate safety to RDA. While this lacks the logic of causality, it has the merits of being a generalised and simple statement as well as having an emotional appeal related to perceived need. In practice it is difficult to quantify because there is no universally-accepted quantification of RDAs for populations and no experimentally-based rationale from which a multiplier of any RDA can be chosen to serve as an index of safety.

A different approach is that which is used in the medical and veterinary fields and is based on the concept of contraindication. In this, each vitamin is considered individually. Its case history in respect of usage is reviewed and any objective and critically assessed report of adverse effect is recorded. The lowest intake associated with the mildest recorded side effect is taken as the first level of consumption for which there is a contraindication. This provides a starting point for an approach to safety which is based on specific observed facts.

An advantage of this approach, that it is specific to each vitamin and relates to individuals rather than to means of populations, can also be seen as a disadvantage, in that it is based on case history and lacks the objectivity of a generalised hypothesis. However, in the current dynamic situation of research on the role of vitamins in intermediary metabolism and health, there is no general hypothesis.

Irrespective of which approach to safety is taken, there are the issues of (i) need for consumption in amounts that are greater than the RDA and (ii) the consumption of foods on a regular basis that are particularly rich in one or more vitamin, so that again the RDA can be exceeded, sometimes substantially.

3.3.4 Need

The definition and quantification of need has been one of the most intransigent problems in nutrition during the second half of this century. At the statistical level there is the issue of the composition of the normal distribution curve which, it is assumed, best describes the range of requirement for a nutrient in a population. For 97.5% of the population the requirement is mathematically defined as being within two standard deviations of the median which, in the assumed symmetrical unimodal distribution is coincident with the mean. However, by the same assumption, 5% of the population differs by more than two standard deviations from the mean and almost 1% will differ by more than three standard deviations. The complete spread of requirement is thus substantial in statistical terms, but because there is so little reliable data on the vitamin requirements of individuals within populations it is not possible to give firm estimates of the size of the spread in terms of multiples of an RDA. Indeed, as has been described in the previous section, even the mean RDA for a population cannot be determined with precision for any vitamin.

Therefore, statistically it is possible to be certain that there is a range of need between different individuals within a population, but its magnitude cannot be reliably measured at this time. The causes of variation are well known (Girdwood, 1971) and are multifactorial, involving availability within a food and physiological characterisics in a consumer as well as sociological factors.

Many have taken it as axiomatic that 'good' nutrition will result in a lower incidence of illness, as distinct from overt deficiency diseases. However it is only in the past two decades that a growing body of evidence has started to be identified to support this long-held opinion. In essence it concerns two mechanisms: immuno-defence systems and oxido-reduction systems.

3.3.4.1 Immuno-defence systems. A report of WHO (Chandra *et al.*, 1978) suggested that 'undernourishment' was the most common cause of secondary immunodeficiency in man. While it is premature to declare that there is now complete understanding of the nutritional and metabolic systems involved, it is becoming clear that certain micronutrients play a key role. In 1989 the New York Academy of Sciences held a symposium relating solely to the role of vitamin E in maintaining health (Diplock *et al.*, 1989). The following year a symposium on micronutrients in health was held in York, England where the roles of vitamin E (Kelleher, 1991) and vitamin A (West *et al.*, 1991) in the immune response were reviewed.

The research is not yet complete, but already it is clear that there is a quantifiable requirement for these two vitamins that is both different from and greater than that which was previously thought to be necessary. Indeed, in the previous review of RDAs in the UK (DHSS, 1979) it was not considered to be necessary to include vitamin E in the considerations of the reviewing body.

Because it is not yet known how to predict the precise levels that are required to optimise the immune system in any one individual, the issue of safety is immediately raised. Is there an upper limit that should not be exceeded or can consumption be at any level over any period?

3.3.4.2 Oxido-reduction systems. In recent years interest has focussed on a particular set of reactions which involve free radicals. A radical is any species containing one or more unpaired electrons, a condition which is usually associated with increased activity. Their role in biology and medicine has been described by Halliwell and Gutteridge (1985) and specifically in nutrition by Halliwell (1987).

In some quarters there has been misunderstanding about the significance of the presence of free radicals. Far from being indicators of disease, they are a normal part of the metabolic process. However, if their concentration in any one metabolic site becomes unusually high for that site in relation to its normal function, then highly reactive species of free radicals can be formed and these can cause biological damage.

There is intensive interest in and a growing body of research on the proposition that an excess of free radicals is a causative agent in some chronic diseases, and that these diseases may be prevented, or their progress delayed, by the consumption of vitamins with an antioxidant capability (Diplock, 1991). This particularly relates

Table 3.2 Foods containing vitamins consumed in large quantities compared with their RDA value

Food	Vitamin	Amount / 100 g	Multiple of RDA
sardines in oil	B_{12}	28 μg	18
peanuts and raisins	E	5.65 mg	0.8
popcorn	E	11 mg	1.6
strawberries	C	77 mg	1.9
calf liver (fried)	A	39 780 μg	56.8
calf liver (fried)	B_2	4.2 mg	3.2

to muscle damage (Jackson, 1987), inflammatory diseases (Merry *et al.*, 1991), cancers (Weisburger, 1991) and coronary heart disease (Ferrari *et al.*, 1991). The vitamins that are primarily concerned are vitamin C and vitamin E. Other micronutrients that may be involved include some of the carotenoids, selenium and folate.

3.3.5 Adventitious acquisition

The extent to which foods are selected and consumed primarily because they are liked rather than because of their content of protein, carbohydrate and fats, results in micronutrients being acquired adventitiously by the consumer. Consequently some vitamins are consumed *inter alia* in amounts that are large compared with their RDA.

Examples are given in Table 3.2 (Holland *et al.*, 1991).

It should be noted that 100 g is a modest helping and that 'peanuts and raisins' is typical of a snack served with drinks and to that extent is consumed without reference to nutrient value.

The examples given are of single common foods, but there is also evidence that a minority of the population consumes vitamins in amounts that are greatly in excess of their RDA (Gregory *et al.*, 1991). The following examples are taken from the results of this survey (Table 3.3), which is the first of its kind in the UK.

3.3.6 Safety and RDAs (DRVs)

The data quoted is for adult men, but comparable data exists for adult women and in all the age groups of both sexes that were recorded (Gregory *et al.*, 1991). Furthermore, the 2.5 percentile of the population over sixteen years of age comprises several hundred thousand individuals. Because, by implication, the consumption of foods and hence of nutrients that was recorded was habitual, it must be concluded from the wide ranges of consumption that safety of vitamins and their RDAs are unrelated. Certainly there is no theoretical reason to lead one to expect that they could be.

3.3.7 Need and consumption

If the antioxidant theory is proven, then it will become desirable for certain vitamins and micronutrients to be consumed in amounts that are significantly greater than

Table 3.3 Vitamins consumed in amounts greatly in excess of their RDA value

Vitamin	Units	Daily intake of men (16–64)		
		Mean	Upper 2.5 percentile	
A (i)	µg	1277	6671	(9.5)
B_1	mg	2.01	3.29	(3.3)
B_2	mg	2.29	4.32	(3.3)
B_3	mg	40.9	67.4	(4.0)
B_6	mg	2.68	5.35	(3.8)
B_{12}	µg	7.3	23.0	(15.3)
Folate	mg	312.0	562.0	(2.8)
Biotin	µg	39.1	71.4	(—)
Pantothenate	mg	6.6	11.2	(—)
C	mg	74.6	227.3	(5.7)
D (ii)	µg	3.78	12.72	(1.3)
E (iii)	mg	11.7	23.4	(5.9)

Numbers in brackets are derived by dividing the value for the upper 2.5 percentile by the RDA so that the intake may be expressed as a multiple of the RDA.
(i) As preformed retinol.
(ii) The main source is from the action of sunlight on the skin; the RDA relates to those who are confined indoors.
(iii) the requirement for vitamin E is influenced by the dietary intake of polyunsaturated fatty acids (PUFA), and 4 mg was proposed as adequate for men.
(—) No value available.

the current estimates of their RDAs (DRVs). The possibility also exists that 'optimal health' will become capable of definition and quantification and involve more vitamins than just vitamins C and E.

Meanwhile, segments of the population are anticipating this situation and are consuming selected vitamins in amounts that are significantly in excess of their RDAs. Hence guidelines on safety are desirable.

3.3.8 Possible guidelines for safety

A linkage to RDA has already been discarded, but two other approaches are possible.

3.3.8.1 Toxicological. The toxicological data can be assembled on each vitamin, using animal data in the absence of appropriate human data. Conventional margins of safety are then achieved if the experimentally determined toxic dose is divided by ten when the data is acquired with humans and 100 when it is of animal origin.

Unfortunately this approach is difficult to apply because truly toxic effects have been recorded for so few vitamins. Where it has been attempted the result has been one of confusion because of the absence of a single and comparable measure of toxicity between vitamins (MAFF, 1991).

3.3.8.2 Clinical. In this approach the data is also assembled for each vitamin, but the record made is that of the lowest level to cause a clinically observed effect

Table 3.4 Daily intakes of vitamins for which no contraindication[†] has been published

Vitamin	Units	Intake	Reference
A*	µg	2250	Bendich and Langseth, 1989
B_1	mg	100	Unna, 1968
B_2	mg	200	Danford and Munro, 1980
Niacin (acid)	mg	100	DiPalma and Ritchie, 1977
(amide)		1000	Coronary Drug Res. Gp., 1975
B_6	mg	100	Coeling Bennink et al., 1975
B_{12}	µg	100	Herbert, 1987a
Folate	µg	500	Herbert, 1987b
Biotin	µg	1000	Zak and D'Ambrosio, 1985
Pantothenate	mg	1000	Stockley, 1981
C	mg	1000	Olson, 1987a
D	µg	10	Marx et al., 1983
E	mg	800	Bendich and Machlin, 1988
K	µg	500	Olson, 1987b

* During pregnancy, foods containing high levels of vitamin A should be avoided (Chief Medical Officer of the United Kingdom 1990).

[†] Contraindication is not synonymous with toxicity. Therapeutic doses will often be in excess of those indicated here and substantially higher levels of the majority of the vitamins may be consumed over short periods (Department of Health, 1991b).

that is undesirable. While the concept is to identify the consumption associated with a very mild disorder, it must be related to consumption over a long period since the concept is to be applied to a nutrient, not a medicine. Table 3.4 has been prepared on this basis. It is suggested that it is an appropriate concept for determining the safety of vitamins when used as foods and, at the time of writing, is under discussion in the Commission of the EC.

It is necessary to draw attention to the note to Table 3.4 relating to vitamin A. The Chief Medical Officer of the UK recommended that, during pregnancy or at times when pregnancy is expected, women should not allow their intake of vitamin A to exceed the current RDA in the UK for adults, namely 750 µg. To accomplish this they are advised not to eat liver or its products and not to take supplements of vitamin A or fish liver oils, except under medical supervision. It is emphasised that this constraint is not relevant to any other class of individual and that no positive evidence for birth defects relating to excessive consumptions of vitamin A has arisen in the UK.

References

Armstrong, B.K., Davies, R.E., Nicol, D.J., van Merwyk, A.J. and Larwood, C.J. (1974) Hematological vitamin B_{12} and folate studies on Seventh Day Adventist vegetarians. *Am. J. Clin. Nutr.* **27**, 712–718.
Arneil, G.C. (1975) Nutritional rickets in children in Glasgow. *Proc. Nutr. Soc.* **34**, 101–109.
Bates, C.J. (1988) Vitamin studies and supplementation of rural Gambian women. In: *Vitamins and Minerals in Pregnancy and Lactation* (ed. H. Berger) Raven Press, New York, pp. 261–263.
Bendich, A. and Langseth, L. (1989) Safety of vitamin A. *Am. J. Clin. Nutr.* **49**, 358–371.
Bendich, A and Machlin, L.J. (1988) Safety of oral intake of vitamin E. *Am. J. Clin. Nutr.* **48**, 612–619.

Bessey, O.A., Adams, D.J.D. and Hansen, A.E. (1957) Intake of vitamin B6 and infantile convulsions: a first approximation of requirements in infants. *Paediatrics* **20**, 33–44.
Blaxter, K.L. (1986a) *People, Food and Resources*. Cambridge University Press, Cambridge, pp. 48–58.
Blaxter, K.L. (1986b) *People, Food and Resources*. Cambridge University Press, Cambridge, p. 33.
Bray, G.W. (1928) Vitamin-B deficiency in infants: its possibility, prevalence and prophylaxis. *Soc. Tropical Med. Trans.* **22**, 9–42.
Bryant, A. (1933) *Samuel Pepys: the Saviour of the Navy*. William Collins and Sons, London.
Buss, D.H. (1986) Variations in recommended nutrient intakes. *Proc. Nutr. Soc.* **45**, 345–350.
Carpenter, K.J. (1986) *The History of Scurvy and Vitamin C*. Cambridge University Press, Cambridge, p. 253.
Chandra, R.K., Cooper, M.D., Hitzig, W.H., Rosen, F.S., Seligmen, M., Soothill, J.F. and Terry, R.J. (1978) *Immunodeficiency*. WHO Technical Report. Series **630**, p. 61.
Coelingh Bennink, H.S.T. and Schreurs, W.H.P. (1975) *Brit. Med. J.* **3**, 13.
Cook, J. (1776) The method taken for preserving the health of the crew of His Majesty's ship the Resolution during her late voyage round the world. *Phil. Trans. R. Soc. (London)* **66**, 402–406.
Coronary Drug *Res. Gp.* (1975) Clofibrate and niacin in coronary artery disease. *J. Am. Med. Assoc.* **231**, 360–381.
Coursin, D.B. (1964) Vitamin B6 metabolism in infants and children. *Vitamins and Hormones* **22**, 755–786.
Davidson, S and Passmore, R. (1965) *Human Nutrition and Dietetics*, 2nd edition. E.S. Livingstone, Edinburgh, p. 253.
DHSS. (1979) Recommended daily amounts of food energy and nutrients for groups of people in the UK. *Report on Health and Social Subjects* **15**. HMSO, London.
DHSS. (1988) Present day practice in infant feeding: third report. *Report on Health and Social Subjects* **32**, HMSO, London, pp. 35–36.
Department of Health (UK) (1991a) *The Health of the Nation*. HMSO, London.
Department of Health (UK) (1991b) Dietary reference values for food energy and nutrients for the United Kingdom. *Report on Health and Social Subjects*, **41**. HMSO, London.
Di Mascio, P, Murphy, M.E. and Sies, H. (1991) Antioxidant and defense systems: the role of carotenoids, tocopherols and thiols. *Am. J. Clin. Nutr.* **53**, 194S–200S.
DiPalma, J.R. and Ritchie, D.M. (1977) Vitamin toxicity. *Ann. Rev. Pharmacol Toxicol.* **17**, 133.
Diplock, A.J. (1991) Antioxidant nutrients and disease prevention: an overview. *Am. J. Clin. Nutr.* **53**, 189S–193S.
Diplock, A.T., Machlin, L.J., Packer, L. and Pryor, W.A. (1989) Vitamin E: Biochemistry and health implications. *Annals N.Y. Acad. Sci.* **570**, 1–555.
Drummond, J.C. (1920) The nomenclature of the so-called accessory food factors (vitamins). *Biochem. J.* **14**, 660.
Dudley, S. (1953) James Lind: Laudatory address. *Proc. Nutr. Soc.* **12**, 202–209.
Dunnigan, M.G., Glekin, B.M., Henderson, J.B., McIntosh, W.B., Sumner, D. and Sutherland, G.R. (1985) Prevention of rickets in Asian children: assessment of the Glasgow campaign. *Brit. Med. J.* **291**, 239–242.
Eijkman, C. (1890) Polyneuritis in chickens. *Geneeskundig Tijdschrift voor Nederlandsch-Indie* **30**, 295–234. (English translation: Roche, Welwyn Garden City, 1990).
Eijkman, C. (1896) Polyneuritis in chickens: a new contribution to its aetiology. *Geneeskundig Tijdschrift voor Nederlandsch-Indie* **36**, 214–269. (English translation: Roche, Welwyn Garden City, 1990).
FAO (1988) Requirements of vitamin A, iron, folate and vitamin B_{12}: report of a joint FAO / WHO expert consultation. *Food and Nutrition Series* **23**.
Ferrari, R., Ceconi, C., Cargnoni, A., Pasini, E., DeGiuli, F. and Albertini, A. (1991) Role of oxygen free radicals in ischemic and reperfused myocardium. *Am. J. Clin. Nutr.* **53**, 215S–222S.
Fraser, D.R. (1975) Advances in the knowledge of the metabolism of vitamin D. *Proc. Nutr. Soc.* **34**, 139–143.
Funk, C. (1912) The etiology of the deficiency diseases. *J. State Med.* **20**, 341–368. (Reprinted: Goldblith, S.A. and Joslyn M.A. (eds) *Milestones in Nutrition*. Avi Publishing, Westport, Connecticut, pp. 145–172).
Girdwood, R.H. (1971) Problems in the assessment of vitamin deficiency. *Proc. Nutr. Soc.* **30**, 66–73.
Gopolan, G. (1989) Dietary guidelines for the populations of developing countries. In: *Dietary Guidelines: Proceedings of an International Conference* (ed. M. Latham, and M. van Veen,) Cornell Int. Monograph Series **21**.

Gregory, J., Foster, K., Tyler, H. and Wiseman, M. (1991) *The Dietary and Nutritional Survey of British Adults.* HMSO, London pp 128–137.
Halliwell, B. (1987) Free radicals and metal ions in health and disease. *Proc. Nutr. Soc.* **46**, 13–26.
Halliwell, B and Gutteridge, J.M.C. (1985) *Free radicals in Biology and Medicine.* Clarendon Press, Oxford.
Harris S. (1941) *Clinical Pellagra.* Mosby, St Louis, Mo.
Hassan, H. Hashin, S.A., Van Italie, T.B. and Sebrell, W.H. (1966) Syndrome in premature infants associated with low plasma vitamin E levels and high polyunsaturated fatty acid diet. *Am. J. Clin. Nutr.* **19**, 147–157.
Herbert, V. (1987a) RDI of vitamin B12 in humans. *Am. J. Clin. Nutr.* **45**, 661–670.
Herbert, V. (1987b) RDI of folate in humans. *Am. J. Clin. Nutr.* **45**, 661–670.
Holland, B., Welch, A.A., Unwin, I.D., Buss, D.H., Paul, A.A. and Southgate, D.A.T. (1991) *McCance and Widdowson's The composition of foods 5th edition.* Royal Society Chemistry and MAFF, Royal Society Chemistry, Cambridge.
Hopkins, F.G. (1912) Feeding experiments illustrating the importance of accessory factors in normal dietaries. *J. Physiol.* **44**, 425–460.
Hornig, D. (1975) Metabolism of ascorbic acid. *World Rev. Nutr. Diet.* **23**, 225–258.
Jackson, M.J. (1987) Muscle damage during exercise: possible role of free radicals and protective effect of vitamin E. *Proc. Nutr. Soc.* **46**, 77–80.
Jeffrey, G.P., Muller, D.P.R., Burroughs, A.K., Matthews, S., Kemp. C., Epstein, O., Metcalfe, T.A., Southam, E., Tazir-Melboucy, M., Thomas, P.K. and McIntyre, N. (1987) Vitamin E deficiency and its clinical significance in adults with primary biliary cirrhosis and other forms of liver disease. *J. Hepatol.* **4**, 307–317.
Kelleher, J. (1991) Vitamin E and the immune response. *Proc. Nutr. Soc.* **50**, 245–249.
Laguna, J. and Carpenter, K.J. (1951) Raw versus processed corn in niacin-deficient diets. *J. Nutr.* **45**, 21–28.
League of Nations (1938) Technical commission on nutrition. *Bulletin Health Organisation* **7**. Geneva.
Leevey, C.M. and Baker, H. (1968) Vitamins and alcoholism *Am. J. Clin. Nutr.* **21**, 1325–1328.
Leitch, I. (1942) The evolution of dietary standards. *Nutr. Abstr. Rev.* **11**, 509.
Lindenbaum, J., Healton, E.B., Savage, D.G., Brust, J.C.M., Garrett, T.J., Podell, E.R., Marcel, P.D., Stabler, S.P. and Allen, R.H. (1988) Neuropsychiatric disorders caused by cobalamin deficiency in the absence of anemia or macrocytosis. *New England J. Med.* **318**, 1720–1728.
MAFF (1991) *Dietary Supplements and Health Foods.* Report of the Working Group, MAFF Publications, London, Reference PBO 139.
Marx, S.J., Liberman, U.A. and Eif, C. (1983) Calciferols: action and deficiency in action. *Vitamins and Hormones* **40**, 235–308.
McClaughlin, P.J. and Weihrauch, J.L. (1979) Vitamin E content of foods. *J. Am. Diet. Assoc.* **75**, 647–665.
McCord, C.P. (1971) Scurvey as an occupational disease. *J. Occup. Med.* **13**, 441–447.
McKenzie, J. (1968) The dissemination of misinformation: a growing problem. *Proc. Nutr. Soc.* **27**, 117–118.
Merry, P., Grootveld, M., Lunec, J. and Blake, D.R. (1991) Oxidative damage to lipids within the inflamed human joint provides evidence of radical-mediated hypoxic-reperfusion injury. *Am. J. Clin. Nutr.* **53** 632S–369S.
Milunsky, A., Jick, H., Jick, S., Bruell, C.L. MacLaughlin, D.S. Rothman, K.J. and Willet, W. (1989) Multi vitamin / folic acid supplementation in early pregnancy reduces the prevalence of neural tube defects. *JAMA* **262**, 2847–2852.
Moore, T. (1957) *Vitamin A.* Elsevier, Amsterdam.
Muller, D.P.R. (1987) Free radical problems of the newborn. *Proc. Nutr. Soc.* **46**, 69–75.
National Research Council (1943) *Recommended Dietary Allowances.* National Academy of Sciences, Washington D.C.
Newton, H.M.V., Schorah, C.J., Habidzadeh, N., Morgan, D.B. and Hullin, P.R. (1985) The cause and correction of low blood vitamin C concentrations in the elderly. *Am. J. Clin. Nutr.* **42**, 656–659.
Nicholls, L. (1938a) *Tropical Nutrition and Dietetics.* (revised by H.M. Sinclair and D.B. Jelliffe, 1961) Bailliere, Tindall and Cox, London, pp. 135–136.
Nicholls. L. (1938b) *Tropical Nutrition and Dietetics.* Bailliere, Tindall and Cox, London, pp. 138–140.
Olson, J.A. (1987a) RDI of vitamin C in humans. *Am. J. Clin. Nutr.* **45**, 693–703.
Olson, J.A. (1987b) RDI of vitamin K in humans. *Am. J. Clin. Nutr.* **45**, 687–692.

Palm, T.A. (1890) The geographical distribution and aetiology of rickets. *Practitioner* **45**, 270–279.
Rodriguez, M.S. (1978) A conspectus of research on folacin requirements of man. *J. Nutr.* **108**, 1983–2103.
Sabry, J.H., Fisher, K.H. and Dodds, M.L. (1958) Human utilization of dehydroascorbic acid. *J. Nutr.* **64**, 457–466.
Schorah, C.J. (1991) Undernutrition in the U.K.: its impact on health. *Food. Sci. Technol. Today* **5**, 32–35.
Schorah, C.J. and Habidzadeh, N. (1986) Folates in embryonic development and old age. *Proceedings 13th International Congress Nutrition* (ed. T.G. Taylor and N.K. Jenkins) John Libby, London, pp. 457–460.
Shrimpton, D.H. (1993 in press) Diseases of affluence. In: *Encyclopaedia of Food Science, Food Technology and Nutrition* (eds G. Fullerlove, R. Macrae, R. Robinson and M. Sadler). Academic Press, London.
Sinclair, H.M. (1955) Vitamin deficiencies in alcoholism. *Proc. Nutr. Soc.* **14**, 107–115.
Sommer, A. (1982) *Nutritional Blindness: Xerophthalmia and Keratomalacia.* Oxford University Press, Oxford.
Stephen, J.M.L. (1975) Epidemiological and dietary aspects of rickets and osteomalacia. *Proc. Nutr. Soc.* **34**, 131–138.
Stewart, C.P. (1953) Scurvey in the nineteenth century and after. In: *Lind's treatise on Scurvey* (ed. C.P. Stewart and D. Guthrie) Edinburgh University Press, Edinburgh, pp. 404–412.
Stockley, I. (1981) *Drug Interactions.* Blackwell Scientific Publishers, Oxford.
Traber, M.G. and Kayden, H. (1989) Alpha-tocopherol as compared with gamma tocopherol is preferentially secreted in humans. *Annals N.Y. Acad. Sci.* **570**, 95–108.
Trusswell, A.S. 1983 Recommended dietary intakes. *Nutr. Abstr. Rev. Series A* **53**, 94–1015.
Trusswell, A.S., Irwin, T., Beaton, G.H., Suzue, R., Haenel, H., Hejda, S., X-C. Hou., Leveille, G., Morava, E., Pederson, J. and Stephen, J.M.L. (1983) Recommended dietary intakes around the world. *Nutr. Abstr. Rev. Series A* **53** 1075–1119.
Unna, K.R. (1968) In: *The Vitamins, 2nd edition* (ed. W.J. Sebrell and R.S. Harris). Academic Press, London, pp. *150–115.*
Usui, Y., Tanimura, H., Nishimura, N., Kobayashi, N., Okanoue, T and Ozawa, K. (1990) Vitamin K concentration in the plasma and liver of surgical patients. *Am. J. Clin. Nutr.* **51**, 846–852.
Wald, N.J. and Polani, P.E. (1984) Neural tube defects and vitamins: the need for a randomised trial. *Brit. J. Obstetrics Gynaecology* **91**, 516–523.
West, C.E., Rombout, W.M., van der Zijp, A.J. and Sijtsma, S.R. (1991) Vitamin A and immune function. *Proc. Nutr. Soc.* **50**, 251–262.
WHO (1982) Control of vitamin A deficiency and xerophthalmia. *Technical Report Series* **674**. WHO, Geneva.
WHO (1990a) Diet, nutrition and the prevention of chronic diseases. *Technical Report Series* **797**, 626. WHO, Geneva.
WHO (1990b) Diet, nutrition and the prevention of chronic diseases. *Technical Report Series* **797**, 26. WHO, Geneva.
Zak, T.A. and D'Ambrosio, J.F.A. (1985) Nutritional nystigmus in infants. *J. Paediatric Opthal. Strabismus* **22**, 140–142.

4 Industrial production
M. J. O'LEARY

4.1 Introduction

Some excellent reviews of the chemistry, synthetic routes and technology of vitamin production have been produced (Kirk and Othmer, 1984; Isler *et al.*, 1988).

4.1.1 History

The true industrial production of vitamins probably began in 1937 with the plant-scale manufacture of vitamin C by Hoffmann–La Roche in Basle, Switzerland. Prior to this, vitamins were utilised as enriched mixtures such as fish oils and rose hip syrup. The exploitation of natural sources of vitamins, with several exceptions, has largely been replaced by synthetic production over the past fifty years.

Extraction of vitamins from natural sources suffers from a considerable number of disadvantages which limit its economic viability. Some of these are:

1. Variable availability of basic raw materials through seasonal effects or harvesting problems.
2. Variation in the concentration of the vitamin in the material.
3. Loss of yield through instability of the vitamin during the extraction process.
4. Low concentration of the vitamin, necessitating the processing of a large quantity of material and generation of a large volume of waste.
5. Difficulty of purifying the vitamin, resulting in a poor quality, unstable, product with adverse organoleptic properties.

The economic effect of these problems and the increasing need for vitamins for nutrition, particularly during the Second World War, were sufficient incentive for the development of synthetic methods for manufacture. The argument that extracted vitamins are preferable because of their natural sources has very little scientific basis except possibly for tocopherols. In addition, the environmental impact of extraction can often exceed that of the equivalent chemical production.

Production of a number of B vitamins began during the 1940s and by the mid 1950s, when a viable vitamin A synthesis was established, tonne quantities of a range of vitamins were available. During the 1960s, the market grew by about 15% per year, more synthetic vitamins were added to the list and prices were favourable enough to fund investment. At the end of this decade, five or six companies emerged as the main contenders in world markets and they built large dedicated plants to increase output and reduce costs.

Growth slowed down in the 1970s, partly because of the two oil crises, and the profitability diminished. Several less committed companies ceased production during this period giving opportunities to those remaining to gain more market share. Process development was mainly aimed at output scale-up, often utilising continuous methods of operation, and efficiency improvements. However, there was little change in the synthetic routes for the main vitamins.

During the 1980s, a significant reduction in growth was experienced across a wide range of vitamins thus increasing competition and lowering prices. This resulted in attempts by the main contenders to expand their market share by differentiation and the introduction of vitamin premixes or coated forms for various outlets.

Technological innovation was directed at cost saving on chemicals, energies and personnel costs. The last was achieved by use of increasing levels of automation and process control computers. An additional incentive for innovation emerged in this period through the impact of environmental and safety concerns.

The emergence of new techniques in biotechnology gave a boost to the search for improved processes for vitamin production.

4.1.2 Current situation

The average annual growth rate for most vitamins has continued to fall and is now probably less than 5%. About six companies now dominate the world vitamin market (Hoffmann–La Roche, Takeda, BASF, Rhône–Poulenc, Daiichi and Merck), with an additional influence exerted from time to time by China.

Rationalisation of the number of manufacturing sites is still occurring although occasionally a new plant emerges, as with the Takeda plant in the USA in 1989. Table 4.1 gives some data on the present status of world production.

The larger volume vitamin plants have become very complex as technological development has refined all the unit operations for high efficiency, low energy and personnel utilisation and minimum impact on the environment. The net result of this is that the capital cost of new plants is relatively high and incremental expansion of old plants has been preferred.

The operating standards of the main producers are generally very high, particularly as most companies have a large involvement in pharmaceutical activities. Quality and good manufacturing practice standards are thus maintained and many companies are engaged in total quality management initiatives and register for international quality standards such as ISO 9000. The main vitamin production plants are also operated to a high standard of safety and environmental care with a low impact on the environment.

Table 4.1 World production of vitamins

Vitamin	Type of production C = chemical, N = natural sources B = biochemical	Approx. tons/year world output*	Approx. price (1991) $/kg	Producers Firms	Producers Countries
A	C	2500	50	1, 2, 3	A, B, F, I, K
Provitamin A	C or B	100	400	1, 2, 3, 19, 20	A, B, F, R, S
B_1	C	2000	37	1, 3, 4, 17	B, C, E, G, H
B_2	C or B	2000	30	1, 3, 7, 18	B, C, D
Nicotinamide	C	8500	5	1, 12, 13, 14, 15, 16	A, B, C, D
Pantothenates	C	4000	10	1, 3, 5, 33	B, C, E, L
B_6	C	1600	43	1, 3, 4, 5, 6	B, C, H, J, M
Folic acid	C	300	115	3, 4, 17, 18, 21	A, C
B_{12}	B	5–10	5000	2, 10, 19, 20	E, F, P
C	N+C+B	65 000	13	1, 3, 4, 6	B, C, D, E, H, J, L, M
D	N+C	25	350	3, 11	B, Q
E	N+C,C	6800	20	1, 2, 3, 8, 21, 22	A, B, C, D, F
F(pufas)	N or B	2200	50–800	3, 10, 23, 24, 25, 26, 27, 28, 29	C, D, E, F
Biotin	C	3	5500	3, 6, 9, 17	A, B, C, D
K	C	1.3	1700	3, 8, 30, 31	A, C

* Source, Vandamme (1989), page 8)

Key to firms:

1, BASF; 2, Rhône-Poulenc; 3, Hoffmann-La Roche; 4, Takeda; 5, Daiichi; 6, Merck-Darmstadt; 7, Tanabe; 8, Eisai; 9, Sumitomo; 10, Roussel-Uclaf; 11, Philips-Duphar; 12, Lonza; 13, Degussa; 14, Yuki Gosei; 15, Nepera; 16, Reilly Tar and Chemicals; 17, Yodagawa; 18, Kongo; 19, Glaxo; 20, Farmitalia; 21, Eastman Kodak; 22, Henkel; 23, Cyanotech; 24, Dainippon Ink; 25, Efamol; 26, Nippon Oils and Fats; 27, QP Corporation; 28, RMC (Evening Primrose Oil Company); 29, Shiseido Co Ltd; 30, Nisshin Chemical; 31, Teikoku.

Key to countries:

A, Switzerland; B, Germany; C, Japan; D, USA; E, UK; F, France; G, Denmark; H, China; J, India; K, Russia; l, E.Europe; M, Croatia; N, S.Korea; P, Italy; Q, Netherlands; R, Australia; S, Israel.

4.1.3 Future production

Unless the emerging nations accelerate their rate of economic progress it is unlikely that the annual growth of the world vitamin market will reach again the level of former years. Consequently, only a few new plants will be built in the future for capacity expansion. The main incentive for new plants would thus be cost saving, rationalisation or replacement of old plants.

Innovation is mainly focused on biotechnology because of the expected increase in environmental pressures on general energy consumption as well as minimisation of waste and use of harmful chemicals. The imposition of a carbon tax and the eventual inability to dispose of waste will force companies to use biochemical processes wherever possible. Progress has been made in the past five years on vitamin C, β-carotene, vitamin B_2 and biotin biosynthesis but the rate of development of these processes is slow (Vandamme, 1989).

The main factors that cause progress to be slow are:

1. The high efficiency and low cost of chemical synthesis present a very difficult target to beat.
2. Low market growth and potential overcapacity are a disincentive to build new plants.
3. The investment costs for building new plants are very high.
4. There are high research and development costs, with extensive development required to give high conversions and concentrations, with stable genetically engineered organisms.
5. The technological risk of a completely new process is high and cannot completely be eliminated by extensive piloting.

Despite these constraints, it is nevertheless likely that a good proportion of vitamin production will be by biotechnological means within fifteen years.

4.2 Vitamin production

The processes utilised for manufacture of the individual vitamins follow in alphabetical order as in Table 4.1.

4.2.1 Vitamin A

The main commercial form of vitamin A is the acetate which is manufactured on a large scale by three companies, Hoffmann–La Roche, Rhône–Poulenc (AEC) and BASF. The main plants are situated in Germany, Switzerland, France and the USA.

All the processes currently in use proceed via the key C_{13} component β-ionone (Mayer and Isler, 1971, Figure 1.1). In the original Hoffmann–La Roche route (Isler *et al.*, 1947), this was made from citral, isolated from lemon grass oil. Synthetic routes starting from acetone were developed later and form the basis of current

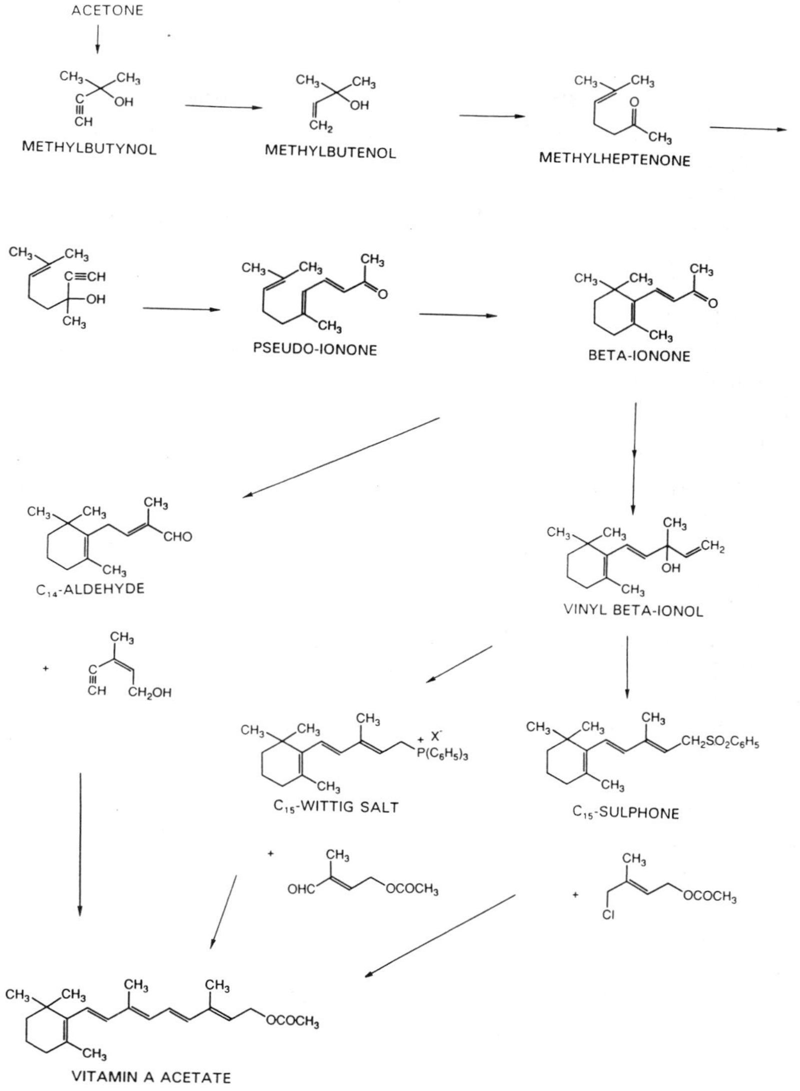

Figure 4.1 Synthetic route to vitamin A from acetone.

manufacturing (Figure 4.1). One route (Mayer, 1971) adds a C_2 unit via an acetylenic Grignard reaction followed by partial hydrogenation to methylbutenol. Reaction with isopropenyl methyl ether followed by a Claisen rearrangement adds a C_3 component to give methylheptenone. Repetition of this process but without the partial hydrogenation yields pseudo-ionone which can be readily cyclised to the β-isomer.

Conversion of β-ionone to vitamin A acetate is carried out by three different routes (Figure 4.1). The Hoffmann–La Roche version employs a $C_{13} + C_1 + C_6$ approach. The C_1 is added to β-ionone by a Darzens glycidic ester condensation yielding a C_{14}-aldehyde. A C_6 component is then added through a Grignard reaction and the product rearranged, isomerised and esterified to give all-*trans* vitamin A acetate. The BASF and Rhone–Poulenc routes use a C_{15} plus C_5 strategy, the former via the Wittig reaction (Pommer, 1960) and the latter via the C_{15}-sulphone (Julia and Arnold, 1973). All three routes are competitive and their relative commercial success depends as much on the scale of manufacture and the extent of refinement of the process as on chemical efficiency.

The Julia reaction has an advantage as the condensing agent can be recycled, whereas in the Wittig case the triphenylphosphine has to be regenerated using highly toxic reagents. All the plants in use by the main manufacturers employ a mixture of batch and continuous operations, and are probably computer controlled with minimum personnel involved.

The existence of large integrated plants and the likely overcapacity in the world market are deterrents to the introduction of new, potentially more efficient routes. However, as the existing plants get older the necessity to reinvest may cause new processes to be established. Any new route that utilises intermediates of application in carotenoid and other manufactures would have an advantage.

The economic characteristics of the current manufactures are all very similar. They are very capital intensive both in terms of the cost and complexity of the chemical plants and in the use of infrastructure such as for energy production and waste treatment. Because the processes utilise basic chemicals for building the terpene chain, the chemical costs do not dominate, and personnel, energy and engineering costs have a significant contribution.

4.2.2 Provitamin A: β-carotene

β-carotene is the most commercially significant of the carotenoids and is both isolated from natural sources and synthesised. The natural source is usually algal. *Dunaliella salina* is a green alga found in high salt concentration natural waters, colouring the brine red. This alga contains a substantial concentration of intracellular β-carotene. For growth, the alga requires a high light intensity and salt concentration, but a low nitrogen level in the medium. The carbon source is carbon dioxide from the atmosphere or carbonate ion in solution.

Isolation of β-carotene from this source has been commercialised in the USA (Klausner, 1986), Israel (Rich, 1978) and Australia (Borowitzka and Borowitzka, 1989). However, this source has had minor commercial significance because of the costs associated with harvesting the low concentration of algae in the brine, and the low, weather-dependent growth rate in large areas of ponds. Nevertheless, a small market is assured because it is a natural source of this pro-vitamin. Industrial production of β-carotene concentrates from the fungus *Blakeslea trispora* has also been reported (Ninet and Renault, 1979).

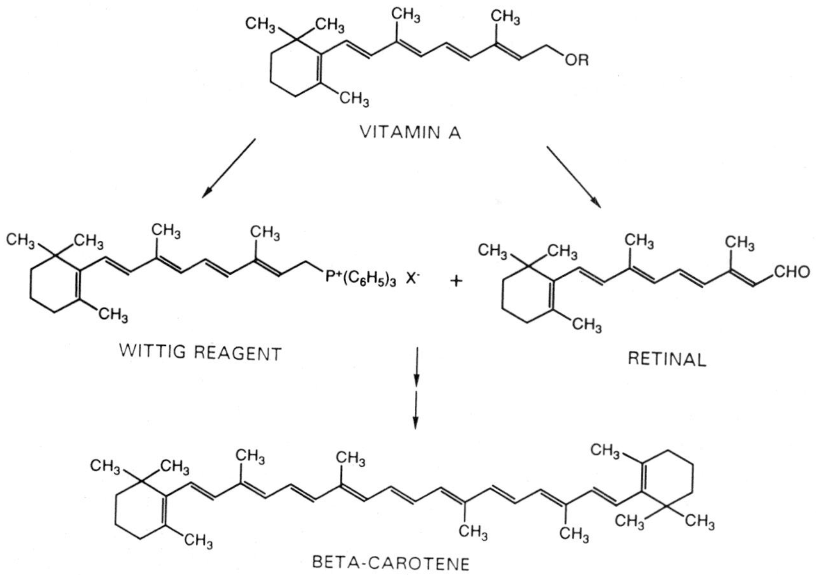

Figure 4.2 Synthesis of β-carotene from vitamin A.

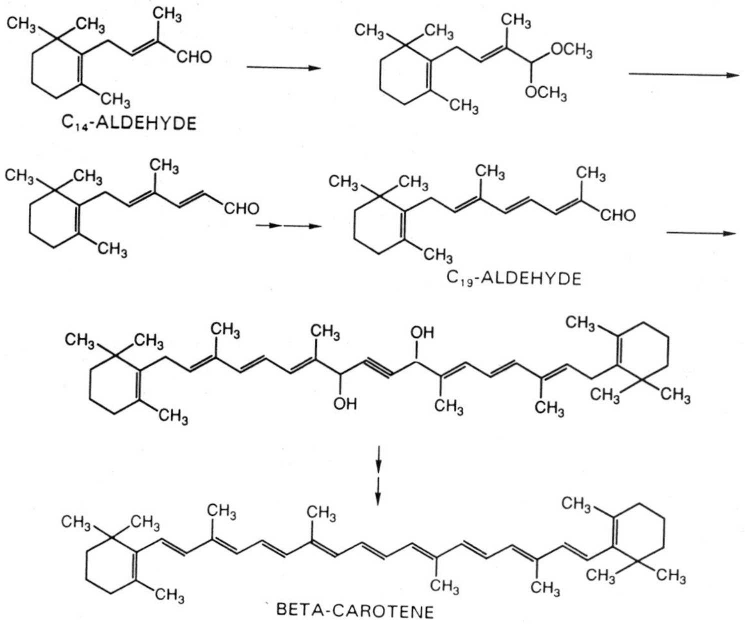

Figure 4.3 Hoffmann-La Roche production of β-carotene utilising a C_{14} aldehyde intermediate from vitamin A process.

The synthesis of β-carotene has been carried out on a commercial scale since the early 1960s. There were initially three processes in operation, by Hoffmann–La Roche, BASF and Rhone–Poulenc (AEC). Vitamin A was the starting material in each case (Figure 4.2).

The processes involve conversion of vitamin A to retinal and then condensing this with another C_{20} vitamin A-derived reagent to give the C_{40} β-carotene molecule. The chain joining reaction is typically through the Wittig reagent (Pommer, 1969).

The $C_{20} + C_{20}$ routes were used because well developed vitamin A productions were already in operation. As β-carotene sales increased it became economic to develop other routes with alternative ways of building up the C_{40} chain. Typical of these is the approach used by Hoffmann–La Roche (Isler *et al.*, 1956) which utilises the C_{14} aldehyde intermediate from the vitamin A process and extends the chain by a cheap C_2 and C_3 homologation to a C_{19} aldehyde (Figure 4.3). Two C_{19} units are then joined through an acetylene Grignard reaction to give the C_{40} carotene, initially in *cis* configuration, which is isomerised and isolated in the mainly *trans* form.

As world consumption of β-carotene increases, other routes will undoubtedly be established that do not utilise vitamin A or its current intermediates.

The β-carotene plants currently in operation are smaller and less well developed than the vitamin A plants. Towards the end of each process, significant volumes of solvent have to be used because of the relatively low solubility of β-carotene. In all routes from vitamin A the chemical cost is the most significant contributor to the overall economics.

4.2.3 Vitamin B_1: thiamine

Only chemical routes are used for the manufacture of the two commercial forms of vitamin B_1, thiamine dihydrochloride and mononitrate. The main suppliers of B_1 are currently Hoffmann–La Roche, Takeda, BASF and Tanabe Seiyaku. Other plants exist in India, Russia and China, the last giving rise to some exports into the western world.

The main routes used at present involve synthesising the pyrimidine ring first, and proceed via the key intermediate Grewe diamine (Figure 4.4) (Isler *et al.*, 1988).

The Roche and BASF processes are believed to start from malononitrile which is made from cyanogen chloride, principally by Lonza in Switzerland. Malononitrile (Figure 4.4) is reacted with trimethylorthoformate to give a methoxymethylene derivative (**1**) which is then reacted with ammonia to yield aminomethylene malononitrile. The base catalysed condensation of this intermediate with acetoiminoether hydrochloride gives the pyrimidine ring (**3**), initially in the nitrile form. Catalytic hydrogenation converts this into Grewe diamine.

The current approach used by Takeda and Tanabe is believed to commence with the catalytic oxidation of acrylonitrile to dimethoxypropionitrile (**4**). This interme-

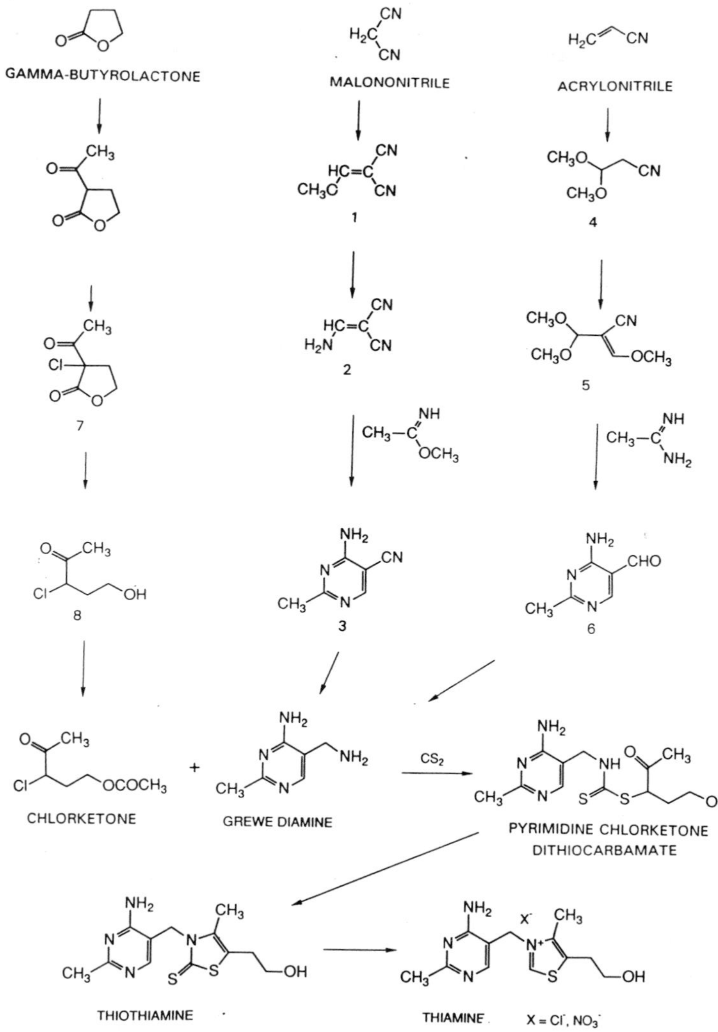

Figure 4.4 Main routes to synthesis of vitamin B₁, thiamine.

diate can be converted into Grewe diamine by a variety of routes. One patent (Migashiro, 1988) describes catalytic carbonylation, followed by *O*-alkylation to give an enol compound (**5**) which can be condensed with acetamidine to give the pyrimidine ring (**6**). Reductive amination of this compound yields Grewe diamine.

The routes from acrylonitrile have the advantage of a much cheaper starting material but the yields are lower than for the malonitrile approach.

In most plants the Grewe diamine is further processed by reaction with carbon disulphide to give a dithiocarbamate followed by addition of a chlorketone acetate. The latter intermediate is usually synthesised from γ-butyrolactone by a Claisen

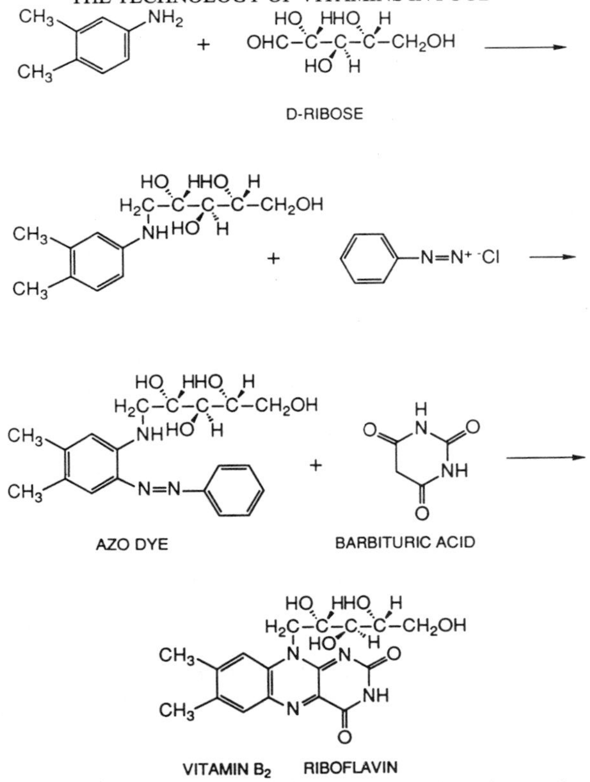

Figure 4.5 Vitamin B$_2$, riboflavin, is synthesised via D-ribose from D-glucose.

condensation with methylacetate followed by chlorination, decarboxylation and acetylation (**7,8**). The pyrimidine chlorketone dithiocarbamate is then cyclised into thiothiamine through acid treatment and the additional sulphur atom then replaced with oxygen by means of hydrogen peroxide. This route originated in Japan (Matsukawa, 1970). An earlier variation involving preparation of chlorinated thiazole and quaternization with a pyrimidine derivative may still be operated in some small plants.

Thiamine plants are equipment intensive because of the large number of steps involved but most of the operations use conventional technology with batch reactions and isolation of intermediates by crystallisation. The individual step yields are high but many reagents are used causing the chemical costs to be more significant than other factors. Some chlorine- and sulphur-containing by-products have to be treated with care.

4.2.4 Vitamin B$_2$: riboflavin

Vitamin B$_2$ is mainly produced by chemical synthesis but fermentation processes are beginning to become more significant. The synthetic route used is the same for all the major manufacturers (Takeda, Hoffmann—La Roche and BASF; see Isler et al., 1988).

The key intermediate D-ribose is currently manufactured from D-glucose by a *Bacillus* fermentation. This process was developed to replace a four-step chemical route from D-glucose that encountered severe environmental problems through the use of sodium amalgam. D-ribose is then condensed with 3, 4-dimethylaniline under reductive conditions to give a secondary amine which is reacted with benzene diazonium chloride (Figure 4.5). The resultant azo dye compound is finally condensed with barbituric acid to give riboflavin. Because of its low solubility riboflavin is difficult to purify so that animal feed and pharmaceutical food grades are produced.

Conventional batch processing equipment is probably used in the chemical steps of the synthesis. Both the azo dye intermediate and riboflavin give small crystals in their purification steps which are difficult to filter and wash efficiently. The intense colour of these two compounds makes good housekeeping on the plants difficult to achieve.

Unlike vitamin B_1, there are many organisms that are capable of producing reasonable concentrations of B_2. Extensive development work has been carried out and a commercial process for feed-grade production has been described (Pfeifer et al., 1950). This utilised *Ashbya gossypu*, but a more recent process with higher efficiency has utilised a mutant strain of *Candida flaveri* (Coors, 1985). Both BASF and Coors Biotech have production plants utilising microbial processes and isolation and purification methods will probably be developed to give a feed-grade product.

4.2.5 Niacin

Two forms of niacin are manufactured and sold, nicotinamide and nicotinic acid. The main suppliers are Lonza, BASF, Degussa and Nepera.

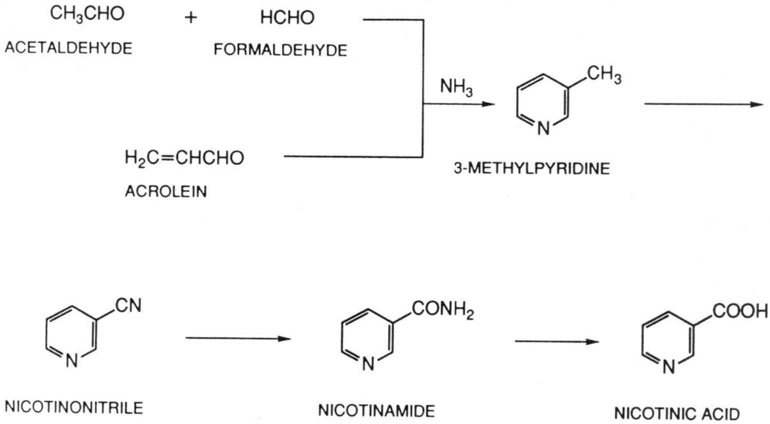

Figure 4.6 Synthesis of niacin as nicotinamide or nicotinic acid.

It is likely that most of the processes in current use (Offermanns et al., 1984) proceed via 3-methylpyridine or 2-methyl-5-ethylpyridine (Figure 4.6). These are prepared by condensing simple components such as acrolein, acetaldehyde and formaldehyde with ammonia. The alkyl pyridine is converted into nicotinonitrile in an ammoxidation reaction carried out in the vapour phase using heterogeneous catalysts comprised of vanadium, titanium and other metal oxides. Nicotinonitrile is partially hydrolysed to nicotinamide which can be further hydrolysed to nicotinic acid.

4.2.6 *Pantothenic acid*

The commercial form of this vitamin is usually calcium pantothenate. Both the active (*R*)-form (Daiichi, BASF and Roche) and the (*R, S*)-racemic mixture (Alps) are produced although the vitamin content of the latter material is only about 45%.

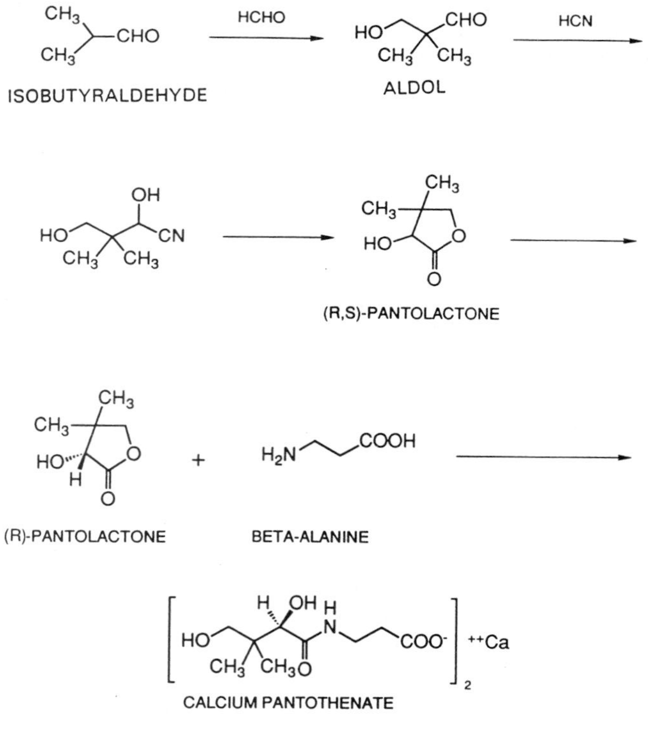

Figure 4.7 Synthetic manufacture of calcium pantothenate.

The first step in the manufacture by all producers involves a base-catalysed aldol reaction between formaldehyde and isobutyraldehyde (Isler et al., 1988). The intermediate aldol is then reacted with cyanide ion and the resulting cyanohydrin hydrolysed and lactonised. In modern plants this step is carried out on a continuous basis and pure (R, S)-pantolactone is isolated.

(R, S)-calcium pantothenate producers then react the lactone with calcium β-alaninate and isolate the product by crystallisation. For (R)-calcium pantothenate, a number of routes have been developed. The potentially simplest route, patented by Daiichi, involves optical resolution of (R, S)-calcium pantothenate by kinetic crystallisation of a methanol / water solvate. The direct yield per crystallisation is only about 25% but the optical purity is high and liquors can be repetitively recycled. Racemisation of the unwanted (S)-calcium pantothenate, however, causes yield loss.

Alternative routes proceed via optical resolution of (R, S)-pantolactone by conventional methods (Isler et al., 1988) or by kinetic crystallisation of the free lactone or its lithium salt. A plant using the latter method developed by Jenapharm was operated for several years in the USA prior to 1982.

Processes involving oxidation of (R, S)-pantolactone to ketopantolactone and stereo-specific reduction to (R)-pantolactone are also described (Ogima et al., 1978). All routes to calcium pantothenate proceeding via (R)-pantolactone have the advantage that this intermediate can also be utilised to make (R)-panthenol, an ingredient of some cosmetics.

Conversion of (R)-pantolactone into calcium pantothenate proceeds via condensation with either sodium β-alaninate, followed by ion-exchange, or directly with calcium β-alaninate (Figure 4.7). The β-alaninate is invariably produced by the reaction of acrylonitrile with ammonia and alkaline hydrolysis of the β-aminopropionitrile to give the calcium or sodium salt.

Biochemical routes for (R)-calcium pantothenate are unlikely to compete with chemical processes unless an efficient stereo-specific enzymatic cyanohydrin reaction is developed to give (R)-pantolactone directly.

Two relatively small new plants constructed in the USA in the early 1980s were not a commercial success, but BASF in Germany commenced manufacture during the same period and the plant is still in operation. The other world-scale plants (Daiichi, Alps and Roche) have been operating in Japan and Scotland for a considerable time.

The processes for (R, S)-pantolactone manufacture tend to be continuous in operation, whereas the diastereomeric resolution to (R)-pantolactone involves batch crystallisations and filtrations. Isolation of calcium pantothenate is achieved either by crystallisation or spray-drying.

4.2.7 Vitamin B_6: pyridoxine

The original production routes for B_6 manufacture (Isler et al., 1988; Coffen, 1984) involved many steps to insert all the substituents onto the pyridine ring. However, the situation was transformed after publication of a Diels–Alder approach by some Russian workers (Karpeiskii and Florenter, 1969). From this original work a number of alternative Diels–Alder production processes were developed and are currently utilised by BASF, Hoffmann–La Roche, Merck–Darmstadt, Daiichi and Takeda.

The components of the Diels–Alder reaction are an oxazole and a butenediol (Figure 4.8).

The substituents have been selected based on ease of manufacture of the components and the efficiency of the Diels–Alder reaction. The butenediol component is prepared by condensation of acetylene with two moles of formaldehyde followed by partial hydrogenation to 2-butene-1, 4 diol. The diol is protected prior to the Diels–Alder reaction by conversion to the dioxepine or diacetate.

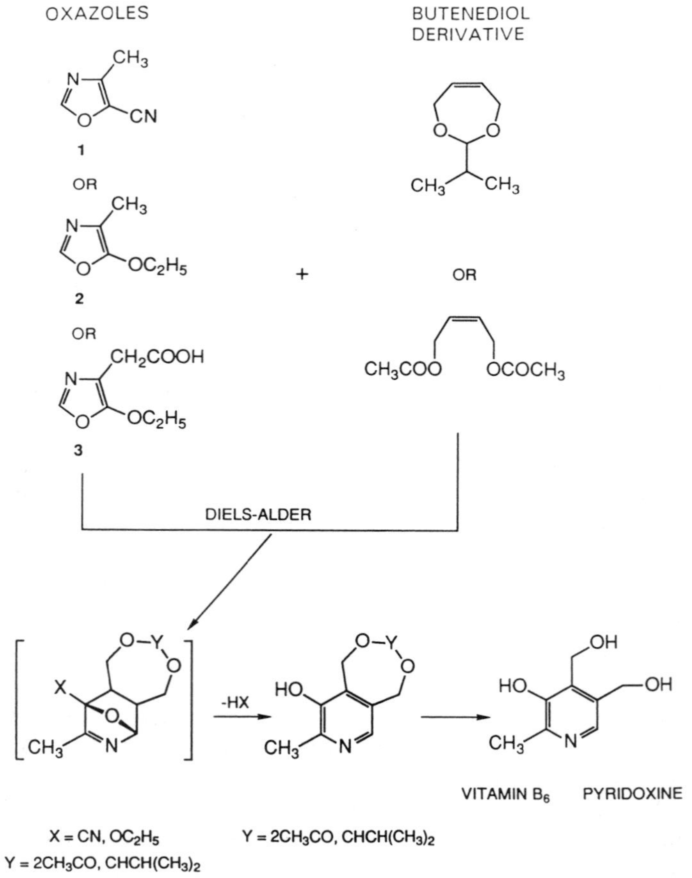

Figure 4.8 Manufacture of vitamin B_6, pyridoxine via a Diels–Alder reaction.

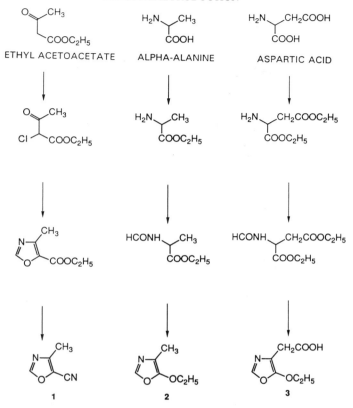

Figure 4.9 Synthesis of three oxazoles for use in the Diels–Alder reaction in the manufacture of vitamin B_6.

Four different oxazoles are described in the patent literature. One route starts from α-alanine (Figure 4.9) which is esterified, formylated and then cyclised typically with phosphorus pentoxide.

An alternative involves chlorination of ethyl acetoacetate and then cyclisation with formamide. The resultant oxazole ester is then converted to a cyanoxazole via the amide.

Aspartic acid is the starting point for the third version which is analogous to the α-alanine route.

The Diels–Alder reaction with these two oxazoles occurs with the elimination of ethanol or hydrogen cyanide respectively (Figure 4.8). Removal of the protective group yields pyridoxine.

BASF derived publications (Boll and Konig, 1979; Konig and Boll, 1977) describe the Diels–Alder reaction of a sulphone derivative of dihydrofuran with 4-methyloxazole (Figure 4.10). However, the protective ether group then must be opened up with hydrogen bromide and the free alcohol groups liberated by displacement with acetate and hydrolysis.

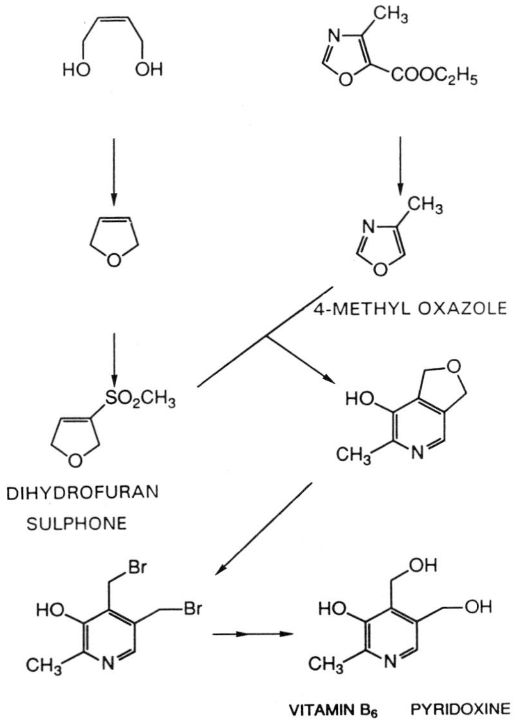

Figure 4.10 Diels–Alder reaction of a sulphone derivative of dihydrofuran with 4-methyloxazole.

Vitamin B_6 plants utilise mainly batch processes and are equipment intensive in view of the number of steps involved.

4.2.8 Folic acid

The main producers of folic acid are Hoffmann–La Roche in Switzerland and Daicel, Takeda, Yodagaura and Kango in Japan. Material for internal consumption is also produced in China and India. All manufacturers use basically the same process which involves reacting three key components in the final step (Isler *et al.*, 1988). These components are either purchased or synthesised by the firms involved (Figure 4.11).

The pyrimidine ring component 2,5,6-triamino-4-oxopyrimidine can readily be produced by condensation of ethylcyanoacetate with guanidine followed by nitrosation and catalytic reduction.

The second component trichloracetone or dichloroacrolein presents more of a problem as the compounds are lachrymatory and chlorinated by-products have to be disposed of. The yield of the desired compound is not particularly high owing to under- and over-chlorination.

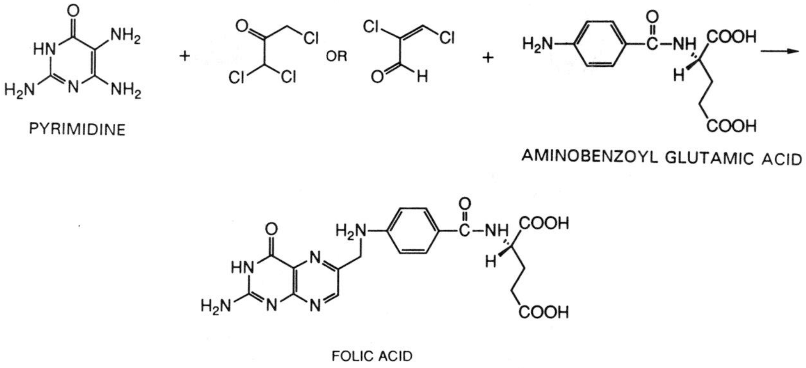

Figure 4.11 Manufacturing route to folic acid.

The aminobenzoylglutamic acid component is manufactured by condensing *p*-nitrobenzoyl chloride with sodium glutamate and reducing the nitro group to amino. The end step condensation of the three components is usually a one-pot process in aqueous hydrochloric acid.

4.2.9 Vitamin B_{12}

In view of the complex structure of vitamin B_{12} it is not surprising that there is no synthetic production. Fermentation processes (Friedrich, 1984) are used by the principle suppliers Glaxo, Rhone–Poulenc, Farmitalia and Roussel–Uclaf, using mutant strains of *Pseudomonas* species or propionibacterium. The fermentations are mainly aerobic and conducted on a fairly large scale in view of the low concentration and the length of time required. A mixture of cobalamins are initially produced and extracted into organic solvents. Treatment with cyanide converts this to cyanocobalamin– vitamin B_{12}.

4.2.10 Vitamin C

The scale of manufacture of vitamin C or L-ascorbic acid is considerably greater than that of any other vitamin. There are several plants in the world that have a reported capacity of over 10 000 tonnes per year (Delic *et al.*, 1989) (Figure 4.12).

The synthetic route used for the greater part of this production is still based on the original Reichstein process (Reichstein and Grussner, 1934) although the route has been extensively developed over the past fifty years. This development was focused on progressive scale-up from the original kilogramme manufacture to the present situation, and on increasing the yield and chemical efficiency. In more recent times the process has undergone considerable refinement to minimise energy usage, eliminate environmental problems and increase the level of automation and control.

Figure 4.12 Hoffmann-La Roche vitamin C plant.

The first two steps in the process (Figure 4.13), the hydrogenation of glucose and the biochemical oxidation to sorbose, have changed the least since Reichstein as they are both very efficient conversions.

A large number of variations on the remaining steps have been published, but few of them are in current use (Jaffe, 1984). The acetonisation reaction to protect hydroxyl groups on carbons 3–6 can be carried out with catalytic quantities of acid catalysts such as perchloric acid (Halder *et al.*, 1971) instead of molar amounts of sulphuric acid.

This avoids the cost of neutralisation and disposal of large quantities of sodium or ammonium sulphate. However, the concentration of the reaction in acetone is much lower and energy costs are high to remove water to displace the equilibrium in favour of 2,3:4,6-diacetone-sorbose (DAS). A plant recently constructed by Takeda in the USA utilises such a process.

The acetone protection step can be avoided altogether by direct catalytic oxidation of sorbose to 2-keto-L-gulonic acid (KGA). The relatively low yield of about 50% and the difficult isolation of pure KGA are major disadvantages of this variant.

The oxidation of the diacetone derivative can be efficiently carried out with alkaline sodium hypochlorite, by catalytic air oxidation (Jaffe and Pleven, 1972) or by direct electrolytic oxidation (Frohlich *et al.*, 1969). The relative economics of these methods depends on electricity costs or proximity to a chlorine producer. The sodium salt of the acid is produced initially and pure diacetone-2-keto-L-gulonic acid (DAG) is obtained by precipitation after mineral acid addition. Rearrangement of DAG after acetone removal can be effected by alkaline or acid catalysis. The alkaline route proceeds via intermediate ester formation and yields initially sodium ascorbate. The cost of sodium removal makes this approach less attractive

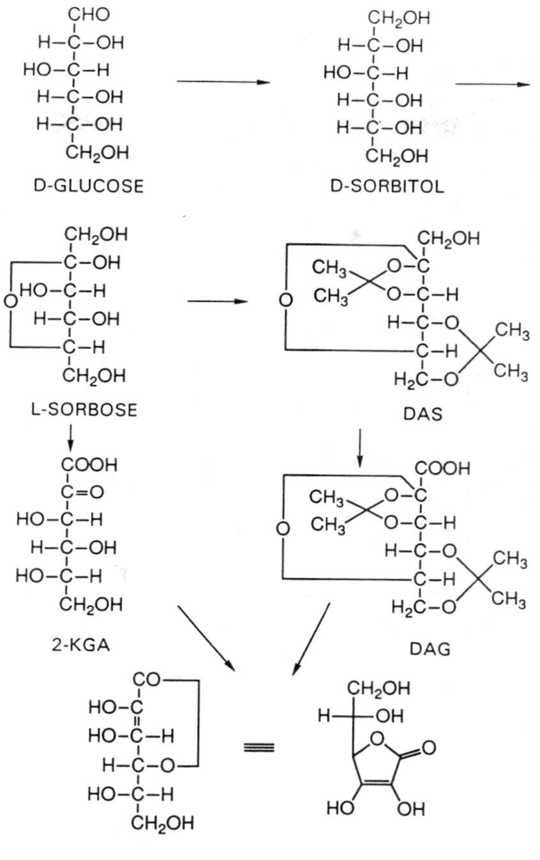

Figure 4.13 Hydrogenation of glucose and biochemical oxidation to sorbose in synthetic route to vitamin C.

than the acid rearrangement but use of recyclable long chain organic amines may be an improvement.

The acid catalysed acetone removal and rearrangement is the main current route to convert the DAG into ascorbic acid. Many patents describe alternative solvent mixtures to effect this transformation, including toluene, chlorinated hydrocarbons, acetone and ethanol. The preferred acid is normally hydrochloric and crystalline ascorbic acid is obtained directly from the mixture. Purification is achieved by crystallisation from water.

A significant proportion of ascorbic acid is further processed and sold as the sodium or calcium salt, the palmitate ester or various phosphate derivatives.

An extensive amount of research and development has been carried out on replacement of some or all of the chemical steps in the synthesis by biochemical processes (Delic et al., 1979). One of the simplest of such processes used on a commercial scale in China involves a double fermentation using two bacteria of the genera *Gluconobacter* and *Bacillus* in a one-pot fermenter to convert sorbose to 2-keto-L-gulonic acid (Ning *et al.*, 1988).

The concentration and yield are probably lower than for the chemical process but the simplicity makes smaller plants competitive. A Japanese process has been described involving conversion of glucose to 2, 5-diketogulonic acid using a mutant strain of *Erwinia*, followed by conversion to 2-keto-L-gulonic acid using a Corynebacterium mutant (Sonoyama *et al.*, 1982).

There are numerous other examples in the literature, including some involving genetically engineered organisms, for replacement of two or more of the chemical steps.

With the economy of scale and efficiency improvements, the production cost in real terms has reduced over the past twenty years but the scope for further improvement is limited. The factors controlling the economics of production and the high capital and development cost mean that it is very difficult for alternative processes to compete and the construction of new plants is thus very infrequent. When further new plants are constructed, however, it is very likely that they will involve replacement of at least two of the Reichstein steps by a biochemical process.

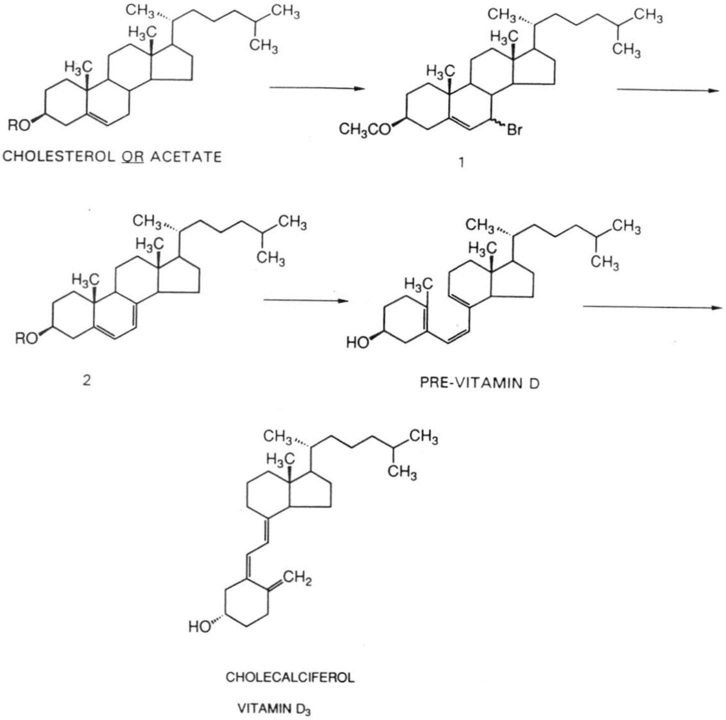

Figure 4.14 Commercial preparation of vitamin D_3 from cholesterol.

4.2.11 Vitamin D

Two vitamin D-active compounds have been commercialised, ergocalciferol (vitamin D_2) and cholecalciferol (vitamin D_3). The latter is by far the more important and is produced principally by Phillips–Duphar and Hoffmann–La Roche.

The starting point for vitamin D_3 is cholesterol (Hirsch, 1984) which is converted to its acetate (Figure 4.14), brominated at C-7 (**1**), typically with *N*-bromosuccinimide, and then subjected to a base-catalysed elimination to introduce additional saturation (**2**). The product is then saponified and photochemically converted to pre-vitamin D which is thermally rearranged to vitamin D_3. The technology of carrying out an organic photochemical reaction on a production scale was developed for this transformation. The other steps are typically batch reactions in conventional equipment.

Vitamin D_2 is prepared in an anologous way but starting from ergosterol.

Biosynthetic methods of vitamin D production, or isolation from natural sources have not achieved commercial significance (Margalith, 1989).

4.2.12 Vitamin E: α-tocopherols

Vitamin E is sold either as a mixture of eight diastereomers (all-racemic-α-tocopherol) by BASF, Rhone–Poulenc, Hoffmann–La Roche and Eisai or as the nature-identical (R,R,R)-α-tocopherol by Eastman Kodak, Henkel and Eisai. The all-racemic material is produced by chemical synthesis, whereas (R, R, R)-tocopherols are normally obtained by extraction from vegetable oils. The vitamin E activity of these forms are similar and both are commercialised mainly as acetate ester.

All synthetic approaches to tocopherols utilise the acid-catalysed condensation of trimethyl hydroquinone (TMHQ; see Figure 4.15) with isophytol which proceeds in high yield (Kasparek, 1980; Isler and Brubacher, 1982). The crude product is then converted into the acetate form prior to purification by high-vacuum fractional distillation.

The trimethyl hydroquinone component is prepared by hydrogenation of the corresponding quinone, which can be obtained by catalytic air oxidation of the trimethyl phenol. Originally coal tar sources were utilised as a source for trimethyl phenol by methylation of dimethyl phenol. Phenol can also be methylated in several steps, but aliphatic sources via condensation of crotonaldehyde with diethylketone are also utilised.

The isophytol component is manufactured by similar isoprenoid chain extension methods to those used for vitamin A and β-carotene.

Thus pseudo-ionone is a common intermediate and is hydrogenated to the saturated ketone before reaction with a metal acetylide. A three-carbon chain extension is then effected by an elegant Claisen rearrangement of an isopropenyl methyl ether adduct. Hydrogenation of the C_{18} ketone and addition of metal acetylide followed by semi-hydrogenation gives isophytol (Figure 4.16).

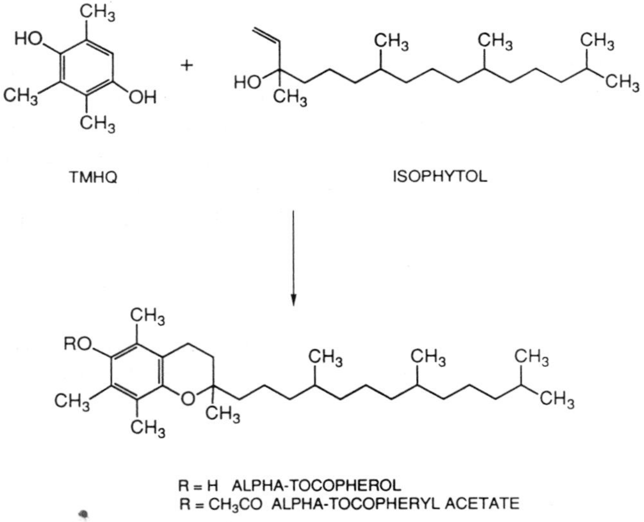

Figure 4.15 Synthetic manufacture of α-tocopherols via TMHQ and isophytol.

Continuous processes have been developed for some of the steps in the isophytol synthesis and the yields are usually high.

The extraction of (R, R, R)-α-tocopherol from natural sources yields first a mixture with β-, γ- and δ-tocopherols. These isomers are converted into α-tocopherol by chemical methylation. The total synthesis of (R, R, R,)-α-tocopherol can be achieved but production is not yet commercially viable.

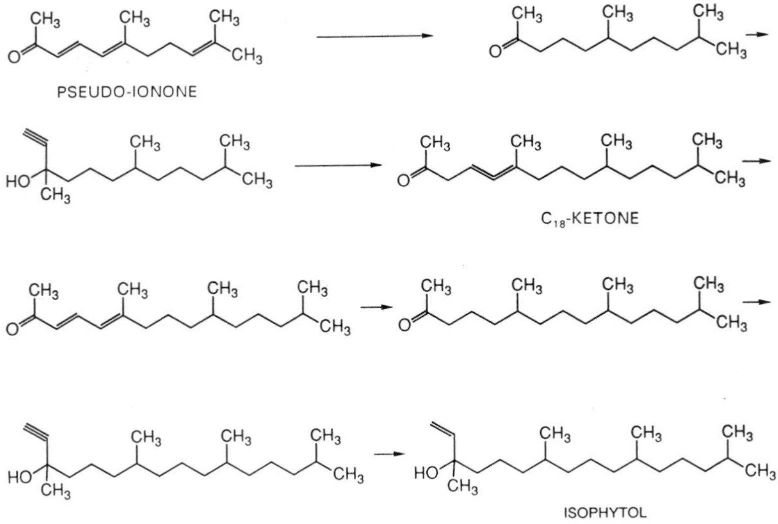

Figure 4.16 Synthetic route to isophytol.

4.2.13 Vitamin F group

Various polyunsaturated fatty acids (PUFAs) have been shown to be essential for human nutrition and are sometimes referred to as the vitamin F group. Concentrates of these compounds are usually prepared by extraction from natural sources. The extraction process usually aims to try to enrich the PUFAs from a mixture of less essential fatty acids. The essential oils are characterised by the first double bond occurring at the C-3 or C-6 from the end of the carbon chain, and they can be C_{18} or C_{20} acids. Typical are γ-linolenic acid (GLA), found in evening primrose seed, and eicosapentaenoic acid (EPA) and docosahexaenoic acid (DHA), found in fish oils.

Microbiological sources of PUFAs have also been investigated and commercial production of γ-linolenic acid using a Mortierella fungus has apparently started in Japan (Shimizu, 1989).

4.2.14 Biotin (vitamin H)

Of all the vitamins currently synthesised on a manufacturing scale (+)-biotin has the most complex process. This is partly due to the fact that it is a chiral compound with three stereogenic centres. The routes used by the major producers, Hoffmann–La Roche, Sumitomo Pharmaceutical, Tanabe Seiyaka and Merck–Darmstadt, probably derive from the original Roche route (Goldberg and Sternbach, 1949). However, much development work has been done on alternative synthesis and improvements to the original route (Uskokovic, 1984; Isler, 1988).

The basic route proceeds via the cycloanhydride (Figure 4.17) which is produced in four steps from fumaric acid. The chemistry is straightforward but the toxic reagents bromine and phosgene are employed in the first and third steps, respectively. This puts constraints on the location for carrying out this part of the

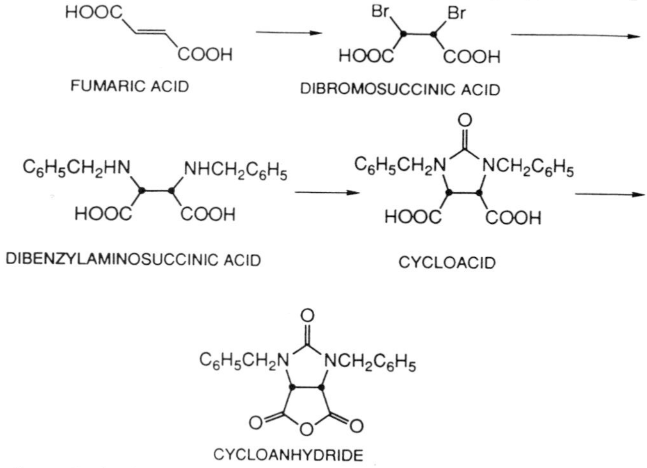

Figure 4.17 Biotin synthesis. (i) Synthetic route from fumaric acid in four steps to the cycloanhydride.

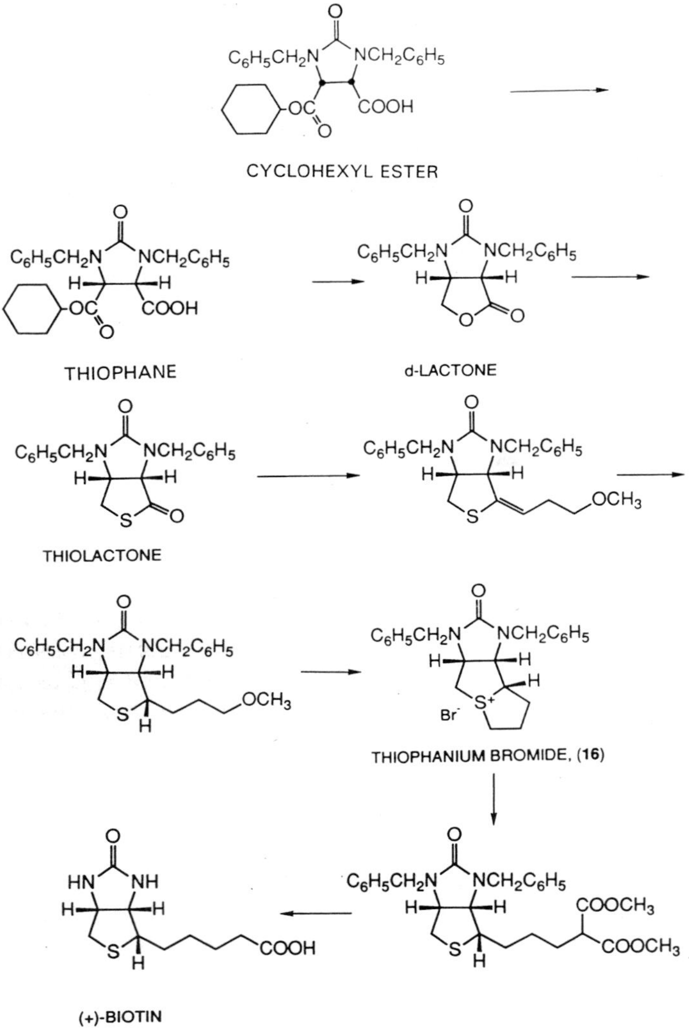

Figure 4.18 Biotin synthesis. (ii) Synthesis from the cyclohexyl ester to biotin.

synthesis. The cycloanhydride is resolved by conversion to the cyclohexyl ester and diastereomeric salt formation with (+)-ephedrine (Figure 4.18). The desired enantiomer is selectively reduced by lithium borohydride yielding the *d*-lactone. The unwanted enantiomer is hydrolysed and the cycloanhydride regenerated for repeated resolution. An alternative method described for the resolution (Hisao *et al.*, 1975) proceeding via a chiral imide is possibly employed by Sumitomo.

Further processing of the *d*-lactone proceeds via the thiolactone and then introduction of the side chain. In the original method this was achieved with a Grignard reagent made from 3-methoxypropyl chloride.

The following steps are dehydration and stereo-selective hydrogenation to give the thiophene. Treatment of this with hydrobromic acid, then reaction with dimethyl malonate anion, hydrolysis and decarboxylation yields (+)-biotin. A Wittig reaction has also been described for adding a 5-carbon side chain directly.

The majority of the reactions and isolation processes are undoubtedly carried out batchwise and the whole synthesis requires a large number of equipment items. However, it is possible to use flexible multi-purpose equipment and produce key intermediates on a campaign basis.

Further development of the chemical synthesis is undoubtedly continuing in an attempt to reduce the number of steps, avoid the use of phosgene and some of the expensive reagents, and improve the yields.

Fermentation processes for (+)-biotin are also under development, but the maximum concentrations produced so far are uneconomical at about 100 mg / litre (Izumi and Yamada, 1989). Some work has also been done on the bioconversion of biotin precursors such as pimelic acid but the economics do not appear favourable yet.

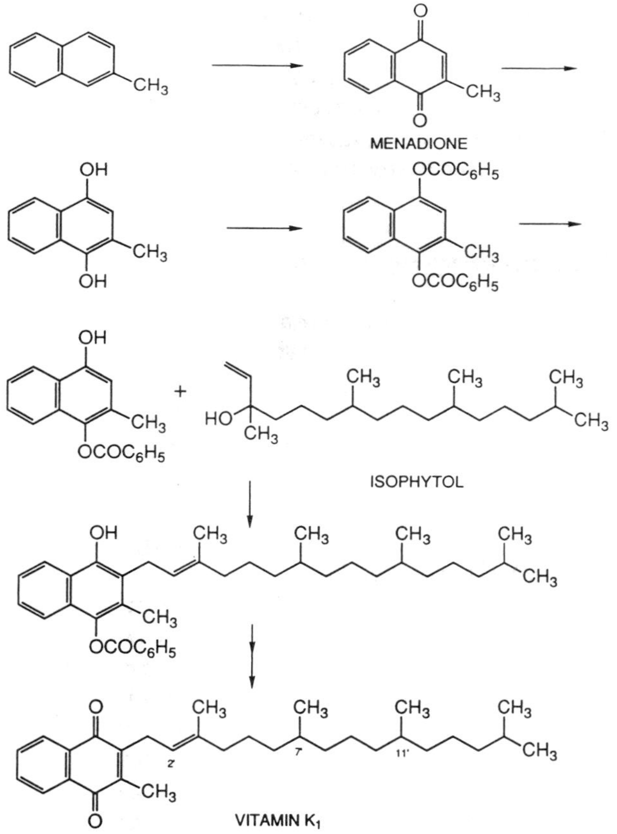

Figure 4.19 Synthetic route to vitamin K.

4.2.15 Vitamin K

Vitamin K compounds are similar in structure to tocopherols and indeed utilise the same chemical intermediate, isophytol, to construct the side chain. The main compound commercialised is phylloquinone, vitamin K_1. This is produced principally by Hoffmann–La Roche, Eisai Nisshin Chemical and Taikoku.

Like tocopherols, vitamin K_1 is also a chiral molecule, possessing two stereogenic centres (Figure 4.19). However, additionally with a double bond in the side chain at C_2 there are Z and E isomers. Naturally occurring vitamin D_3 is one diastereomer only.

Synthesis of the aromatic ring system commences typically with 2-methylnaphthalene which is oxidised to the quinone menadione. Catalytic hydrogenation yields the corresponding hydroquinone which is protected by forming the benzoyl derivative of the hydroxyl group adjacent to the methyl group. Condensation with isophytol is then directed at the required location. Hydrolysis of the benzoate and oxidation to the quinone structure yields vitamin K_1.

No microbiological process has yet been developed for vitamin K.

References

Böll, W. and König, H. (1979) *Liebigs Ann Chem*, 1657.
Borowitzka, L.J. and Borowitzka, M.A. (1989) In: *Biotechnology of Vitamins, Pigments and Growth Factors* (ed. E.J. Vandamme). Elsevier, London, pp. 15–26.
Coffen, D.L. (1984) In: *Encyclopedia of Chemical Technology* (ed. R. Kirk and D.F. Othmer). John Wiley and sons, New York, pp. 94–107.
Coors (1985) *European Patent* No. 231,605.
Delić, V., Šunić, D. and Vlašeć (1989), In: *Biotechnology of Vitamins, Pigments and Growth Factors* (ed. E.J. Vandamme). Elsevier, London, pp. 299–333.
Frohlich, G.J., Kratavil, A.J. and Zrike, E. (1969) *U.S. Patent* 3,453, 191.
Goldberg, M.W. and Sternbach, L.H. (1949) *U.S. Patent* 2,489,234.
Halder, N., Hindley, N.C., Jaffe, G.M., O'Leary, M.J. and Weinert, P. (1971) *U.S. Patent* 3,607,862.
Hirsch, A.L. (1984) In: *Encyclopedia of Chemical Technology* (eds R. Kirk and D.F. Othmer). John Wiley and sons, New York, Volume 24, pp. 186–213.
Hisao, A., Yasuhiko, A., Shigeru, O. and Hiroyuki, S. (1975) *U.S. Patent* 3,876,656.
Isler, O., Huber, W., Ronco, A. and Koffer, M. (1947) *Helv. Chim. Acta* **30**, 1911.
Isler, O., Lindlar, H., Montavon, M., Ruegg, R. and Zeller, P. (1956) *Helv. Chim. Acta* **39**, 249.
Isler, O. and Brubacher, S. (1982) *Vitamine I*. Georg Thieme Verlag, New York, pp. 126–151.
Isler, O Brubacher, S., Ghisla, G. and Kräutler, B. (1988) *Vitamin II*. Georg Thieme Verlag, New York.
Izumi, Y. and Yamada, H. (1989) In: *Biotechnology of Vitamins, Pigments and Growth Factors* (ed. E.J. Vandamme) Elsevier, London, pp. 231–256.
Jaffe, G.M. (1984) In: *Encyclopaedia of Chemical Technology* (eds R. Kirk and D.F. Othmer). John Wiley and sons, New York, Volume 24, pp. 8–40.
Jaffe, G.M. and Pleven, E. (1972) *U.S. Patent* 3,832,355.
Julia, M. and Arnold, D. (1973) *Bull. Soc. Chim. Fr.*, 746.
Karpeiskii M. Ya. and Florent'ev, V.L. (1969) *Russ. Chem. Rev.* **38**, 540.
Kasparek, S. (1980) Chemistry of tocopherols and tocotrienols. In: *Vitamin E: A Comprehensive Treatise* (ed. L.E. Machlin). Marcel Dekker, New York, pp. 7–65.
Kirk, R. and Othmer, D.F. (1984) (eds.) *Encyclopedia of Chemical Technology*. John Wiley and Sons, New York.
Klausner, A. (1986). Algaculture: food for thought. *Biotechnology* **4**, 947–53.
Konig, H. and Boll, W. (1977) *U.S. Patent* 4,039,554 (to BASF).

Margalith, P. (1989) In: *Biotechnology of Vitamins, Pigments and Growth Factors* (ed. E.J. Vandamme). Elsevier, London, pp. 81–93.
Mayer, H. and Isler, O. (1971) *Total Syntheses in Carotenoids* (ed. O. Isler). Birkhauser Verlag, Basle.
Ninet, L. and Renault, J. (1979). Carotenoids. In: *Microbial Technology*, 2nd edition, Volume 1 (eds H.J. Peppler and D. Perlman). Academic Press, New York, pp. 529–44.
Ning, W., Tao, Z., Wang, C., Wang, S., Yan, Z. and Yin, G. (1988) *European Patent* No. 278,447.
Offermanns, H., Klemann, A., Tanner, H., Beschke, H. and Friedrich, H. (1984) Vitamins (nicotinamide and nicotinic acid) In: *Kirk–Othmer Encyclopedia of Chemical Technology*, (ed. M. Grayson). John Wiley and Sons, New York, Volume 24, pp. 59–93.
Ojima, I., Kogure, T., Terasaki, T. and Achiwa, K. (1978) *J. Org. Chem.* **43**, 3444–6.
Pfeifer, V.F., Tanner, F.W., Vojnovich, C. and Traufler, D.H. (1950) Riboflavin by Fermentation with Ashbya gossypii. *Ind. Eng. Chem.* **42**, 1776–81.
Pommer, H. (1960). *Angew. Chem.* **72**, 811.
Reichstein, T. and Grussner, A. (1934) *Helv. Chim. Acta* **17**, 311.
Rich, V. (1978) Israel's place in the sun. *Nature* **275**, 581–2.
Sonoyama, T., Tani, H., Kageyama, B., Matsuda, K., Tanimoto, M., Kobayashi, K., Yagi, S., Kyotani, H. and Mitsushima, K. (1982) *Appl. Environ. Microbiol.* **43**, 1064–9.
Uskokovic, M.R. (1984) In: *Encyclopedia of Chemical Technology*, Volume 24 (eds. R. Kirk and D.F. Othmer). J. Wiley and sons, New York, pp. 41–49.
Vandamme, E.J. (1989) (ed.) *Biotechnology of Vitamins, Pigments and Growth Factors*. Elsevier Science Publishers, London.

5 Stability of vitamins in food
P. BERRY OTTAWAY

5.1 Introduction

Vitamins are one of the few groups of food constituents in which it is possible to demonstrate quantitatively a deterioration in content over a period. The rate of this reduction in vitamin content is influenced by a number of factors:

1. Temperature
2. Moisture
3. Oxygen
4. Light
5. pH
6. Oxidising and reducing agents
7. Presence of metallic ions (e.g. iron, copper)
8. Presence of other vitamins
9. Other components of food such as sulphur dioxide
10. Combinations of the above.

The most important of these factors are heat, moisture, oxygen, pH and light. Vitamin stability can also be affected by the reaction of the vitamins with other ingredients or components of the food. With the increased use of nutritional labelling of food products the vitamin levels in foods have become the subject of label claims that can easily be checked by enforcement authorities.

Chemically the vitamins are a heterogeneous group of compounds with no common structural attributes; some are single compounds (e.g. biotin) while others (e.g. vitamin E) are large groups of compounds; and the stability of the individual vitamins varies widely from the relatively stable, in the case of niacin, to the relatively unstable. An example of the latter is vitamin B_{12} (cyanocobalamin) which is decomposed by oxidising and reducing agents, sensitive to light, unstable in acid or alkaline solutions, and interacts adversely with vitamin B_1, ascorbic acid and niacin.

Vitamin deterioration can take place naturally during the storage of vegetables and fruits as seen in the gradual reduction of the vitamin C content of potatoes (Figure 5.1), and losses can occur during the processing and preparation of ingredients and foods, particularly those subjected to heat treatment. The factors that affect the degradation of vitamins are the same whether the vitamins are naturally occurring or are added to the food from synthetic sources.

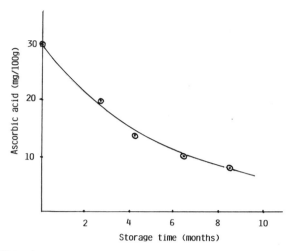

Figure 5.1 Effect of storage on vitamin C content of potatoes (adapted from Marks, 1975).

The use of vitamins for the enrichment or fortification of food products poses a number of technological problems. If more than one vitamin is used it is highly probable that the rates of degradation of each of the vitamins will differ significantly; and, if the amounts of these micronutrients are included in nutritional labelling, the shelf life of the product is determined by the life of the most unstable component.

In order to comply with the legal requirements of maintaining the label claim throughout the life of the product the technologist needs to obtain a reasonably accurate estimation of the stability of the various vitamins added to the formulation. This has to be evaluated in the context of the food system (solid, liquid, etc.), the packaging and the probable storage conditions and is achieved by conducting well designed stability trials. Data obtained from these trials should lead to the determination of a realistic shelf life and expiry date.

Where the rates of degradation of the vitamins in a product are found to vary widely the formulated levels of the more sensitive vitamins are adjusted above the declared value to ensure that all label declarations are met at the end of the stated shelf life of the product. These increases over the declared values are termed 'overages', and in order to determine the required overages the technologist needs to be aware of the factors which will affect the stability of the vitamin.

5.2 Fat soluble vitamins

5.2.1 Vitamin A

Vitamin A is one of the more labile vitamins, with retinol being less stable than the retinyl esters. The presence of double bonds in its structure makes it subject to isomerisation, particularly at acid pH in an aqueous medium. The isomer with the highest biological activity is the all-*trans* vitamin A. The predominant *cis* isomer is 13-*cis* or neovitamin A which only has a biological activity of 75% of the all-*trans* isomer; and 6-*cis* and 2, 6-di-*cis* isomers which may also form during isomerisation have less than 25% of the biological activity of the all-*trans* form of vitamin A. The natural vitamin A preparations usually contain about one third neovitamin A while most synthetic sources generally contain considerably less. For aqueous products where isomerisation is known to occur, mixtures of vitamin A palmitate isomers at the equilibrium ratio have been produced commercially. Vitamin A is relatively stable in alkaline solutions.

Vitamin A is sensitive to atmospheric oxygen with the alcohol form being less stable than the esters. The decomposition is catalysed by the presence of trace minerals. As a consequence of its sensitivity to oxygen, vitamin A is normally available commercially as a preparation which includes an antioxidant and often a protective coating. While butylated hydroxyanisole (BHA) and butylated hydroxytoluene (BHT) are permitted in a number of countries for use as antioxidants in vitamin A preparations, the recent trend has been towards the use of tocopherols (vitamin E).

Both retinol and its esters are inactivated by the ultraviolet component of light.

In general, vitamin A is relatively stable during food processing involving heating, with the palmitate ester more stable to heat than retinol. It is normally regarded as stable during milk processing, and food composition tables give only small differences between the retinol contents of fresh whole milk, sterilised and ultra high temperature (UHT) treated milk (Paul and Southgate, 1979). However, prolonged holding of milk or butter at high temperatures in the presence of air can be shown to result in a significant decrease in the vitamin A activity.

5.2.2 Vitamin E

There is a considerable difference in the stability of the naturally occurring (tocopherol) forms of vitamin E and the tocopheryl esters. While vitamin E is regarded as being one of the more stable vitamins, the unesterified tocopherol is less stable due to the free phenolic hydroxyl group.

Vitamin E is unusual in that it exhibits *reduced* stability at temperatures below freezing. The explanation given for this is that the peroxides formed during fat oxidation are degraded at higher temperatures but are stable at temperatures below 0°C and as a consequence can react with the vitamin E (IFST, 1989). It has also been shown that α-tocopherol may function as a pro-oxidant in the presence of metal ions such as iron.

α-Tocopherol is readily oxidised by air. It is stable to heat in the absence of air but is degraded if heated in the presence of air. α-Tocopherol is readily oxidised during the processing and storage of foods. The most important naturally occurring sources of tocopherols are the vegetable oils, particularly wheat germ and cottonseed oils. While deep frying of the oils may result in a loss of vitamin E of around

10%, it has been found that the storage of fried foods, even at temperatures as low as -12°C, can result in very significant losses.

dl-α-tocopheryl acetate is relatively stable in air but is hydrolysed by moisture in the presence of alkalis or strong acids to free tocopherols.

5.2.3 Vitamin D

The two forms of vitamin D which are commercially available are vitamin D_2 (ergocalciferol) and vitamin D_3 (cholecalciferol). Crystalline vitamin D_2 is sensitive to atmospheric oxygen and will show signs of decomposition after a few days storage in the presence of air at ambient temperatures. Crystalline cholecalciferol (D_3) is also destroyed by atmospheric oxygen but is relatively more stable than D_2, possibly due to the fact that it has one less double bond.

Both vitamins D_2 and D_3 are sensitive to light and can be destroyed relatively rapidly if exposed to light. They are also adversely affected by acids.

Preparations of vitamin D in edible oils are more stable than the crystalline forms, and the vitamin is normally provided for commercial usage as an oil preparation or stabilised powder containing an antioxidant (usually tocopherol). The preparations are usually provided in light proof containers with inert gas flushing.

The presence of double bonds in the structure of both forms of vitamin D can make them susceptible to isomerisation under certain conditions. Studies have shown that the isomerisation rates of ergocalciferol and cholecalciferol are almost equal. Isomerisation in solutions of cholecalciferol resulted in an equilibrium being formed between ergocalciferol and precalciferol with the ratio of the isomers being temperature dependent. The isomerisation of ergocalciferol has been studied in powders prepared with calcium sulphate, calcium phosphate, talc and magnesium trisilicate. It was found that the isomerisation was catalysed by the surface acid of these additives (De Ritter, 1982).

5.2.4 Vitamin K

Vitamin K occurs in a number of forms. Vitamin K_1 (phytomenadione or phylloquinone) is found in green plants and vegetables, potatoes and fruits, while vitamin K_2 (menaquinone) can be found in animal and microbial materials.

Vitamin K is rarely added to food products and the most common commercially available form is K_1 (phytomenadione), which is insoluble in water. A water soluble K_3 is available as menadione sodium bisulphite.

The various forms of vitamin K are relatively stable to heat and are retained after most cooking processes. The vitamin is destroyed by sunlight and is decomposed by alkalis. Vitamin K_1 is only slowly affected by atmospheric oxygen.

The presence of double bonds in both vitamin K_1 and K_2 makes them liable to isomerisation. Vitamin K_1 has only one double bond in the side chain in the 3-position whereas in K_2 double bonds recur regularly in the side chain. Vitamin

K$_1$ exists in the form of both *trans* and *cis* isomers. The *trans* isomer is the naturally occurring form and is the one which is biologically active. The *cis* form has no significant biological activity.

5.2.5 β-Carotene (provitamin A)

Provitamins are compounds which can be converted in the body to vitamins. These compounds are generally found in plant pigments but can also be found in liver, kidney, the spleen and milk.

The provitamin with most commercial and nutritional importance is β-carotene and its derivatives (β-apo-8'-carotenal and the ethyl ester of β-apo-8'-carotenoic acid). These provitamins all have vitamin A activity.

The stability of these carotenoids is similar to vitamin A in that they are sensitive to oxygen, light and acid media.

It has been reported that treatment with sulphur dioxide reduces carotenoid destruction in vegetables during dehydration and storage. A study by Baloch *et al.* (1977) with model systems showed that the stability of β-carotene was greatly enhanced by sulphur dioxide added either as a sulphite solution to cellulose powder prior to β-carotene absorption or as a headspace gas in containers of β-carotene. While it was found that the β-carotene stability was improved by increasing the nitrogen levels in the containers, the stability was even greater when the nitrogen was replaced by sulphur dioxide. Comparative values for the induction period were 19 hours for β-carotene samples stored in oxygen only, 120 hours in nitrogen and 252 hours in sulphur dioxide.

Investigations into the effect of sulphur dioxide treatment on the β-carotene stability in dehydrated vegetables have given varying results and it has been postulated that the effects of the drying and storage conditions on the stability of the sulphur dioxide has a consequential effect on the stability of the β-carotene in dehydrated products (Baloch, 1976).

Products containing β-carotene should be protected from light and headspace air kept to the minimum.

5.3 Water soluble vitamins

5.3.1 Thiamin (vitamin B$_1$)

While thiamin (vitamin B$_1$) is widely distributed in living tissues, it is only commercially available for addition to food in the form of thiamin chloridehydrochloride (commonly termed thiamin hydrochloride) and thiamin mononitrate. Both salts have specific areas of application as the hydrochloride is considerably more soluble in water than the mononitrate while the mononitrate is very much less hygroscopic and therefore is preferentially used in dry powder products which are required to have a long shelf life.

Both the thiamin salts are relatively stable to atmospheric oxygen in the absence of light and moisture, and both are normally considered to be very stable when used in dry products with light and moisture-proof packaging.

Thiamin is unstable in alkaline solutions and becomes increasingly unstable as the pH increases. The stability of the vitamin in low pH solutions such as fortified fruit drinks is very good.

A considerable amount of research has been carried out on the heat stability of thiamin and its salts, particularly in the context of cooking losses. Early work on thiamin losses during bread making showed an initial cleavage of the thiamin to pyrimidine and thiazole (Dwivedi and Arnold, 1973). The destruction of thiamin by heat is more rapid in alkaline media. Vitamin B_1 losses in milk, which has an average fresh content of 0.04 mg thiamin per 100 g, are normally less than 10% for pasteurised milk, between 10 and 20% for UHT milk and about 20% for sterilised milks. Between 30 and 50% of the vitamin B_1 activity can be lost during the production of evaporated milk.

Losses of thiamin during the commercial baking of white bread are between 15 and 20%. Part of this loss is due to the yeast fermentation which can convert thiamin to cocarboxylase which is less stable than thiamin.

Thiamin is very sensitive to sulphites and bisulphites as it is cleaved by sulphite. This reaction is rapid at high pH, and is the cause of large losses of the vitamin in vegetables blanched with sulphite, and in meat products such as comminuted meats where sulphites and bisulphites are used as preservatives. Where the pH is low, such as in citrus fruit juices, the bisulphite occurs mainly as the unionised acid, and thiamin losses in such systems are not significantly different from those in products not containing bisulphite (Borenstein, 1981).

Studies on the rate of sulphite-induced cleavage of thiamin during the preparation and storage of minced meat showed that losses of thiamin were linear with sulphur dioxide concentrates up to 0.1%. The storage temperature did not have a significant effect on the losses. It has also been reported that thiamin is cleaved by aromatic aldehydes.

Thiamin is decomposed by both oxidising and reducing agents. If it is allowed to stand in alkaline solution in air it is oxidised to the disulphide and small amounts of thiothiazolone. It is unstable in dilute solutions containing rutin or *para*-aminobenzoic acid which accelerate the oxidation of the vitamin by dissolved oxygen in neutral or alkaline solutions. The addition of sodium tartrate to the solution prevents oxidation.

A range of food ingredients have been shown to have an effect on the stability of thiamin. In general, proteins are protective of the vitamin, particularly food proteins such as egg albumin and casein. When heated with glucose, either as a dry mixture or in solution, a browning analogous to a Maillard reaction can occur. This reaction is similar to the reaction between sugars and amino acids and may be important in the loss of thiamin during heat processing. Work has shown that fructose, invertase, mannitol and inositol can actually retard the rate of destruction of thiamin (Dwivedi and Arnold, 1973).

The stability of thiamin is adversely affected by the presence of copper ions. This effect can be reduced by the addition of metal-chelating compounds such as calcium disodium ethylenediamine tetra-acetate (EDTA). The heavy metals only appear to influence thiamin stability when they are capable of forming complex anions with constituents of the medium.

The enzymes, thiaminases, which are present in small concentrations in a number of animal and vegetable food sources can degrade thiamin. These enzymes are most commonly found in a range of sea foods such as shrimps, clams and raw fish, but are also found in some varieties of beans, mustard seed and rice polishings. Two types of thiaminases are known and these are designated thiaminase I and thiaminase II. The former catalyses the decomposition of the thiamin by a base-exchange reaction involving a nucleophilic displacement of the methylene group of the pyrimidine moiety. Thiaminase II catalyses a simple hydrolysis of thiamin.

A problem associated with the addition of vitamin B_1 to food products is the unpleasant flavour and odour of the thiamin salts. The breakdown of thiamin, particularly during heating, may give rise to off-flavours, and the compounds derived from the degradation of the vitamins are believed to contribute to the 'cooked' flavours in a number of foods.

5.3.2 Riboflavin (vitamin B_2)

Like vitamin B_1, riboflavin (vitamin B_2) is widely found in both animal and plant tissue. While riboflavin is available commercially as a crystalline powder it is only sparingly soluble in water. As a consequence the sodium salt of riboflavin-5'-phosphate which is more soluble in water is used for liquid preparations.

Riboflavin and riboflavin phosphate are both stable to heat and atmospheric oxygen, particularly in an acid medium. In this respect riboflavin is regarded as being one of the more stable vitamins. It is degraded by reducing agents, and becomes increasingly unstable with increasing pH.

However, the most important factor influencing the stability of vitamin B_2 is light, with the greatest effect being caused by light in the 420 to 560 µm range. Fluorescent light is less harmful than direct sunlight, but products in transparent packaging can be affected by strip lighting in retail outlets.

While riboflavin is stable to the heat processing of milk, one of the main causes of loss in milk and milk products is from exposure to light. Liquid milk exposed to light can lose between 20 and 80% of its riboflavin content in two hours with the rate and extent of loss being dependent upon the light intensity, the temperature and the surface area of the container exposed. Although vitamin B_2 is sensitive to light, particularly in a liquid medium such as milk, it remains stable in white bread wrapped in transparent packaging and kept in a lit retail area.

5.3.3 Niacin

Niacin (nicotinic acid) and niacinamide (nicotinamide) are both available commercially and possess the same vitamin activity. Niacin occurs naturally in the meat and liver of hoofed animals and also in some plants. The form in which niacin is present in maize and some other cereals is not one which can be utilised by man unless treated with a mild alkali.

Both niacin and niacinamide are normally very stable in foods. They are stable to atmospheric oxygen, heat and light in both the dry state and in aqueous solution. Niacinamide can be hydrolysed to the acid by heating in strongly acid or alkaline solutions.

5.3.4 Pantothenic acid

Pantothenic acid is rarely found in the free state in animal and vegetable tissue although it is widely distributed in nature as a component of coenzyme A, where it occurs in animals in muscle tissue, liver, kidney, and in egg yolk and milk. It is also found in yeast, some cereals and legumes.

Free pantothenic acid is an unstable and very hygroscopic oil. Commercial preparations are normally provided as calcium or sodium salts. The alcohol form, panthenol, is available as a stable liquid but it is not widely used in foods.

Pantothenic acid is optically active and only its dextro-rotatory forms have vitamin activity. Losses of pantothenic acid during the preparation and cooking of foods are normally not very large. Milk generally loses less than 10% during processing, and meat losses during cooking are not excessive when compared to the other B vitamins.

The three commercial forms, calcium and sodium D-pantothenate and D-panthenol, are moderately stable to atmospheric oxygen and light when protected from moisture. All three compounds are hygroscopic with sodium pantothenate being the worst.

Aqueous solutions of both the salts and the alcohol form are thermolabile and will undergo hydrolytic cleavage, particularly at high or low pH. The compounds are unstable in both acid and alkaline solutions and maximum stability is in the pH range of 6 to 7. Aqueous solutions of D-panthenol are more stable than the salts, particularly in the pH range 3 to 5.

5.3.5 Folic acid

Compounds with folic acid activity contain one or more linked molecules of glutamic acid and this is an essential part of the biological activity. Folic acid is normally found in nature as a conjugate and is widely distributed throughout most living cells. The folic acid synthesised for commercial usage has only one glutamic group.

Commercially available crystalline folic acid is moderately stable to heat and atmospheric oxygen. It is stable in solution around pH 7 but becomes increasingly unstable in acid or alkali media, particularly at pH less than 5. Folic acid is decomposed by oxidising and reducing agents. Sunlight, and particularly ultraviolet radiation, has a serious effect on the stability of folic acid. Cleavage by light is more rapid in the presence of riboflavin. This reaction can be retarded by the addition of the antioxidant BHA to solutions containing folic acid and riboflavin (Tansey and Schneller, 1955).

Folic acid loss during the pasteurisation of milk is normally less than 5%. Losses in the region of 20% can occur during UHT treatment and about 30% loss is found after sterilisation. UHT milk stored for three months can lose over 50% of its folic acid. The extra heat treatment involved in boiling pasteurised milk can decrease the folic acid content by 20%. Losses of around 10% are found in boiled eggs, while other forms of cooking (fried, poached, scrambled) give between 30 and 35% loss. Total folic acid losses from vegetables as a result of heating and cooking processes can be as high as 50%.

5.3.6 Pyridoxine (vitamin B_6)

Vitamin B_6 activity is shown by three compounds, pyridoxol, pyridoxal and pyridoxamine. These are often considered together as pyridoxine. The commercial form is the salt, pyridoxine hydrochloride. Vitamin B_6 is found in red meat, liver, cod roe and liver, milk and green vegetables.

Pyridoxine hydrochloride is normally stable to atmospheric oxygen and heat. Decomposition is catalysed by metal ions.

Pyridoxine is sensitive to light, particularly in neutral and alkaline solutions. One of the main causes of loss of this vitamin in milk is sunlight with a 21% loss being reported after eight hours exposure (Borenstein, 1981).

Pyridoxine is stable in milk during pasteurisation but about 20% can be lost during sterilisation. Losses during UHT processing are around 27% (Scott and Bishop, 1986), but UHT milk stored for three months can lose 35% of this vitamin. Average losses as a result of roasting or grilling of meat are 20% with higher losses (30 to 60%) in stewed and boiled meat. Cooking or canning of vegetables results in losses of 20 to 40%.

5.3.7 Vitamin B_{12}

The most important compound with vitamin B_{12} activity is cyanocobalamin. This has a complicated chemical structure and occurs only in animal tissue and as a metabolite of certain microorganisms. The other compounds showing this vitamin activity differ only slightly from the cyanocobalamin structure. The central ring structure of the molecule is a 'corrin' ring with a central cobalt atom. In its natural form vitamin B_{12} is probably bound to peptides or protein.

Vitamin B_{12} is commercially available as crystalline cyanocobalamin which is a dark red powder. As human requirements of vitamin B_{12} are very low (about 2 μg a day), it is often supplied as a standardised dilution on a carrier.

Cyanocobalamin is decomposed by both oxidising and reducing agents. In neutral and weakly acid solutions it is relatively stable to both atmospheric oxygen and heat. It is only slightly stable in alkaline solutions and strong acids. It is sensitive to light and ultraviolet radiation, and controlled studies on the effect of light on cyanocobalamin in neutral aqueous solutions showed that sunlight at a brightness of 8000 foot candles caused a 10% loss for each thirty minutes of exposure, but exposure to levels of brightness below 300 foot candles had little effect (De Ritter, 1982).

Vitamin B_{12} is normally stable during pasteurisation of milk but up to 20% can be lost during sterilisation, and losses of 20 and 35% can occur after spray drying of milk.

The stability of vitamin B_{12} is significantly influenced by the presence of other vitamins.

5.3.8 Biotin

The chemical structure of biotin is such that eight different isomers are possible and of these only the dextro-rotatory or D-biotin possesses vitamin activity. D-biotin is widely distributed, but in small concentrations, in animal and plant tissues. It can occur both in the free state (milk, fruit and some vegetables) or in a form bound to protein (animal tissues and yeast). It is commercially available as a white crystalline powder.

Biotin is generally regarded as having a good stability, being fairly stable in air, heat and daylight. It can, however, be gradually decomposed by ultraviolet radiation.

Biotin in aqueous solutions is relatively stable if the solutions are either weakly acid or weakly alkaline. In strong acid or alkaline solutions the biological activity can be destroyed by heating.

Avidin, a protein complex, which is found in raw egg white, can react with biotin and bind it in such a way that the biotin is inactivated. Avidin is denatured by heat and biotin inactivation does not occur with cooked eggs.

5.3.9 Vitamin C

A number of compounds possess vitamin C activity and the most important is L-ascorbic acid. L-ascorbic acid is widely distributed in nature and can occur at high levels in some fruits and vegetables. The D isomer does not have vitamin C activity.

Ascorbic acid is the enolic form of 3-keto-1-gulofuranolactone. The endiol groups at C-2 and C-3 are sensitive to oxidation and can easily convert into a diketo group. The resultant compound, dehydro-L-ascorbic acid also has vitamin C activity.

Commercially, vitamin C is available as ascorbic acid and its calcium and sodium salts (calcium and sodium ascorbates).

Crystalline ascorbic acid is relatively stable in dry air but is unstable in the presence of moisture. It is readily oxidised in aqueous solutions, first forming dehydro-L-ascorbic acid which is then further and rapidly oxidised. Conversion to dehydroascorbic acid is reversible but the products of the latter stages of oxidation are irreversible.

Ascorbic acid is widely used in soft drinks and to restore manufacturing losses in fruit juices, particularly citrus juices. Research has shown that its stability in these products varies widely according to the composition and oxygen content of the solution. It is very unstable in apple juice but stability in blackcurrant juice is good possibly as a result of the protective effects of phenolic substances with antioxidant properties.

The effect of dissolved oxygen is very significant. As 11.2 mg of ascorbic acid is oxidised by 1.0 mg of oxygen, 75 to 100 mg of ascorbic acid can be destroyed by one litre of juice. Vacuum treatment stages are normally added to the process to deaerate the solution to reduce the problem. It is also important to avoid significant head-spaces in containers of liquids with added ascorbic acid as 3.3 mg of ascorbic acid can be destroyed by the oxygen in 1 cm^3 of air (Bender, 1958).

Traces of heavy-metal ions act as catalysts to the degradation of ascorbic acid. Studies on the stability of pharmaceutical solutions of ascorbic acid showed that the order of the effectiveness of the metallic ions was $Cu^{+2} > Fe^{+2} > Zn^{+2}$ (De Ritter, 1982). A Cu^{+2} – ascorbate complex has been identified as being intermediate in the oxidation of the ascorbic acid in the presence of Cu^{+2} ions. Other work on model systems has shown that copper ion levels as low as 0.85 ppm was sufficient to catalyse oxidation, and that the reaction rate was approximately proportional to the square root of the copper concentration.

Work with sequestrants has shown that ethylenediamine tetra-acetate (EDTA) has a significant effect on the reduction of ascorbic acid oxidation, with the optimal level of EDTA required to inhibit the oxidation of vitamin C in blackcurrant juice being a mole ratio of EDTA to [Cu + Fe] of approximately 2.3 (Timberlake 1960 a, b). Unfortunately, EDTA is not a permitted sequestrant for fruit juices in many countries. The amino acid cysteine has also been found effectively to inhibit ascorbic acid oxidation.

Cu and Fe ions play such a significant part in metal catalysed oxidation of ascorbic acid that the selection of process equipment can have a marked effect on the stability of vitamin C in food and drink products. Contact of product with bronze, brass, cold rolled steel or black iron surfaces or equipment should be avoided and only stainless steel, aluminium or plastic should be used.

The rate of ascorbic acid degradation in aqueous solutions is pH dependent with the maximum rate at about pH 4. Vitamin C losses can occur during the frozen storage of foods, and work has shown that oxidation of ascorbic acid is faster in ice than in the liquid water. Frozen orange concentrates can lose about 10% of their vitamin C content during twelve months' storage at -23°C (-10°F) (Grant and Alburn, 1965).

Light, either in the form of sunlight or white fluorescent light, can have an effect on the stability of vitamin C in milk with the extent of the losses being dependent on the translucency and permeability of the container and the length and conditions of exposure. Bottled orange drinks exposed to light have been found to lose up to 35% vitamin C in three months (Borenstein, 1981).

The destruction of vitamin C during processing or cooking of foods can be quite considerable, with losses during pasteurisation being around 25%, sterilisation about 60% and up to 100% in UHT milk stored for three months. Milk boiled from pasteurised can show losses of between 30 and 70%.

Large losses of vitamin C are also found after cooking or hot storage of vegetables and fruits. The commercial dehydration of potatoes can cause losses of between 35 and 45%. Destruction of vitamin C during the processing of vegetables depends on the physical processing used and the surface area of product exposed to oxygen. Slicing and dicing of vegetables will increase the rate of vitamin loss. Blanching of cabbage can produce losses of up to 20% of the vitamin C, whilst subsequent dehydration can account for a further 30%.

5.4 Vitamin–vitamin interactions

The fortification or enrichment of products using more than one added vitamin can lead to the possibility of mutual interactions of the vitamins, with the consequent reduction in the shelf stability of the product.

Most of the work in this area has been carried out in the pharmaceutical industry, and has tended to concentrate on the effects of vitamin combinations on both the stability and solubility of the vitamins in aqueous multivitamin solutions.

A number of studies have shown that the vitamins responsible for producing deleterious effects in combinations are ascorbic acid and four from the B group; thiamin, riboflavin, folic acid and cyanocobalamin. These five vitamins have been reported as having a number of mutual interactions, the principal ones are described below.

5.4.1 Ascorbic acid–folic acid

Cleavage of folic acid can occur in solutions containing both ascorbic and folic acids due to the reducing effect of ascorbic acid. The breakdown of the folic acid is most rapid in the pH range 3.0 to 3.3, and slowest at pH 6.5 to 6.7.

5.4.2 Ascorbic acid–vitamin B_{12}

Vitamin B_{12} (as cyanocobalamin) has been shown to be unstable in the presence of ascorbic acid in aqueous solution. This instability is pH dependent being greatest at pH 7 and least below pH 1. Cyanocobalamin has been found to be more stable in the presence of ascorbic acid than hydroxycobalamin.

It was originally thought that the vitamin B_{12} content of meals was seriously affected by the presence of vitamin C. Early reports indicated the destruction of between 50 and 90% of the vitamin B_{12} content of food by 0.5 g of vitamin C. More recent studies have shown that the low vitamin B_{12} values reported in the early work were entirely due to inappropriate extraction methods and that intakes of up to 1 g of vitamin C with a meal did not significantly decrease the vitamin B_{12} content (Klaui, 1979).

5.4.3 Thiamin–folic acid

Thiamin has been shown to have a significant effect on the stability of folic acid, particularly in the pH range 5.9 to 7. The decomposition of thiamin in solutions can affect the rate of breakdown of folic acid which is accelerated in the presence of the degradation products of thiamin, particularly hydrogen sulphide.

5.4.4 Thiamin–vitamin B_{12}

The decomposition of thiamin can also increase the rate of breakdown of cyanocobalamin with the main cause being attributed to 4-methyl-5-(β-hydroxyethyl) thiazole which is formed during thiamin cleavage.

5.4.5 Riboflavin–thiamin

Riboflavin can have an oxidative action on thiamin leading to the formation and precipitation of thiochrome. This reaction appears to be specific to solutions containing only the B-vitamins and is not seen in solutions containing thiamin, riboflavin and ascorbic acid.

5.4.6 Riboflavin–folic acid

The stability of folic acid is affected by the combined actions of riboflavin and light which produce an oxidative reaction resulting in the cleavage of the folic acid. This effect occurs more rapidly at pH 6.5 and can be reduced, but not eliminated, by deaeration.

5.4.7 Riboflavin–ascorbic acid

The oxidation of ascorbic acid exposed to light can be catalysed by the presence of riboflavin which acts as a light-energy receptor. This reaction is commonly found to be the cause of accelerated vitamin C loss in milk exposed to sunlight. Light-induced vitamin C oxidation is greatly reduced in milk which has had the riboflavin removed.

5.4.8 Other interactions

In addition to the above interactions which decrease vitamin stability, interactions have been reported which can increase the solubility of some of the less soluble vitamins in aqueous solutions.

Niacinamide acts as a solubiliser for both riboflavin and folic acid with the concentration reported to have an effect on riboflavin being a 1% solution.

Ascorbic acid is reactive as an acid and a reducing agent and is involved in many browning reactions. Vitamin B_6 may take part in Maillard reactions, and thiamin (vitamin B_1) can react with polyunsaturated fatty acids to produce the typical 'meat' flavour and odour.

5.5 Processing losses

The relative instability of vitamins as a group means that the naturally occurring vitamins in foods are susceptible to destruction during processing and storage. While the specific effects of some processing and storage conditions on the stability of vitamins has already been considered in an earlier part of this chapter, it is important that the treatments are reviewed in the context of their effects on vitamin stability in general. As many factors can influence the amounts of vitamins retained in processed foods it is only possible to give general indications of stability.

5.5.1 Vegetables and fruits

The vitamins in vegetables and fruits can be subjected to considerable temperatures during blanching and canning. Most of the research on this subject was carried out during the 1940s and 1950s, and although there have been substantial improvements in assay techniques, some of the general findings can still be considered to be valid. Guerrant *et al.* (1947) found that a high temperature–short time water blanch gave a better vitamin retention than a low temperature–long time blanch and that steam blanching was superior to water blanching. β-carotene was found to be the most stable of the vitamins assayed with 100% retention under the range of blanching conditions evaluated. Thiamin and riboflavin both had retentions in the range 80 to 95% while niacin was slightly more unstable at 75 to 90%. Vitamin C was in the range 70 to 90%. The addition of sulphite to the blanching water can significantly affect the thiamin levels in vegetables and fruits. Mallette *et al.* (1946) found a thiamin loss in cabbage of 45% in sulphite-treated blanching water compared with 15% in untreated water.

A comprehensive study of heat processing in tin and glass containers showed significant losses of vitamin C and thiamin. Ascorbic acid levels immediately after processing were between 15 and 45% of the fresh product and they reduced further during storage. Thiamin retained about 50% after processing and the levels reduced to between 15 and 40% after twelve months' storage. β-Carotene was found to be very stable. Riboflavin losses averaged about 12 to 15% immediately after processing but losses of around 50% were observed after twelve months' storage. Niacin showed initial losses of 15 to 25% but storage losses over the twelve months were not as great as riboflavin. Reduction of vitamin B_6 content during canning of vegetables is normally in the range of 20 to 30% (Guerrant and O'Hara, 1953).

Dehydration of blanched vegetables can produce additional vitamin loss. For example, dehydration of blanched cabbage can result in an additional 30% loss of vitamin C, 5 to 15% niacin and about 15% thiamin when prepared from an unsulphited blanch. Riboflavin is normally more stable than thiamin during dehydration.

While most vitamins are stable in frozen fruits and vegetables for periods up to a year, losses of vitamin C can occur at temperatures as low as -23°C.

Losses of B-vitamins and vitamin E during the milling and processing of cereals can be large. Up to 90% of vitamin B_6 can be lost during the milling of wheat, and extrusion cooking of cereals can result in the loss of up to 50% of the thiamin content.

Studies on folic acid content of flour show that it is fairly stable. After twelve months' storage at 29°C (84°F) there was a retention of 86% (Keagy *et al.*, 1973).

Vitamin losses during the baking of bread are about 20% for thiamin, up to 17% for vitamin B_6 and up to one third of the natural folic acid content. Pantothenic acid and niacin are relatively stable.

5.5.2 Meat

The reports of B-vitamin losses during the cooking and processing of meats vary widely. Cooking conditions can have a marked effect and it has been shown that the roasting temperatures can have a significant effect on the retention of thiamin in beef and pork (Cover *et al.*, 1949).

In general, riboflavin, niacin and vitamin B_{12} are stable during the cooking of meat if the vitamin content of the drippings is taken into consideration. Pantothenic acid losses are usually less than 10%. Folate losses (both free and total) of over 50% in beef, pork and chicken which have been boiled for 15 minutes have been reported (Taguchi *et al.*, 1973). Vitamin B_6 losses during cooking of meat can vary from 20 to 55%.

Although little work has been carried out, biotin is believed to be relatively stable in meat processing.

Vitamin A losses during frying and cooking of livers are normally within the range of 10 to 20%. While niacin losses of up to 30% in seven days have been reported from the post-mortem ageing of beef (Meyer *et al.*, 1960), ageing for 21 and 42 days caused a small increase in pantothenic acid content (Meyer *et al.*, 1966).

5.5.3 Milk

The fat soluble vitamins A and D are relatively stable to the heat treatment used in the processing of milk. Of the water soluble vitamins, riboflavin, niacin, pantothenic acid and biotin are all relatively stable.

Vitamin C, thiamin, vitamin B_6, vitamin B_{12} and folic acid are all affected by milk processing and the more severe the process the greater the loss. In general, and with the exception of vitamin C, vitamin losses are less than 10% after pasteurisation of milk and between 10 and 20% after UHT treatment. Average losses on sterilisation are 20% for thiamin, vitamin B_6 and vitamin B_{12} and 30% for folic acid.

The stability of vitamin C is also affected by the oxygen content of the milk. Average losses for vitamin C are 25% after pasteurisation, 30% after UHT and 60% after sterilisation. Vitamin C appears to be particularly well preserved in condensed full cream milk (Scott and Bishop, 1986).

5.6 Irradiation

The use of irradiation has become an accepted method of food preservation in many countries, although in some, such as those in the European Community, its use is restricted to a small number of foods and food ingredients. In common with other food processing techniques, vitamin levels in foods can be affected by irradiation, and also in common with other food processing, the vitamins exhibit a range of sensitivities with some being considerably more stable than others.

The nutritional changes during irradiation are in general related to dose; at low doses of up to 1 kGy most losses are not significant. At doses in the range 1 to 10 kGy, vitamin loss may occur in food exposed to air during irradiation and storage. With high irradiation doses, the food may have to be subjected to protective techniques such as exclusion of air during processing or carrying out the irradiation at low temperatures.

Niacin, riboflavin and vitamin D have all been reported to be relatively stable to ionising radiation, while thiamin and vitamins A, E and K are more sensitive. The evidence for vitamin C is contradictory with some studies showing significant losses after irradiation and others none.

Studies on the effects of irradiation on wheat and maize showed virtually no change in riboflavin content at doses up to 2.5 kGy and just over 2% reduction at 5.0 kGy. Niacin showed a 2% reduction in wheat and 4% reduction in maize at 5.0 kGy. Thiamin is considerably more unstable with losses of 9% in wheat and 11% in maize at 5.0 kGy under the same conditions (Hudson, 1991).

There is some evidence that irradiation, particularly in combination with heat treatment such as cooking or baking, can increase the nutritive value of B-vitamins in foods. In one study it was shown that although the niacin content of flour decreased after irradiation, the niacin level in bread made from this irradiated flour was 17% higher than in control samples of bread made from non-irradiated flour. It has been postulated that the irradiation combined with the heat from the baking releases some niacin from a bound form (Diehl, 1991). An increase in niacin content of pork chops by 24% and riboflavin by 15% after irradiation with a dose of 3.34 kGy at 0°C has also been reported (Fox et al., 1989).

Of the oil soluble vitamins both vitamin E and A have been shown to be sensitive to irradiation. Important aids in reducing the effects of irradiation on these vitamins particularly are the exclusion of oxygen and irradiating at very low temperatures. Studies on tocopherol levels in rolled oats irradiated at 1 kGy showed that packaging the oats under both vacuum and with nitrogen had a significant effect on the stability of the vitamin E when compared with those irradiated and stored in air. Vitamin A in cream cheese and egg powders was also found to be protected by both nitrogen and vacuum packing (Diehl, 1981, 1991).

The beneficial effects of very low temperatures during irradiation may be due to the drastic reduction in diffusion rates on freezing. Studies of the reaction of solvated electrons with various solutes in polycrystalline systems at -40°C have shown G-values for products of only 1 to 10% of the values found in fluid systems. G-values indicate the yield of chemical changes in an irradiated substance in terms of the number of molecules produced or destroyed per 100 eV of energy produced (Taube *et al.*, 1979). Under frozen conditions, the constituents of the medium are protected, but it should be noted that while the use of low temperature during irradiation protects the vitamins and other food constituents, it can also protect bacterial cells. Consequently, a higher dose of irradiation may be required.

Most research on the effects of irradiation on vitamin C has been concerned with losses in vegetables and fruits. In a wide range of these (e.g. bananas, tomatoes, carrots, lychees and mangoes) it was found that there was no significant effect, and in many cases losses were less than those caused by conventional handling and processing such as blanching and freezing.

Reports of the effect of irradiation on the vitamin C content of potatoes are contradictory, with some workers reporting significant losses and others no losses at similar doses of irradiation. It is possible that some of the ascorbic acid in the potato is converted to dehydroascorbic acid as a result of radiation-enhanced reactions, and this may have influenced the analytical results.

5.7 Food product shelf life

The increased awareness of healthy eating and the widespread use of nutritional labelling means that many food manufacturers now claim the vitamin content of their products on the labels. As a group, the vitamins are the only nutrients normally declared on a label which can exhibit significant quantitative changes on storage. This is the case both for vitamins naturally occurring in foods and for those added for enrichment or fortification.

Considerable research has been carried out on the kinetics of vitamin degradation in foods in an attempt to predict losses during storage, but it has not been widely published. As the composition of each food is different, and even similar foods can differ in moisture and mineral content, it is not possible to rely on general predictions of shelf life of vitamins, and properly conducted and controlled storage tests have to be carried out on the product in the proposed packaging (or range of packaging) under a range of storage conditions.

The vitamins in a product will degrade at a rate determined by the factors previously referred to in this chapter and each vitamin will deteriorate at a different rate. The varying rates of degradation in food, and particularly in food products containing a number of added vitamins, can pose considerable development problems.

To meet label claims during a realistic shelf life, the amounts of the vitamins in the product need to be above the amounts stated on the label. The difference between formulated and declared levels, known as the 'overage', will vary according to the inherent stability of the vitamin, the conditions under which the food is prepared and packed and the anticipated shelf life of the product.

Overages are normally expressed as a percentage of the declared level:

$$\frac{\text{Amount of vitamin present in product} - \text{amount declared}}{\text{Amount declared on product label}} \times 100$$

Thus an input level of 45 mg of vitamin C and a declared level of 30 mg would give an overage of 50%.

The overages for fortified foods and food supplements, where the added vitamins are the only significant source of those nutrients, are usually calculated as a percentage of the amount required in the product at the end of the required shelf life. In foods where vitamins are added for restoration purposes to replace those lost during processing, or for enrichment to increase the levels of nutrients already present, the overage calculation must take into consideration the amount of the vitamin already in the food.

The stability of the vitamin in the food and the length of required shelf life will govern the amount of overage selected. Experience has shown that the more unstable vitamins such as vitamins C, A and B_{12} generally require high overages whilst those that are inherently more stable such as niacin and vitamin E will only need small overages. Examples of the amounts of overages used in different types of products to ensure that declared levels are met at the end of the shelf life are given in Table 5.1.

Table 5.1 Examples of vitamin overages used in three different products

	Product / Required shelf life		
	Milk based fortified drink powder (12 months) %	Fortified meal replacement bar (12 months) %	Multivitamin tablet (30 months) %
Vitamin A	25	45	60
Vitamin D	25	30	30
Vitamin E	10	10	10
Vitamin B_1	15	15	20
Vitamin B_2	15	20	15
Niacin	10	15	10
Vitamin B_6	20	30	15
Pantothenic acid	15	30	20
Folic acid	20	25	20
Biotin	20	20	10
Vitamin B_{12}	20	20	25
Vitamin C	30	35	45

Table 5.1 clearly shows that, when compared with the drink powder, the higher moisture content of the meal replacement bar and the longer shelf life required by the vitamin tablets both increase the overage levels of more sensitive vitamins. While the overages shown in the table are typical, certain of them may need to be increased in specific circumstances. For example, if sulphur dioxide is likely to be present in a product as carry over from dried fruits or vegetables (e.g. in a fruit bar), higher levels of thiamin will be required. In a liquid milk formula an overage of 40 to 50% may be required for vitamin C to obtain a six month shelf life in cans at ambient conditions.

The shelf life of a product is often dictated by commercial pressures which take into account the time taken to reach the consumer, the range of temperatures that it is likely to be subjected to between production and consumption and the rate at which it is likely to be consumed. This information can be obtained by studying the distribution patterns of similar products already on the market, and should include estimates of the longest distribution chain and the rate of sale through the smaller retailer or caterer. Data on the ambient temperatures and seasonal fluctuations in the countries in which the product is to be sold should also be taken into consideration, as shelf lives based on vitamin levels can vary significantly between products kept under tropical conditions and those stored in more temperate climes.

Once the basic criteria have been established for the shelf life, the overages have to be assessed. The estimation of the overages during product development is difficult and has to be based initially on historical information from products of a similar composition. One factor which can affect the calculations is the legal situation on label claims in the countries in which the product is to be sold. In some countries the legal requirement is for all vitamin levels to be at or slightly above the declared amount on the label on the last day of the stated shelf life ('Best before end' date in the European Community). In other countries the authorities will accept one or more vitamins to be between 90 and 100% of claim at the end of the shelf life. It is accepted by most authorities that there will be some variation between analytical results from different laboratories. Foods subjected to abnormal storage conditions, particularly high temperature and high humidity, which can be proved to be significantly different from those recommended on pack, are normally treated more tolerantly by the authorities.

While the lower levels of vitamins are defined or implied in the legislation of most countries as they affect the accuracy of claims, there is no general agreement on the maximum amount of overage that can be added to a product. Worldwide, only a small number of countries specifically regulate overages for vitamins while most rely on the general principles of good manufacturing practice.

In terms of maximum overages the French law (passed on 4th August 1988) has been very generous and allows a maximum of three times the label claim for all vitamins except vitamin A to be present in the product on the day it is shipped from the factory. Vitamin A is restricted to double the label claim.

In Switzerland, there is legislation published in March 1957 which allows vitamins, other than vitamin D, to be added to a product to amounts of up to three times the Swiss recommended daily amount (RDA) to compensate for losses during storage. No overage is allowed on vitamin D. Under Swiss law products with added vitamin content greater than three times the RDA are considered as pharmaceuticals.

In the United Kingdom and Germany no specific regulations exist on overages. In neither country are the amounts allowed to go below label claim at any time during the declared shelf life. In theory, British and German law requires that the amount declared is the same as the amount added or already present in the food, and there is no provision for declaration of levels below the actual content to allow for storage loss. In practice, manufacturers add overages which are not declared and this is accepted by the authorities on the basis that the declared level is met.

In the USA reliance is placed on the principles of good manufacturing practice and overages for vitamins in food are not regulated, but in Canada very specific guidelines are laid down for the addition of vitamins to foods which are allowed to be fortified.

An Australian working party in the early 1970s developed a set of guidelines on overages which were published by the National Health and Medical Research Council. In general this allowed an overage of 50%. These guidelines were incorporated into the Therapeutics Goods Order in 1990 which allows a maximum of 50% overage on multivitamin preparations and can include products sold as food supplements (Australian Government, 1990).

As already discussed, the shelf life of a product is determined by the stability of its most unstable ingredient in the context of the packaging used and anticipated storage conditions. In a product containing a number of vitamins, the shelf life of vitamins C, A and B_{12} tends to be the controlling factor on the shelf life of the product as a whole. The vitamins in a product will degrade at a rate determined by the factors previously referred to, but each vitamin will deteriorate at a different rate.

In terms of physical chemistry the degradation of most of the vitamins follows 'first order' or 'zero order' kinetics (Labuza, 1982) and a classical Arrhenius model can be developed (Labuza and Riboh, 1982). This allows predictions to be made of both shelf life and overages on the assumptions that the model holds for all the reactions being studied, the same reaction mechanism occurs throughout the temperature range of the study, the energy of activation is between 10 and 20 kcal /mole and that the effects of moisture at ambient temperature are equivalent to maintaining the same relative humidity at the higher temperatures.

Labuza and Riboh (1982) have studied the kinetics of nutrient loss and show that the thiamin loss in pasta predicted by an Arrhenius model suggested a shorter shelf life than actually measured. However, an analysis of the data shows that the errors fall within statistical limits, and as the thiamin assay had a precision at best of ± 5% it was concluded that the application of the Arrhenius equation could produce useful information. The authors concluded that the precision of the technique could be considerably improved if more temperatures were used for the

Table 5.2 Comparison of predicted and actual vitamin losses in a multivitamin preparation stored for 6 months

	Percentage loss of vitamins at 20°C / 75% relative humidity	
	Predicted from Arrhenius model	Actual, assayed at 6 months' storage
Ascorbic acid (vitamin C)	24.0	23.0
Vitamin A preparation	15.0	10.0
Folic acid	8.1	7.4
Vitamin B_{12}	9.2	7.7

study and more frequent samples taken from each temperature. They acknowledge that in a commercial situation, increasing the temperatures and number of samples can add significantly to development costs.

An ideal stability protocol for foods requires storage of replicate samples at at least three temperatures (typically, 0°C, 25°C, 35°C, 40 or 45°C, and 50 or 55°C, and 75% relative humidity). The selection of the two higher temperatures depends upon the product under test as phase changes (i.e. solid to liquid) during storage must be avoided. The tests run for up to 24 weeks with samples from each temperature being removed at predetermined intervals and stored at 0°C. All samples are held at 0°C until the entire study is completed, and all the samples are then assayed at the same time. The data is analysed using the Arrhenius equations. Whilst the technique has some limitations, studies have shown that if all the experimental controls are maintained, useful predictions of product stability can be made (Table 5.2).

The method can be used to estimate overage levels to allow each vitamin in the product to meet a given expiry date, it allows comparison of different packaging options, and it enables the technologist to identify potential stability problems.

Kinetic models have also been developed for the prediction of nutrient losses during the thermal processing of products. Lenz and Lund (1980) have reviewed the techniques used to obtain data on destruction kinetics and show that these can be applied to the study of vitamin losses during the processing of canned meat and vegetable products.

5.8 Protection of vitamins

In order to reduce the large differences between the rates of degradation of vitamins, the vitamin manufacturers have developed various methods for coating or protecting the vitamins. These are very useful for low moisture products and vitamin tablets but are less effective in products with higher moisture levels.

Coated or encapsulated forms of the more sensitive vitamins are commercially available, and coating agents include gelatin, edible fats and their derivatives, starches and sugars. A product formulator needs to check out the composition of

Table 5.3 Summary of vitamin stability

	Heat	Light (UV)	Atmospheric oxygen	pH <7	pH >7	Oxidising agents	Reducing agents	Metallic ions	Ionising radiation	Additional information
Vitamin A		u	vu	u				u	u	Susceptible to isomerisation
β-Carotene		u	u	u						Susceptible to isomerisation
Vitamin D		u	u							
Vitamin E (tocopherols)	u (in presence of air)	u	u (slowly)						u	Vitamin E esters very stable
Vitamin K		u			u					Susceptible to isomerisation
Vitamin B$_1$ (thiamin)	u (at high pH)				u	u			u	Cleaved by sulphites
Vitamin B$_2$ (riboflavin)		vu			u (high pH)		u			
Niacin										Normally very stable
Vitamin B$_6$ (pyridoxine)		u								
Pantothenic acid (as pantothenates)				u (low pH)	u (high pH)					Free pantothenic acid very unstable
Folic acid		vu		u (below pH 5)		u	u			
Vitamin B$_{12}$ (cyanocobalamin)		u		u (low pH)	u (high pH)	u	u			
Biotin										Inactivated by avidin
Vitamin C		u	u	u (vu at pH 4.0–4.3)				vu (Cu and Fe)	u?	Unstable with dissolved oxygen in solutions

Key: u = unstable, vu = very unstable.

the coating as it has been found in the past that some which were originally developed for pharmaceutical use were not technically legal in foods. Problems have also been encountered with gelatin coatings and kosher certification. It is also important to be aware of the vitamin activity of the coated preparations as it can be as low as 10% of the preparation. Table 5.3 shows a summary of vitamin stability.

References

Australian Government (1990) *Therapeutic Goods* Order No 36. Commonwealth Dept. of Community Services and Health, Australian Govt. Pub. Serv., Canberra.

Baloch, A.K. (1976) The stability of beta carotene in model systems. *PhD Thesis*, University of New South Wales, Kensington, Australia.

Baloch, A.K., Buckle, K.A., Edwards, R.A. (1977) Stability of beta carotene in model systems containing sulphate. *J. Food Technol.* **12**, 309–316.

Bauernfeind, J.C. and Pinkert, D.M. (1970) Ascorbic acid as an added nutrient to beverages. *Adv. Food Res.* **18**, 219–235.

Bender, A.E. (1958) The stability of vitamin C in a commercial fruit squash. *J. Sci. Food Agric.* **9**, 754–760.

Bondelin, F.J. and Tuschhof, (1955) The stability of ascorbic acid in various liquid media. *J. Am. Pharm. Assoc. (Sci. Ed.)* **44**, (4), 241–244.

Borenstein, B. (1981) Vitamins and amino acids. In: *Handbook of Food Additives* (ed. T. Furia) Volume I, CPC Press, Boca Raton, Florida, USA, pp. 85–114.

Carstensen, J.T. (1964) Stability patterns of vitamin A in various pharmaceutical dosage forms. *J. Pharm. Sci.* **53**, (7), 839–840.

Cover, S., Dilsaver, E.M., Hays, R.M. and Smith, W.H. (1949) Retention of B vitamins after large scale cooking of meat II. Roasting by two methods. *J. Am. Diet. Assoc.* **25**, 949–951.

De Ritter, M. (1982) Vitamins in pharmaceutical formulations *J. Pharm. Sci.* **71** (10), 1073–1096.

Diehl, J.F. (1981) Effects of combination processes on the nutritive value of food. In: *Combination Processes in Food Irradiation*. International Atomic Energy Agency, Vienna, pp. 349–366.

Diehl, J.F. (1991) Nutritional effects of combining irradiation with other treatments. *Food Control* **2** (1), 20–25.

Dwivedi, B.K. and Arnold, R.G. (1973) Chemistry of thiamin degradation in food products and model systems: a review. *J. Agric. Food Chem.* **21** (1), 54–60.

Feller, B.A. and Macek, T.J. (1955) Effect of thiamine hydrochloride on the stability of solutions of crystalline vitamin B_{12}. *J. Am. Pharm. Assoc. (Sci. Ed.)* **44** (11), 662–665.

Fox, J.B., Thayer, D.W., Jenkins, R.K. et al (1989) Effect of gamma irradiation on the B vitamins of pork chops and chicken breasts. *Internat. J. Radiat. Biol.* **55** 689–703.

Grant, N.H. and Alburn, H.E. (1965) Fast reactions of ascorbic acid and hydrogen peroxide in ice. *Science* **150**, 1589–1590.

Guerrant, N.B. and O'Hara, M.B. (1953) Vitamin retention in peas and lima beans after blanching, freezing and processing in tin and in glass, after storage and after cooking. *Food Technol.* **7**, 473–477.

Guerrant, N.B., Vavich, M.G., Fardig, O.B., Ellenberger, H.A., Stern, R.M. and Coonen, N.H. (1947) Effect of duration and temperature of blanch on vitamin retention of certain vegetables. *Ind. Eng. Chem.* **39**, 1000–1007.

Hudson, D. (1991) Irradiation—its effects on grain, milling and flour. *Internat. Milling Flour and Feed* **184** (9), 20–23.

Institute of Food Science & Technology (IFST) (1989) *Nutritional Enhancement of Food*—Technical Monograph N° 5, IFST, London.

Keagy, P.M., Stokstad, E.I.R. and Fellers, D.A. (1973). Folacin stability during bread processing and family flour storage. *Cereal Chem.* **52**, 348–351.

Killeit, U. (1988) *The Stability of Vitamins— a Selection of Current Literature*. Hoffmann-La Roche A G, Grenzach-Wyhlen, Germany

Kläui, F. (1979) Inactivation of vitamins. *Proc. Nutr. Soc.* **38**, 135–141.

Labuza, T.P. (1979) A theoretical comparison of losses in foods under fluctuating temperature sequences. *J. Food Sci.* **44**, 389–393.

Labuza, T.P. (1982) *Open Shelf Life Dating of Foods*. Food and Nutrition Press, Westport, Conn, USA.

Labuza, T.P. and Riboh, D. (1982) Theory and application of Arrhenius kinetics to the prediction of nutrient losses in foods. *Food Technol.* **36** (10) 66–74.

Lenz, M.K. and Lund, D.B. (1980) Experimental procedures for determining destruction kinetics of food components. *Food Technol.* **34** (2), 51–55.

Mallette, M.F., Dawson, C.R., Nelson, W.L. and Gortner, W.A. (1946) Commercially dehydrated vegetables, oxidative enzymes, vitamin content and other factors. *Ind. Eng. Chem.* **38**, 437–441.

Marks, J. (1975) *A Guide to the Vitamins—their role in health and Disease*. Medical & Technical, Lancaster, UK.

Meyer, B.H., Thomas, J. and Buckley, R (1960). The effect of ripening on the thiamin, riboflavin and niacin content of beef. *Food Technol.* **14**, 190–192.

Meyer, B.H., Mysinger, M.A. and Cole, J.W. (1966) Effect of finish and ripening on B_6 and pantothenic acid content of beef. *J. Agric. Food Chem.* **14**, 485–486.

Paul, A.A. and Southgate, D.A.T. (1979) *McCance and Widdowson's—The Composition of Foods*. HMSO, London.

Slater, G., Stone, H.A., Palermo, B.T. and Duvall, R.N. (1979) Reliability of Arrhenius Equation in predicting vitamin A stability in multivitamin tablets. *J. Pharm. Sci.* **68** (1), 49–52.

Taguchi, H., Hara, K., Hasei, T. and Sanada, H. (1973) Study of folic acid content of foods. II Loss of folate from foods by boiling. *Vitamin* **47**, 513–516.

Tansey, R. P. and Schneller, G. H. (1955) Studies in the stabilization of folic acid in liquid pharmaceutical preparations. *J. Am. Pharm. Assoc. (Sci. Ed.)* **44** (1), 35–37.

Taub, I. A., Halliday, J. W. and Sevilla, M. D. (1979) Chemical reactions in proteins irradiated at sub-freezing temperatures. *Adv. Chem. Ser.* **180**, 109–140

Timberlake, C. F. (1960a) Metallic components of fruit juices. III Oxidation and stability of ascorbic acid in model systems. *J. Sci. Food Agric.* **11**, 258–268.

Timberlake, C. F. (1960 b) Metallic components of fruit juices IV. Oxidation and stability of ascorbic acid in blackcurrant juice. *J. Sci. Food Agric.* **11**, 268–273.

6 Vitamin fortification of foods (specific applications)
A. O'BRIEN and D. ROBERTON

6.1 Addition of vitamins to foods

6.1.1 Introduction

There are thirteen generally recognised vitamins, grouped together by their essential nutritional role rather than any chemical similarity. Thus while vitamins are often classified as water or fat soluble, or referred to as a 'group' in the case of the B-vitamins, they are not chemically related and as such the food technologist needs to be aware of significant stability differences between all thirteen vitamins when considering their addition to a food. A guide to the sensitivity of the vitamins towards various factors is shown in Table 5.3.

The manufacturers of vitamins are obviously aware of these stability characteristics and are therefore able to assist the food technologist by supplying some of the vitamins in protected forms as described in more detail in chapter 9.

For example, the commercial forms of vitamin A are supplied as either dilutions in high quality vegetable oils containing added vitamin E as an antioxidant, or as dry powders in which the vitamin A is dispersed in a coating of gelatine and sucrose or gum acacia and sucrose, thus protecting the vitamin A from oxidation.

When considering which vitamins to add to a food, decisions have first to be made as to the purpose. Is it for vitaminisation, restoration, standardisation or fortification? Restoration is the term used when vitamins and other nutrients lost in processing are restored in comparable amounts. Standardisation describes fortifying a food to a standard within its class. For example the addition of vitamin C to apple juice to bring it up to the levels contained in orange juice. Fortification involves the addition of vitamins to a food to be given a specific nutritional profile, be it a processed or formulated food, promoted as a substitute for a traditional product, for example margarine.

The answer to this question will then determine the required level of each vitamin in the finished food and the consequences for ingredient listing and labelling claims.

Once the required vitamin levels have been decided, the real challenge begins. Can all the necessary vitamins be added at the same stage of the process? What 'overage', that is to say, what quantity of the vitamin above that to be claimed will be required in order to compensate for losses during manufacture and subsequent storage of the food? Which form of the vitamin is most appropriate, for example, should ascorbic acid or sodium ascorbate be the source of vitamin C? Will they be added to the food as a dry powder or in solution?

The answers to these questions can only be obtained through a combination of chemical, manufacturing and analytical skills. In reality, a knowledge of the stability characteristics of the vitamins, together with an understanding of their possible interaction with other ingredients under the anticipated processing conditions, enable the food technologist to make a best guess as to the rate of addition of each vitamin to be added to the food. The only way then of confirming the accuracy of this best guess is to add the required vitamins to the food and then assay for them immediately after processing and then at regular intervals over a period of time equivalent to their effective 'use by' date. This analytical work will then enable the necessary addition rate of each vitamin to be fine tuned.

When two or more vitamins are added to a food at the same manufacturing stage, it is becoming increasingly common for food manufacturers to purchase a specific homogeneous blend of vitamins, known as a vitamin premix. Each premix is tailor made for an individual customer and product after the detailed analytical work described in the previous paragraph has been completed. The advantages of using a vitamin premix include:

- the addition of up to thirteen vitamins as a single ingredient;
- inventory levels are reduced;
- quality control, production and assay procedures are simplified giving marked savings in labour costs and time.

6.2 Beverages

The term beverages covers a wide spectrum of products from dry instant drinks through to concentrates, squashes, fruit juices and nectars.

Vitamin C has been associated with fruit drinks since 1753 when James Lindt showed that supplementation of sailors' diets with oranges and lemons prevented scurvy. In the 1930s syrups based upon rosehips and blackcurrants were made available to the public as a rich source of vitamin C containing 60–100 mg ascorbic acid per fluid ounce (Counsell and Hornig, 1981), while during the 1939–45 war British schoolchildren received a daily dose of orange juice to swallow with their cod liver oil. More recently the trend towards multivitamin drinks has increased in both developed and developing countries for different reasons; in Germany, USA and the UK a range of vitamins are now added to fruit juice-based products to help consumers towards the goal of 'optimal health', while in countries such as Nigeria and South Africa these nutrients are added to selected drinks to help prevent the occurrence of deficiency diseases.

The principles of vitamin addition are similar for all products and stability is

116   THE TECHNOLOGY OF VITAMINS IN FOOD

dependent upon factors including process conditions, shelf life, packaging materials and presence of other ingredients. The combination of vitamins added and claims made are dependent on the manufacturer but there are some limitations with regard to vitamin stability in certain drinks.

6.2.1   Vitaminisation of instant beverages

This is a simple operation involving mixing of the dry vitamins with the remaining ingredients making up the powder.

Cold water dispersible forms of the vitamins should be used to avoid delayed surface dispersion on make up. The mixer used must ensure complete dispersion and avoid over mixing which can lead to segregation. A slight overage for the vitamins should be added to compensate for any distribution variations on mixing. Whether make up is hot or cold, losses should be minimal with consumption being within a short period after preparation.

6.2.2   Vitaminisation of concentrates, nectars and juice drinks

In all liquid beverages it is important to ensure that the vitamins are well dispersed and processing parameters are taken into account in assessing suitable overages.

Of all the vitamins those being more heat labile are thiamin (vitamin B_1), folic acid and vitamin C. Maximum temperatures reached are those of pasteurisation being 90°C for fifteen seconds. Tunnel pasteurisation will be more severe than batch processes as the liquid is subjected to heat treatment for longer periods of time.

Acidity is a major drawback as far as vitamins A, folic acid and calcium pantothenate are concerned. At pHs below 5.0 these vitamins are very unstable and greater losses occur. Most liquid beverages fall below a pH of 3.0. An added disadvantage of folic acid is that it has very low solubility in water. The higher the total solids content, the more stable the vitamins. Carbonation gives a higher degree of stability due to expulsion of oxygen.

6.2.3   Vitamin stability

The results shown in Table 6.1 are taken from a trial carried out on a pasteurised multivitamin orange juice.

The vitamin mix was predispersed in some juice before addition to the bulk to ensure adequate mixing. Pasteurisation was at 90°C for fifteen seconds, the product then being packed in a six layer laminate of polythene and aluminium foil and stored at ambient temperature for six months.

The results show that all of the vitamins tested showed a good degree of stability throughout the heat process including the more heat labile vitamins namely thiamin and vitamin C.

Vitamin A showed a slight degradation on storage during the first two months,

Table 6.1 Trial carried out to find vitamin stability in pasteurised multivitamin orange juice

	Added	Initial	2 months	6 months
Vitamin A				
Added as vitamin A				
iu / 100 ml retinol	219	232	168	163
Added as β-carotene				
iu / 100 ml retinol	175	191	203	180
Vitamin B_1				
Thiamin				
mg / 100 ml	0.168	0.192	0.177	0.162
Vitamin B_2				
Riboflavin				
mg / 100 ml	0.20	0.24	0.17	0.20
Vitamin C				
Ascorbic acid				
mg / 100 ml	12.0	50.9	44.7	39.4

stability was much improved when added as β-carotene. As expected there was a steady loss of vitamin C on storage, being higher in initial stages when the ascorbic acid is acting as an oxygen scavenger. Subsequent losses indicate that the packaging was not airtight and it is likely that oxygen had permeated into the pack along the seal.

All of the B-group vitamins tested, namely B_1, B_2, B_6 and niacin, showed good recovery over the six months storage period.

Apart from a consideration of the potential losses due to heat processing and subsequent storage, some other general facts must also be recognised.

Ascorbic acid is unstable at high oxygen concentrations, a factor to be taken into account when considering packaging and shelf life. Initial levels present will decrease with any head-space and dissolved oxygen in the product. Subsequent to this, barrier packaging plays an important role in reducing further losses.

Ascorbic acid is added as an antioxidant at levels up to around 300 ppm in squashes and concentrates and 100 ppm in 'ready-to-drink' (RTD). Over and above this, a suitable overage should be added to compensate for any processing losses.

Vitamin B_2 (riboflavin) is very unstable to UV light and must be protected when in solution. Any beverages incorporating vitamin B_2 should be packed in brown bottles, or UV barrier materials.

The presence of sulphur dioxide can have a detrimental effect on the thiamin (vitamin B_1) causing instant degradation. Some fruit juices have a carry over of sulphur dioxide and levels should be determined prior to fortification.

6.2.4 Vitamin incorporation (see Figure 6.1)

To ensure adequate mixing, the vitamins should be predispersed in an aliquot of the juice or drink before being added to the bulk. The term generally used is the brew, where a concentrate is made of the minor ingredients. This is added to the

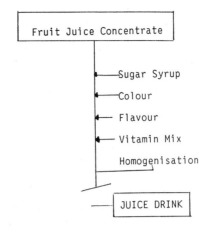

Figure 6.1 Vitaminisation of a juice drink (Roche, 1970).

bulk prior to homogenisation ensuring a thorough mixing. In initial pilot plant trials a series of samples should be taken at intervals on exit from the pasteuriser to determine vitamin distribution and stability.

In conclusion, taking into account the processing conditions it is feasible to add a range of vitamins to a long-life liquid beverage and achieve label declarations by applying suitable overages.

6.3 Cereal products

The cereal crops including wheat, maize and rice are staple foods of both developed and third world countries and the addition of vitamins to bread flour, maize meal and rice grains is well established. Similarly the fortification of 'ready-to-eat' breakfast cereals produced from these basic foods is widely practiced in recognition of the significance of these processed foods in many modern diets.

6.3.1 Breakfast cereals

Breakfast cereals are traditionally divided into two types:

- Ready-to-eat (RTE) products which include rice-, wheat- and maize-based cereals.
- Hot products, which are prepared by adding hot milk or water, for example porridge.

There is a constant stream of new products onto the market a majority of which are aimed at children. Fortification technology must, above all else, be kept pragmatic. After the decisions have been made regarding which nutrients are required in the final product, the food technologist is faced with the challenge of not only getting the job done, but getting it done so that:

- the product is not negatively affected in odour, flavour or colour;
- the added nutrients are acceptably stable with sufficient overage to compensate for losses on processing and storage;
- production of the end product remains practical and economically viable.

The more complex and abusive the processing conditions, the more difficult it is to achieve these ends. In this respect, the industry's ability routinely to fortify many million tonnes of breakfast cereals in the UK alone with a wide spectrum of nutrients places it in the forefront of fortification technology (Johnson *et al.*, 1988)

The range of breakfast cereals is achieved by modification of existing and new processes. Of all the processes used for RTE cereal manufacture, the most widely applied are extrusion, gun puffing, flaking, toasting and dry blending (Steele, 1976). It is more than likely that one or more of these processes will be used.

The manufacturer has several options available when it comes to fortification technology. The method selected must be based on sound manufacturing practices as well as a good knowledge of nutrient chemistry and of their own product and process. Cereal product characteristics that must be considered are; desired nutrient claim, pH, moisture content, processing temperatures, holding times (if any), storage temperature, storage times, packaging, product formulation and others (Johnson *et al.*, 1988).

Processing factors that need to be considered are temperature ranges (121–204°C), moisture ranges (10–60%), toasting temperatures (148–204°C), flaking pressures, extrusion pressures, coating systems etc. As with all other processes it is advisable to add the vitamins at as late a stage as possible but this is not always practical.

With the various processes and conditions in mind it is apparent that certain vitamins can be added to the basic cook whereas the more heat labile can be applied either just before coating or just after coating, prior to the final drying.

6.3.1.1 Extrusion cooking. Extrusion cooking is a high-temperature short-time (HTST) process with many applications. The operation at high temperature, pressure and shear, and at low or intermediate water content distinguishes extrusion cooking from alternative processes, and provides conditions for reactions of nutrients.

Apart from heat, humidity and oxygen there are additional parameters in extrusion cooking that influence the stability of vitamins (Schlude, 1987):

- Raw materials
- Pressure
- Flow rate
- Screw speed
- Energy input
- Die diameter.

120   THE TECHNOLOGY OF VITAMINS IN FOOD

6.3.1.2 Vitamin addition. For ease of operation the most convenient way of adding vitamins is in the dry mix prior to extrusion. Dependent on the processing conditions there will be losses of the more heat labile vitamins from both the temperatures applied and the mechanical energy which builds up in the barrel. If extrusion is followed by expansion there is another point of high temperature and pressure, on emergence from the die.

A number of studies have been carried out which look at the influence of individual processing parameters on the stability of vitamins in extrusion cooking.

On a co-rotating twin-screw extruder Killeit and Weidman (1984) investigated

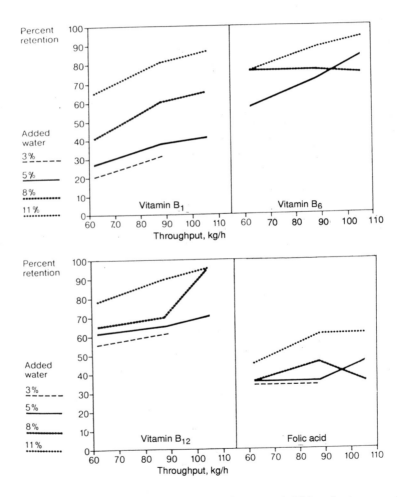

Figure 6.2 Effect of extrusion parameter flow rate and addition of water.

Table 6.2 Effect of die diameter and temperature in dehulled triticale

Nozzle diameter (mm)	°C	Water (%)	Thiamin retention	Riboflavin retention
15.9	177	22	0.60	0.39
	204	22	0.33	0.41
	232	22	0.08	0.62
31.8	177	22	1.02	0.19
	204	22	0.86	0.34
	232	22	0.28	0.63

the effect of flow rate, addition of water and energy input on vitamins B_1, B_6, B_{12} and folic acid in a flat wheat bread.

The retention of vitamins B_1, B_6, B_{12} improved with increasing flow rate although the extrusion pressure increased nearly proportionally. In this case the shorter residence time seems to have a predominating influence.

As illustrated in Figure 6.2 a higher water content in the mixture improved the retention of all four vitamins, probably due to the fact that with increasing moisture, extrusion temperature and residence time decrease. Maga and Sizer (1978) found similar results for thiamin in extruded potato flakes.

Looking at energy input Killeit and Weidman (1984) found a negative correlation between mechanical energy input and vitamin stability for B_1, B_6, B_{12} and folic acid.

Comparing die size and temperatures Beetner *et al.* (1976) looked at stability of natural levels of vitamins B_1 and B_2 in extruded triticale (a cross between wheat and rye). The results showed that for vitamin B_1 there is an increased retention for increasing die size and a decreased retention for increasing temperature. The results are illustrated in Table 6.2.

The vitamins can also be injected into the barrel via the water inlet or the oil inlet although this method is less practiced.

More often than not on emergence from the extruder the cereal product will undergo another process, e.g. drying (passage through a drying chamber), toasting and or coating, being in most cases a sugar syrup or a straight flavour coating.

If the product is coated it is an ideal route for addition of the more heat labile vitamins which can be added to the coating slurry. This is an ideal route for vitamins C, A and B_1.

The critical factors in designing the spray system are to ensure uniformity in the spray coverage and optimising protection of those vitamins which pose unique stability problems namely vitamins A and D. These vitamins are both oxygen and temperature sensitive, oxidative degradation being maximised by spraying. Johnson *et al.* (1988) show that by building in adequate protection into the spray, oxidation can be minimised. The simplest and most often used oxygen barrier is sucrose, which should constitute at least 10% of the spray formula, but is generally in the 15–25% range. Their work showed that vitamin A stability on RTE breakfast

Table 6.3 Vitamin stability in two cereals with different moisture content sprayed with two sugar solutions

	Vitamin A (iu / oz)	Vitamin C (mg / oz)
Cereal 1		
Initial	2180	27.6 (15–20% sugar in spray)
6 months	2230	18.9
12 months	2230	11.2
	no loss	60 % loss
Cereal 2		
Initial	1430	18.8 (1% sugar in spray)
6 months	1170	17.1
12 months	1040	13.4
	27% loss	29% loss

Moisture contents of Cereal 1 is 6% and Cereal 2 is 2–3%.

cereals is directly related to the method of addition and the 'coating' protection of the sugar, while moisture was the most important factor affecting vitamin C stability. The results given in Table 6.3 compare a spray solution of 15–20% sugar against a spray of less than 1% sugar in two cereals with different mositure contents.

The figures confirm higher vitamin A stability with increased sugar content in the spray compared to almost 30% loss where the only antioxidant was butylated hydroxytoluene (BHT). With vitamin C an almost 60% loss occurred at high moisture content against half that level in the low moisture cereal.

A summary of trials carried out over a period of two years shows a range of vitamin losses during extrusion cooking (Table 6.4) (Schlude, 1987).

The ranges given highlight the dependence on processing conditions and raw materials.

Table 6.4 Vitamin losses during extrusion cooking

Fat soluble vitamins	Percentage loss (%)
Vitamin A	12–88
Vitamin E	7–86
Water soluble vitamins	Percentage loss (%)
Vitamin C	0–87
Vitamin B_1	6–62
Vitamin B_2	0–40
Vitamin B_6	4–44
Vitamin B_{12}	1–40
Niacin	0–40
Folic acid	8–65
Calcium pantothenate	0–10
Biotin	3–26

(Documentation Roche Basle.)

Table 6.5 Vitamin addition pre- and post- extrusion

	Vitamin B_1	Vitamin B_2	Vitamin B_6	Vitamin B_{12}*	Niacin	Folic acid	Vitamin C
Pre-extrusion addition							
Added (mg, *µg)	1.73	2.0	3.5	3.2	17.3	530	48.3
% Processing loss**	15.6	13.5	—	—	—	—	16.4
% Loss after 2 months**	28.9	11.5	—	—	—	—	37.1
% Loss after 3 months**	39.9	21.0	19.1	15.6	0	37.8	44.7
Post-extrusion addition							
Added (mg, *µg)	1.27	1.7	2.9	2.2	18.5	365	38.1
% Processing loss**	9.4	9.4	—	—	—	—	0
% Loss after 2 months**	9.4	0	—	—	—	—	9.2
% Loss after 3 months**	21.3	0	0	3.6	10.8	28.8	15.2

** Based on the added vitamin quantity

Trials carried out by Roche UK compare vitamin stability of pre- and post-extrusion (Table 6.5).

In summary the most sensitive vitamins to extrusion are A, E, C, B_1, and folic acid. Other vitamins of the B-group are shown to be very stable. In this case it would be ideal to add the stable vitamins to the cook and spray the labile vitamins post-extrusion.

6.3.1.3 Roller drying.

In this process it is possible to achieve a high rate of drying with economic use of heat Brennan *et al.* (1990). It is used for liquid cereal slurries which are able to withstand relatively high temperatures for short periods of time (of the order of 2–30 seconds).

The vitamins would be dry blended with the remaining dry ingredients and made up to a slurry. Because of the nature of the process the material reaches very high temperatures as it forms a film over the drum surface. As the drum rotates the material is dried and then scraped from the surface.

Vitamin losses on roller drying have been shown to be lower than for extrusion cooking, as shown in Table 6.6.

Table 6.6 Vitamin losses: extrusion vs roller drying

		Vitamin per 100 g dry material			
		A	C	B_1	B_2
Quantity added		2932 IU	71.8 mg	3.7 mg	2.2 mg
Extrusion					
Process loss %		62.4	8.2	32.2	18.2
Storage loss %	after 6 months	75.0	31.1	46.0	30.7
	after 12 months	85.3	45.8	43.2	37.6
Roller drying					
Process loss %		26.2	9.2	33.2	9.1
Storage loss %	after 6 months	39.5	24.0	41.3	29.5
	after 12 months	60.6	37.0	40.6	33.5

Grebaut *et al.* (1983).

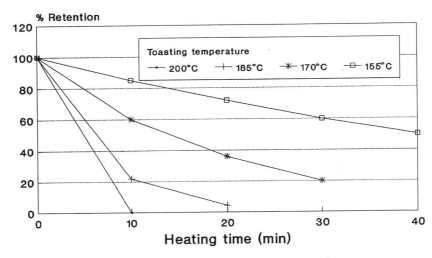

Figure 6.3 Vitamin B_6 stability during toasting of breakfast cereals (B_6 level added: 33 or 66 mg / kg).

As with extrusion cooking it is the heat labile vitamins B_1, C, folic acid, E and A which show greater losses. Losses on storage are slightly lower than extruded products due to the low moisture content and the nature of the product.

6.3.1.4 Additional processes for breakfast cereals. As mentioned earlier one or more processes are generally applied to produce the end product. Other processes involved include baking, toasting, puffing and browning. In all cases it is a combination of temperature reached, exposure time and moisture content which dictates the fate of the added vitamins. As an example Figure 6.3 shows the stability of vitamin B_6 during toasting of breakfast cereals (personal communication).

Once combined in the matrix of the cereal the vitamins will be more stable against secondary processes whereas if they have been surface sprayed they are more susceptible. The total process from start to finish must be assessed along with expected shelf life and packaging details before the vitamin inclusion rate is calculated.

6.3.2 Bread

The demand for white bread is satisfied by the miller producing flour from which the husk and germ has been removed. Unfortunately this process also results in the removal of essential vitamins, particularly thiamin (vitamin B_1), riboflavin (vitamin B_2), niacin and vitamin E.

The graph in Figure 6.4 shows the relation between extraction rate and the proportion of total vitamins of the grain retained in flour (Aykroyd and Doughty, 1970).

Bread and cereals are major contributors of niacin, thiamin and riboflavin to the diet (Dietz and Erdman, 1989). Fortification of bread is an area of growing interest.

VITAMIN FORTIFICATION OF FOODS (SPECIFIC APPLICATION)

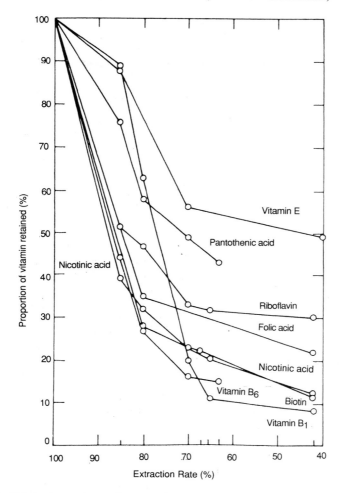

Figure 6.4 Relation between extraction rate and proportion of total vitamins of the grain retained in flour (Aykroyd and Doughty, 1970).

Bread has been a staple part of a diet for hundreds of years with the flour used naturally containing small amounts of many of the B-group vitamins. In addition, in many countries regulations already require the addition of calcium carbonate, iron, thiamin and niacin. Bread is an ideal product for providing essential nutrients to the diet being a regular food substance consumed by all groups.

The most common method used for producing bread in plant bakeries is the Chorleywood Bread Process (CBP). Baking temperature is in the region of 220–270°C for 22–25 minutes. Vitamins can easily be incorporated along with the other dry ingredients at the initial mixing stages. It is essential to ensure a thorough distribution of the vitamins prior to adding the water. A suggested route would be predilution with a small quantity of the flour taken from the bulk.

Table 6.7 Vitamin recoveries in bread

	% Recovered			Unsupplemented (mg per 100 g)
Soft grain white bread				
Vitamin A	84	94	87	ND
Vitamin B_1	110	107	106	0.05
Vitamin B_2	100	105	98	0.02
Nicotinamide	70	77	74	ND
Vitamin B_6	114	116	105	0.02
Wholemeal bread				
Vitamin A	88	84	71	ND
Vitamin B_1	93	77	85	0.07
Vitamin B_2	82	95	83	0.025
Nicotinamide	76	84	66	ND
Vitamin B_6	108	107	104	0.05

ND: not determined
Taken from trials carried out by Roche Product Limited at the Flour Milling Baking Research Association (FMBRA) using the Chorleywood Process.

6.3.2.1 Vitamin stability. Trials carried out to date have shown very good recovery of added vitamins.

Table 6.7 shows results of analyses carried out on vitamins in soft grain and wholemeal breads prepared by the Chorleywood Process. Of all of the vitamins added vitamin A is shown to be the most labile although recoveries are above 80%.

Even vitamin B_1, the most heat sensitive of the B-group vitamins shows excellent stability. The results show levels of nicotinamide to be lower than expected due to difficulties in recovery by the analytical methods used.

Baking parameters: batch size 800 g loaf; 7 kg flour batches; baking temperature 270 °C; baking time 22 min (softgrain), 25 min (wholemeal).

Further trials reported by Cort *et al.* (1975) as shown in Table 6.8 show the stability of vitamins during baking and after five days of storage at room temperature. The vitamins were added via a premix at the rate of 140 mg / lb flour. The results confirm the findings of Borenstein (1966) and Cort and Scheiner (1970) that vitamin B_6, tocopheryl acetate and folic acid are stable in flour and the final bread.

Table 6.8 Stability of bread* made with flour fortified with vitamins

Nutrient	Label claim per loaf	Level found per loaf	
		Bread after baking	Bread after 5 days at room temperature
Vit A (IU)	7500	8280	8300
Vit E (IU)	15	16.4	16.7
Pyridoxine (mg)	2	2.4	2.5
Folic acid (mg)	0.3	0.34	0.36

* 9% moisture bread

Alkaline leavening agents such as baking powder significantly increase the pH of bread dough. Up to 84% of thiamin has been shown to be destroyed when large amounts of alkaline leavening agents are used (Briant and Klosterman, 1950). It is recommended that a pH of 6.0 or lower be maintained in bread dough to favour stability of thiamin during baking (Adams and Erdman, 1988).

A wide range of vitamins can and are added to bread.

At the vitamin levels added (being a percentage of the RDA per serving, e.g. 2 slices) there appear to be no flavour problems arising. However dependent on the levels added, riboflavin can impart a slight yellow colouring to the crumb giving it a 'golden' appearance. The stability of riboflavin in the bread is dependent on exposure times of the dough to UV light in its wet phase although the heat process should not be a factor of concern.

6.3.3 Pasta

The outline of the manufacturing process for most pasta products is very simple. Water is added to semolina flour to obtain about 31% moisture. The mixture is then kneaded to obtain a very stiff dough, which is extruded through a die, cut, dried and packed.

The processing parameter most likely to affect vitamin stability is the drying process, which can range from temperatures of 50°C for 9–12 hours to very high temperatures of 95°C for 3–5 hours. Factors affecting stability are the raw material type, process, temperature and the vitamin form used.

The drying temperature has been shown to have no influence on vitamins B_1, B_2, B_6 and niacin, whereas vitamin A losses increase with increasing drying temperatures. β-Carotene which may be applied as the source of the vitamin A is shown to be far more stable than the straight vitamin. Storage losses have been shown to be similar for instant and normal pasta.

Work carried out on commercially produced spaghetti, noodle and macaroni products showed the following cooking losses; thiamin 41.6–54.4%, riboflavin 29.5–41.3% and niacin from 38.5–50.0% (Ranhotra et al., 1983) On enrichment of instant products it is possible to add the vitamin mix to the spice mix which would be included in the noodle pack.

Table 6.9 Vitamin retention in spaghetti (durum wheat) after drying and after cooking (%)

	Drying at 50°C (9–12 hours)		Drying at 95°C (3–5 hours)	
	Dry	Cooked	Dry	Cooked
Vitamin A	86	72	77	66
Vitamin B_1	98	67	100	71
Vitamin B_6	93	73	94	68
Niacin	100	73	100	73
β-Carotene 1% CWS	95	86	92	52

Table 6.10 Vitamin retention in instant pasta (steamed and dried), on processing and storage (%)

	Drying at 50°C (9–12 h) (months)			Drying at 95°C (3–5 h) (months)		
	0	3	6	0	3	6
Vitamin A	73	76	66	67	64	56
Vitamin B_1	112	—	81	—	104	85
Vitamin B_2	55	—	42	53	—	36

The following results (Tables 6.9 and 6.10) are taken from a trial carried out with Buhler Brothers and Roche Products (1989).

6.3.3.1 Vitamin addition. Vitamins can be added to pasta either by dry addition or by wet addition.

Dry addition. In this method the vitamins should be prediluted with 2–50 parts semolina / flour before being added to the bulk. Addition to the bulk may be either a batch process or via a metering device. Good mixing equipment is essential to ensure a good distribution of the vitamins.

Wet addition. With this method the vitamins are pre-dispersed in a small quantity of water ensuring complete dissolution. This is then added by direct batch system or a continuous metering pump to the flour.

In practise both methods are used, and good distribution and stability of the vitamins can be achieved. The main parameter to take into account when calculating overages are the combined drying temperature and time.

6.4 Dairy products

6.4.1 Milk

Milk is a nutritious food and therefore the concept of increasing the levels of vitamins in dairy products may seem unreasonable. However the fortification of cow's milk with vitamin D in the USA, initiated in the 1940s, was largely credited with the marked decline in rickets by the early 1960s (Roche, 19xx) while the modern trend in developed countries towards more healthy living has created a consumer demand for reduced calorie foods. The consumption of liquid skimmed milk or of products manufactured from it has therefore increased significantly. Unfortunately one consequence of separating off the cream is the simultaneous removal of the fat soluble vitamins A and D. It is therefore logical to add these two vitamins back to skimmed milk in order to restore the nutritional value. Similarly many dieters eat yoghurt as a meal replacement and the addition of a wide range of vitamins is justified.

Milk processing is divided into three main categories:

- Pasteurisation
- Ultra high temperature (UHT)
- Spray drying

All three categories are broad in respect to ranges of processes used and steps involved. The major factors to be considered when adding vitamins to milk undergoing one or more of the mentioned processes are temperature (maximum and time held), degree of aeration, total processing time and product packaging.

6.4.1.1 Pasteurisation. Pasteurisation temperatures do not exceed 100°C and times of processing depend on the temperature reached (the standard is 72 °C for 15 s). There are two main methods of pasteurisation:

- Batch: milk is pasteurised in individual batches in stirred, jacketed stainless steel vessels.
- Continuous: milk is pasteurised by passage through plate heat exchangers involving four stages — preheating (regeneration), heating, holding and cooling.

6.4.1.2 Ultra high temperature (UHT). This treatment utilises two different methods, 'live' steam injection or plate heat exchanger followed by aseptic cooling and filling. The maximum temperature reached is higher than pasteurisation but for a shorter time (e.g. 137°C for 1–3 s). The packaging is more often than not a composite pack made from laminates of polythene / paperboard / aluminium foil / polythene.

6.4.1.3 Spray drying. In this process the milk is introduced into the drying chamber in the form of a fine spray where it is brought into intimate contact with a stream of hot air, enabling rapid drying. The process involves very short drying times of 1–10 s giving low product temperatures.

In all three processes every factor of the process should be taken into account in deciding which vitamin form to add, how much to add, where to add it and how to store the final product.

6.4.1.4 Vitamin addition to liquid milk. In all processes it is essential to ensure a good distribution of the vitamins in the bulk milk prior to any heat treatment. Both oily and dry forms of the vitamins can be used but the following factors should be considered.

Water dispersible forms are easy to incorporate especially if they are cold water dispersible but these forms are less stable in the final product. This is due to the carrier becoming hydrated causing the vitamin to be exposed to the processing parameters.

Oily forms are more favourable to add but the process of addition must ensure adequate dispersion. The oily vitamin forms should be prediluted in a suitable oil to give manageable quantities to handle but should not be added directly to the bulk

130   THE TECHNOLOGY OF VITAMINS IN FOOD

Table 6.11 Percentage loss of vitamin A in UHT milk during processing and storage: (indirect plate-heat exchanger)

Losses on processing	Losses on storage					
	1 month		3 months		6 months	
	5°C	Amb	5°C	Amb	5°C	Amb
19%	5 %	6 %	7 %	10 %	7 %	11 %

Results are from a small scale batch UHT, 200 litres (skimmed milk), carried out on pilot plant (packaging - Tetrabrik).

as it will form a slick on top adhering to the sides of the vessel on emptying from the base. Prehomogenisation of the vitamin mix into an aliquot of the milk prior to addition to the bulk will lead to a better distribution. This suggestion is not always possible, in which case the vitamin mix should be added prior to homogenisation ensuring adequate stirring on exit from the homogeniser. Injection via a metering pump upstream to the homogeniser is a method used in continuous processes. In all cases the operator must ensure a minimum degree of aeration controlled by paddle design and speed. Less oxygen will be incorporated if the vessel also has a lid.

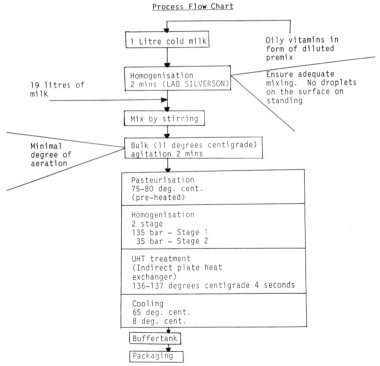

Figure 6.5 Flow diagram of method of vitamin incorporation and processing.

VITAMIN FORTIFICATION OF FOODS (SPECIFIC APPLICATION) 131

6.4.1.5 Factors affecting vitamin stability. Vitamins A and D_3 become unstable in the presence of oxygen, the rate of destruction being accelerated at high temperatures. During pasteurisation and UHT processing, losses have been shown to be minimal, ranging from 0% to about 25%. Subsequent storage losses range from 0% up to about 18% (Table 6.11).

Figure 6.5 shows the flow diagram for the method of vitamin incorporation and processing.

6.4.1.6 Fortification of dried milk. The simplest and most efficient way of adding vitamins to dry milk is to dry blend the vitamin powders into an aliquot of the milk powder and in turn blend this into the bulk. However, this is not always possible as it requires extra mixing equipment. If this method is used it is important to select vitamin forms which are compatible in particle size to the milk powder as distribution may lead to segregation and a heterogeneous mix (Figure 6.6).

Alternative methods are:

- Dispersing the dry vitamins in warm milk or warm water immediately before use and blending into the bulk or concentrate before spray drying.
- Homogenising oily vitamin forms into a small portion of the milk then adding to the bulk followed by spray drying.
- Injecting the homogenised vitamins via a metering pump into the bulk on route to the spray drier.

6.4.1.7 Storage of fortified milk. Despite the milk process and mode of addition of the vitamins, the final product must be packaged and stored in optimal conditions to ensure top quality product and minimal vitamin losses on subsequent storage.

- Dry milk should be stored in a cool / dry environment in moisture proof packaging.
- Long-life milk must be stored in airtight, lightproof barrier material.
- Pasteurised short-life milk should be refrigerated and protected from UV light if possible.

In all development work it is essential to carry out pilot plant trials sampling the

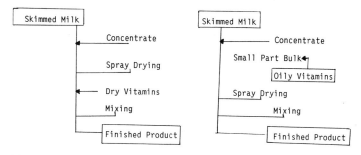

Figure 6.6 Most widely applied methods of adding vitamins to spray dried milk (Roche, Pub. 1056).

product both on the exit prior to filling as well as sampling throughout the shelf life period. Overages should be added to ensure that claimed levels are reached after processing and at the end of the use by date.

6.4.2 Yoghurt

The yoghurt market is a growing area with innovative new products being added to ranges at a constant rate. The following section looks at adding vitamins to standard stirred and set yoghurt and does not go into detail on newer product concepts, although the principles described would apply. Vitamins should ideally be introduced into yoghurt in such a way that processing parameters do not need to be changed, while the flavour and appearance of the finished product are not altered.

6.4.2.1 Vitamin addition. Dependent on the method of manufacture and preferred stage for the addition of the vitamins, there are a number of possible routes which are shown on the flow diagram (Figure 6.7). Temperatures and times of processing are not given as these vary from process to process.

The two basic methods of addition are:

- addition as a dry premix to the base ingredient at initial mixing stages
- addition to the fruit conserve via the fruit supplier.

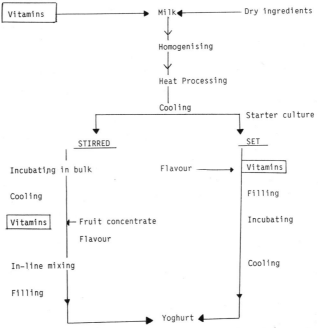

Figure 6.7 Possible points of addition for vitamins in yoghurt manufacture.

6.4.2.2 *Factors affecting stability.*

6.4.2.2.1 Addition to the base: In this instance the vitamins would be added as a dry mix. To ensure a good distribution throughout the mix it is advisable to preblend the vitamin mix with the other dry ingredients, e.g. sugar or skim-milk powder, before addition to the bulk milk. On addition the mix should be stirred adequately, minimising excessive aeration to ensure good distribution.

The heat process, usually pasteurisation, will cause slight losses of the more heat labile vitamins, e.g. vitamin B_1, folic acid and vitamin C, but not to any extreme. The incubation period has been shown to have minimal effect on the vitamin stability, however certain microorganisms have been shown to metabolise some vitamins. Nicotinamide is metabolised by standard yoghurt cultures; this area has not been explored fully in trials carried out to date.

6.4.2.2.2 Addition to the fruit conserve: Temperatures reached in fruit processing cause a slight loss of the more heat labile vitamins but a greater problem is caused by the combined effect of high temperature and low pH. These parameters have severe effects on vitamin A which is unstable at pH below 3.7. The problem can be reduced by rapid cooling of the conserve but this is usually impractical with the volumes processed.

The point of addition of the vitamins is an important consideration. As the gel thickens, distribution becomes more difficult. The suggested route is addition of vitamin mix via the starch slurry once the conserve has been brought up to maximum temperature. The shorter the cooling time, the less severe are the effects of the processing parameters.

A further disadvantage of adding the vitamins to the fruit conserve is the expected shelf life prior to addition to the yoghurt base. The manufacturer must ensure that the required vitamin levels are provided at the maximum storage period of the conserve and subsequent shelf life of the yoghurt.

6.4.2.3 *Effect of packaging material on vitamin stability.*

Of all vitamins added to yoghurt, vitamin C is the most labile when it comes to product storage. With the combined oxygen content of the head space and that dissolved in the product, vitamin C levels decrease through acting as an oxygen scavenger. The vitamin C level therefore shows a dramatic drop followed by gradual loss on storage. The polystyrol pots used for yoghurts allow oxygen penetration which in turn is mopped up by the ascorbic acid.

This can be dramatically reduced by using a totally oxygen-impermeable barrier plastic or glass.

Sufficient vitamin overages can be added to compensate for the losses but it can become a major problem in long-life yoghurts. There are limits to the level of ascorbic acid added above which taste and textural problems arise together with browning problems.

The graph in Figure 6.8 and the results shown in Table 6.12 are from a laboratory

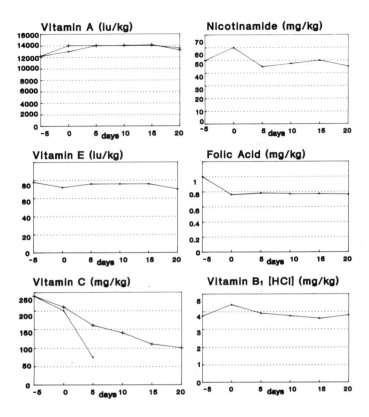

Figure 6.8 Vitamin stability in yoghurt. Processing at –4 to 0, storage at 5°C. ●——● polystyrol pot, ✶——✶ glass pot.

trial carried out comparing polystyrol plastic cups to glass jars, the product being stored at 5°C for 21 days (Roche, 1989). The vitamins were added in the dry form to the base at the beginning of the process. The yoghurt was prepared using a

Table 6.12 Losses (of added levels) of vitamins on processing and storage

Heat process losses	Vitamin E	5%
	Folic acid	22–30%
	Vitamin C	20–27%
	Vitamins A, B$_1$	stable
	Nicotinamide	stable
Storage losses	Vitamins A, B$_1$,	stable
	Folic acid	stable
	Vitamin E	stable
	Nicotinamide	stable
	Vitamin C	Highly sensitive to oxygen

standard culture of *L. bulgaricus* and *S. thermophilus*. Glass jars were sealed with screw-cap lids, the polystyrol pots having a 25 ml head space sealed with an aluminium foil lid.

In summary it can be said that the stability of vitamins in yoghurt is dependent on a number of factors:

- processing conditions
- type of culture media
- incubation time
- packaging material
- head-space of pots

In developing any new product requiring vitamin addition, it is essential that trials are initially carried out and all processing parameters are set as for large-scale manufacture.

6.4.2.4 Taste and odour related to vitamins. Some of the B-group vitamins, namely thiamin (vitamin B_1) and riboflavin (vitamin B_2) have been shown to impart slight off-flavours to yoghurt base. This can be overcome by using protected forms of the vitamins. Riboflavin can impart a slight yellow coloration dependent on the level added, in some cases this is used as an advantageous point in imparting a creamy appearance, in other cases it is not so favourable but again can be avoided by using the protected forms.

In general yoghurts are not fortified to give the total daily requirement of the vitamins, and levels added do not contribute to any detrimental changes to the standard product.

6.4.3 Ice cream

Ice cream is defined as being 'a frozen dairy food made by freezing a pasteurised mix under agitation to incorporate air, ensuring uniform consistency. The mix contains milk solids, sugar, water, flavouring and stabiliser. Physically, it is a partly frozen foam with an air content of about 40 to 50% by volume.'

This statement highlights the three main processes applied in ice cream manufacture, namely, pasteurisation, homogenisation and freezing.

In the manufacture of ice cream the ingredients are mixed together, pasteurised and homogenised. The emulsion then passes into a freezer in which the temperature is reduced sufficiently partly to freeze the mixture, at the same time air is whipped in. A solid film forms on the walls of the freezer and this is continuously scrapped off and taken to a room where the temperature is reduced further. This causes the rest of the water to freeze and the product to harden.

It is interesting to note the mild conditions to which the vitamins would be subjected if incorporated into the recipe. Freezing has not been shown to have a detrimental effect on vitamin stability and with such a product there are no thawing or warming requirements prior to consumption.

The method of pasteurisation is dependent on the size of operation. In small volume operations, pasteurisation is often by the batch method. Those ice cream makers who produce more than 25 000 litres of mix per day usually employ HTST continuous pasteurisers. In such processes the temperature–time relationship is about 80°C for 25 seconds. Suitable overages would be included for the more heat labile vitamins added, these being thiamin (vitamin B_1), folic acid, and vitamin C.

Aeration is a very important stage of ice cream making determining the final over-run and texture of the product. The fat soluble vitamins and vitamin C are readily oxidised in the presence of oxygen, but due to the material being partly frozen at the point of aeration, the vitamin losses will be minimal, and suitable overages can be applied.

6.4.3.1 Vitamin addition. In small scale operations the mix ingredients including the vitamins are usually blended in the pasteurised vat. The mix is usually blended warm enabling a thorough mixing and hydration of the stabilisers.

In larger scale operations, metering pumps are used for adding the dry ingredients and the blending tank would have a turbine agitator, forming a deep vortex into which the dry ingredients can fall. This assures rapid wetting and dispersion of the vitamins.

6.4.3.2 Flavour and colour considerations. At the vitamin levels added, being in the range of 17–33% UK recommended daily allowances (RDA) per serving, the vitamins added will not have any effect on flavour.

Riboflavin (vitamin B_2) may be added with a dual purpose, to give the required vitamin claim and to impart a yellow colour to the mix. Very advantageous in vanilla ice cream, the colour achieved is a clear pale to bright yellow dependent on the level added. Its use is limited in many food products due to instability when subjected to ultraviolet light, although this is not a problem in the storage of an ice cream-type product.

β-Carotene (provitamin A) gives an ideal vanilla shade to ice cream and can contribute to a part of the vitamin A claim with the necessary calculations being made.

In summary, ice cream is an ideal food to add vitamins to, requiring no modifications to the standard method of manufacture and thereby enabling a range of these essential nutrients to be added dependent only upon on the concept employed aimed at specific end users.

6.4.4 *Margarine*

Margarine is an economical, nutritious and palatable substitute for butter. It resembles butter in appearance, form and composition with the fat being vegetable based.

Naturally butter's nutritional value lies in its content of easily absorbable fatty

VITAMIN FORTIFICATION OF FOODS (SPECIFIC APPLICATION)

acids and the fat soluble vitamins. On the other hand, margarine too provides fatty acids but is enriched with fat soluble vitamins.

Enrichment of margarine with vitamins A and D is common practice in a large number of countries, be it on a voluntary or compulsory basis. On average, about 30 000 international units of vitamin A and 3 000 international units of vitamin D are added per kilogram of margarine. Margarines with a particularly high content of polyunsaturated fatty acids (PUFA), are generally also enriched with vitamin E, levels of which should be increased when the level of PUFA is increased.

6.4.4.1 Methods of enrichment. The methods of manufacture of margarine can be divided into two broad categories, batch and continuous, both having similar processing steps (see Figures 6.9 and 6.10). In both processes, the oily vitamins should be mixed thoroughly in five to ten times the quantity of warm oil being mixed thoroughly to produce a uniform solution which is added to the fat blend before the emulsification process.

In some countries water soluble vitamins may also be added to margarine. In this case the vitamins would be dissolved in about twenty times the quantity of pasteurised milk and added to the aqueous phase before pumping into the emulsification churn.

Because of the mild processing conditions applied, only a small overage would

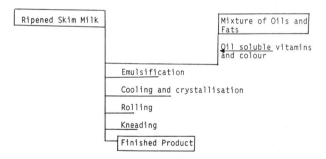

Figure 6.9 Batch process for manufacture of margarine.

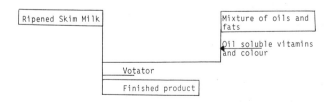

Figure 6.10 Continuous process for manufacture of margarine (Roche, 1986)

be required, more to compensate for distribution variations rather than heat degradation. For vitamins A and D3, and β-carotene, overages of only 10% are more than sufficient for six months at normal storage temperatures. Vitamin E, in the form of tocopheryl acetate, is extremely stable and well covered for the usual shelf life of the product, with an overage of about 5–10% (Goldrick, 1976).

As shown in the flow diagrams, the colour is often added as a combined ingredient with the oily vitamins. β-carotene (provitamin A) not only provides a natural colour hue to margarine, but is also a significant source of vitamin A. Using the appropriate legislative conversion factor it can be added to contribute a part of the vitamin A declaration. The β-carotene can be added as a premix with the oily vitamins or as an oily stock solution prior to emulsification.

As a result of the nature of the margarine and the storage conditions employed vitamin losses on shelf life have been shown to be minimal. Losses occurring would be due to oxidation of the oily vitamins, a process which would at the same time cause rancidity of the fats themselves.

6.5 Confectionery

Fortification of confectionery is an area of growth for those demanding more than just calories and flavour in sweets. The most promising products to focus on are those which also combine the use of sugar substitutes, e.g. calorie reduction, diabetic and tooth-saving products. Examples of such products are chocolate covered dietetic bars, multivitamin lollipops, high level vitamin C-containing hard candies, multivitamin hard candies, chewy sweets and pastilles.

Vitamin addition during confectionery making is a very simple operation. Inappropriate or excessive addition of vitamins to a food is not desirable but foods essentially carbohydrate in character should contain an appropriate amount of vitamins associated with carbohydrate utilisation. Vitamin losses in processing in different types of confections vary considerably but the overages plus natural vitamin content of some of the ingredients will usually cover the losses in processing and storage.

The almost universal way for incorporating vitamins into sugar confectionery involves the preparation of a suitable premix. The carrier for the premix should be either a part of the ingredients of the formula or some substance that is miscible with it.

Although some types of confectionery require careful handling and proper timing in the vitamin addition to prevent excessive losses, no unusual difficulties should be encountered. The vitamin premix becomes only another simple ingredient in the formula.

6.5.1 Hard boiled candies

Traditionally hard candies are either deposited or drop rolled in their manufacture.

In the deposited method, the sugar mass is heated to 140–145°C and then the colour, flavour and citric acid or other acidulants are added prior to depositing. If vitamins are to be added they would be dispersed in the colour and flavour or in a very small quantity of water and added at the last stages. With rapid cooling and no significant exposure to high temperatures, losses of the vitamins are minimal.

Evidence in work carried out to date has shown that problems of browning can occur when ascorbic acid is added at temperatures of around 145°C. The browning is accompanied by the production of carbon dioxide with foaming and furfural production. The problem appears to be caused by the acidity, with time and temperature being the determining factors.

A critical control of the temperature should be applied when fortifying with ascorbic acid and it should not be added with the citric acid or other acidulant. Slight overages should be added to compensate for possible distribution variation and losses of the more heat labile vitamins. Retention of approximately 90% of added vitamin B_1 and B_2 can be achieved, according to Siemers.

The second method of production involves cooling of the sugar mass on the slab after boiling. Here, a premix of the vitamins will be added directly onto the slab along with the citric acid. The flavouring and colouring is then added over the premix. The entire mass is then folded continuously to ensure a thorough mixing of the minor ingredients. The resultant cooled mass (now 85–90°C) is passed through rollers and formed.

Losses of no more than 15–20% may occur over a storage period of six months to a year.

The addition of vitamin C to hard candy also benefits from the acidity of the ascorbic acid which will add to the zestful effect of the fruit acids that are normally added when this type of confection is fruit flavoured.

6.5.2 Chocolate

Chocolate is said to be one of the easiest confectionery products to fortify. It seems to exercise a protective effect on added vitamins and retard losses on storage (Leighton, 1951). The accurately weighed vitamins are mixed with the correct quantity of molten chocolate at 30–40°C and then milled. After milling, the semi-fluid mixture is cast into moulds and coated. This is then used when required to fortify large batches of chocolate simply by melting and blending with the bulk before shaping (Lees and Jackson, 1983).

6.5.3 Fondant

Since the B-group vitamins are susceptible to destruction by heat, it is always advisable to add the vitamin mix as late as possible in the manufacturing process but in sufficient time to insure good distribution throughout the batch. Caramel, fudge and other pan-worked goods can be fortified by adding a premix composed of the vitamins and powdered sugar. For convenience approximately 25 g of a premix should provide the desired quantities of vitamins when added to 4.5 kg of bulk ingredients. The premix is added after the sugar cook has reached the proper temperature and the heat has been turned off. In the cases where a small quantity of egg albumin or fondant (or both) must be added to the cook at the end of the heating period, the vitamin mix can be incorporated into one of these before adding to the batch in the pan.

In large scale production, a retention of approximately 90% of added vitamins B_1 and B_2 can be anticipated when operating under careful control.

6.5.4 Marshmallows

On a general basis the fortification of marshmallows does not present any problems. However, they may not be suitable for the addition of vitamins which are known to oxidise because aeration will be expected to bring about losses of these vitamins. The same applies for nougat type products. If fortified with other vitamins it is usually achieved by adding a powder premix with a sugar base to the liquid ingredients of the formula.

6.5.5 Pectin jellies

In this type of product the vitamins would be added with the colour, flavour and citric acid in a minimum amount of water once the mass has reached the correct solids content. In the case of pectin jellies this is around 78%. Pectin jellies may not be a good vehicle for vitamin C, not because the vitamin would be harmful but because addition of the acid may be detrimental to the setting properties of the jellies, the pH of which runs within narrow limits.

A concentrated premix can be made of the water soluble vitamins together with water dispersible forms of the fat soluble vitamins. If necessary the premix can be diluted with sucrose or dextrose. The premix is added at the same time as the flavour, allowing sufficient time for blending the sucrose or dextrose base, and always before the final addition of the citric acid solution.

Due to the moisture content of a jelly there will be slightly higher losses on storage compared with other types of sugar confectionery. However, this can be compensated by higher overages.

6.5.6 Starch jellies

The pH of starch jellies is not as critical as that of pectin jellies. There will be some losses due to the cooking temperature which can be compensated for by adding overages for the heat and oxygen sensitive vitamins. Losses of vitamins B_1, C and A have been shown to be approximately 10% (Lees and Jackson, 1983). The vitamins, flavour, colour and citric acid can be added at one time after boiling has been completed.

6.5.7 Chewing gum

The majority of work carried out to date has looked at vitamin C in chewing gum. It is a suitable food for the regular intake of supplemental vitamin C enabling slow release from the polymer matrix.

In regular stick gum where the vitamin has been mixed into the mass, shelf life is shortened due to the yellowing of the sticks caused by a reaction between ascorbic acid and an ingredient of the gum base. Water activity of packaged gum is sufficient to turn the sticks yellow to brown in the course of a few months.

Regular crystalline ascorbic acid is too coarse for use in stick gum, ascorbic acid fine powder has to be used. Sodium ascorbate has been shown to cause excessive browning, however despite the colour development, vitamin C retention is very good.

The stability tests show good vitamin C retention (with no overage) and slight discoloration after six months. After twelve months there is definite browning caused by the action of ascorbic acid on the gum base constituents (Table 6.13).

In a standard sugar-containing stick gum recipe, ascorbic acid fine powder is added with the powdered sucrose. Maximum processing time and temperature in the kneader are 10 minutes and 50°C, respectively.

In the case of coated chewing gum, similar problems have been reported when vitamin C was mixed into the mass of gum centres to be used for coating. The easiest way to avoid gum discoloration is to incorporate ascorbic acid in the coating syrup. 15 to 60 milligrams of ascorbic acid is recommended per coated product.

A 15% overage has proven sufficient to compensate for manufacturing losses. Storage losses are expected to be as low as in stick gum (6% after one year). A precaution to note is that inversion of the hot sugar syrup by the ascorbic acid must be avoided by maintaining the sugar syrup temperature at or below 50°C. 'Sugarless' gum coatings with vitamin C are also possible. The use of polyalcohols in this process avoids inversion and the syrup temperature can be higher.

The packaging of fortified confectionery products is of upmost importance. Lozengers and gums in foiled blister packs or individually wrapped would give the ideal protection. Wrapping of chocolate in foil would give added protection. Some of the sugar substitutes (e.g. Lycasin) have the property of being very hygroscopic which should be taken into account when choosing suitable packaging material.

Table 6.13 Vitamin C retention and colour development (see Siemers)

Months storage colour at room temperature	Storage (months) at room temperature	Vitamin C retention (%)
White	0	100
White	1	100
Yellowish	6	94
Light brown	12	93

Fortification of other confectionery products not covered above, for example fruit chews, gums and pan coated sweets will follow similar lines to the recommendations given. Major factors to consider are pH range, moisture content, processing temperatures and mixing variables. Once these have been determined the best suitable method can be adopted.

References

Adams, C.E. and Erdman, J.W. Jr. (1988) Home food preparation practices. In Harris R.S. and Karmen, (eds) Nutritional Evaluation of Food Processing, AVI Publishing Co., Westport CT, 557–605.
Aykroyd, W.R. and Doughty, J. (1970) *Wheat in Human Nutrition*. Food and Agricultural Organisation of the UN, Rome.
Beetner, G., Tsao, T., Frey, A. and Lorenz, U. (1976) Stability of Thiamin and Riboflavin during extrusion processing of triticale. *J. Milk Food Technol.* **39**, 244–245.
Brennan, J.G., Butters, J.R., Cowell, N.D. and Lilley, A.E.V. (1990) *Food Engineering Operations*, 3rd Edition. Applied Science Publishers, London.
Briant, A.M. and Klosterman, A.M. (1950) Influence of ingredients on thiamin and riboflavin retention and quality of plain muffins. *Trans. Am. Assoc. Cereal Chem.* **8**, 69.
Cort, W.M., Borenstein, B., Harley, J.H., Osaca, M. and Scheiner, J. (1975) Nutrient stability of fortified cereal products. A paper presented at *Micronutrient Stability Symposium* at the 35th Annual Meeting of the Institute of Food Technologists, Chicago III, June 8–11.
Counsell, J.N. and Hornig, D.H. (1981) (eds) *Vitamin C, Ascorbic Acid*. Applied Science Publishers, London.
Dietz, J.M. and Erdman, J.W. (1989) Effects of thermal processing upon vitamins and proteins in foods. *Nutrition Today*, July/August.
Goldrick, E.A. (1976) The application of vitamins in food fortification. In: *Food Technology in New Zealand*, Volume 11, No 8.
Grebaut, J., Luquet, F.M., Mareschi, J.P. and Mercier, C.H. (1983) Influence du traitement d'une farine de ble sur cylindre ou par cuisson extrusion sur sa valeur nutritionelle. *Cab Nutr Diet XVIII* **6**, 327–334.
Johnson, L.E., Gordon, H.T. and Borenstein, B. (1988) Technology of breakfast cereal fortification. *Cereal World* **33** (3) 278–330.
Killeit, U. and Wiedman, W.M. (1984) Einfluss der kochextrusion auf die stabilitat von B-vitaminen. *Getreide Brot und Mehl* **39**, 299–302.
Lees, R. and Jackson, E.B. (1983) *Sugar Confectionery and Chocolate Manufacture*. Leonard Hill, Glasgow.
Leighton, A.E. (1951) How to vitaminise candies. In: *Food Engineering*. McGraw-Hill, New York, p. 18.
Magna, J.A. and Sizer, C.E. (1978) Ascorbic acid and thiamin retention during extrusion of potato flakes. *Lebensm. Wiss. Technol.* **11**, 192–194.
Ranhotra, G.S., Gelroth, J.A., Novak, F.A. and Bock, M.A. (1983) Losses of enrichment vitamins during the cooking of pasta products. In: *Nutrition Reports International*, Volume 28, No 2.
Roche Publications 1056, *Enrichment and Fortification of Dairy Products and Margarine*
Roche Publication 1970/8 (1985) *Vitiminisation of Concentrates, Nectars and Juice Drinks*.
Roche Publication 2038 (1986) *Enrichment of Margarine*.
Roche Publication (1988) *Food Marketing News*, **2**.
Roche Publication 2217 (1989) *Vitamins and Carotenoids in Pasta and Noodles*.
Roche Publication 2218 (1989) *Yoghurt with Vitamins: for a clear positioning of your products*.
Schlude, M. (1987) The stability of vitamins in extrusion cooking. In: *Extrusion Technology for the Food Industry* (ed. Colm O'Conner). Elsevier Applied Science, London.
Siemers, G.F., Adding Vitamins to Candies. Hoffmann La Roche Inc, Nutley, N.J.
Steele, C.J. (1976) Cereal fortification – technological problems. *Cereal Foods World* **21**, (No 10), 538–540.

7 Vitamins as food additives
J. N. COUNSELL

In the fields of physiology and nutrition there have been many attempts to classify the vitamins but they have failed to progress very far beyond the two very simple groups; 'oil soluble' and 'water soluble'. Indeed, the only claims that the vitamins have to being regarded as a group of substances is that they are all essential, dietary, trace, organic chemicals, and even that falls down if the substance is synthesised *in vivo*. Their molecular structures and physical, chemical and biological functions are all quite different. There is, however, one thing they do have in common: they are very reactive substances. That, combined with their obvious physiological acceptability, makes at least some of them interesting as possible food additives. The development of industrial scale vitamin manufacture has provided materials on a large enough scale to enable this potential to be realised.

Not all vitamins are suitable for use as food additives: some, indeed, are quite unsuitable. Of the thirteen recognised vitamins, five have had some applications in food technology. The premier position among them must be taken by ascorbic acid (vitamin C) of which it has been said that its applications are limited only by the imagination of the technologist. Close behind it, although in narrower fields, are the carotenoids (provitamins A) and dl-α-tocopherol (vitamin E). Of lesser significance are nicotinic acid and its amide (sometimes called vitamin PP), and riboflavin and its phosphate sodium salt (vitamin B_2).

7.1 Ascorbic acid (vitamin C)

The industrial production of vitamin C began in Basel, Switzerland, in 1934, one year after Reichstein first achieved his synthesis. Since then production has increased steadily to the point where it is now a commodity chemical with a number of manufacturers and pricing subject to strong competitive pressures. At least half of the tonnage produced is destined for food industry use; a relatively small proportion for its nutritional value and by far the larger amount for its use as a food additive.

In practice, of course, the two uses are not so easy to separate. Vitamin C added, for example, to a fruit beverage for its nutritional benefit cannot be expected not to act as an antioxidant should the circumstances require it. Similarly, ascorbic acid added as an antioxidant to the same beverage may have a residual amount present at the time of consumption and so may contribute nutritionally.

The fact of the practical uses being limited only by the imagination of the technologist have been mentioned. Such uses outside the food industry include photography, plastics manufacture, water treatment (for removing excess chlorine), stain removers, hair preparations and skin treatment. These are outside the scope of this chapter but are mentioned to illustrate the huge range of applications of this versatile chemical.

The food additive uses are no less wide in scope. The principal applications are in soft drinks, meat products and bread manufacture.

At first sight it is difficult to understand why a vitamin, especially such an unstable compound, should have achieved such importance as a food additive. Two aspects are useful to gain an understanding. First is the quite powerful reducing properties together with its obvious physiological acceptance and safety and second is the worldwide legislative acceptance of ascorbic acid and many of its technical uses. The availability of crystalline vitamin C to very high standards of purity in industrial quantities has undoubtedly been of assistance in the growth of the market.

In its technical applications, ascorbic acid is used in the forms of ascorbic acid itself, the sodium salt (and to a lesser extent, the calcium salt) and the palmitic acid ester. Each has its own particular strengths and special applications.

7.1.1 Properties

Some of the more important properties of ascorbic acid and the salts and ester used in food additive applications are given in Table 7.1.

7.1.2 Fruit, vegetables and fruit juices

In this context there are two kinds of fruit, those that show discoloration on cutting or bruising (for example, apples and bananas) and those that do not (examples are citrus fruits). Table 7.2 gives examples of the ascorbic acid content of some common fruits.

Table 7.1 Properties of ascorbic acid, salts and ester used as food additives

	Ascorbic acid	Sodium ascorbate	Calcium ascorbate	Ascorbyl palmitate
EC No	E300	E301	E302	E304
Solubilities (approx per 100 ml at room-temperature)				
Water	30 g	90 g	55 g	ns
Ethanol	2 g	ns	ns	12.5 g
Ethyl ether	ns	ns	ns	0.76 g
Arachis oil	ns	ns	ns	0.03 g
Sunflower oil	ns	ns	ns	0.03 g
Olive oil	ns	ns	ns	0.03 g
Vitamin C equivalence	1	0.889	0.826	0.425

ns = not soluble

VITAMINS AS FOOD ADDITIVES

Table 7.2 Ascorbic acid content of some common fruits

Fruit	Ascorbic acid (mg/100 g edible portion)
Apple	2–10
Apricot	7–10
Bananas	10
Blackberry	15
Blackcurrant	90–360
Damson	3
Grapefruit	37–50
Lemon	30–55
Mango	30
Orange	30–65
Peach	7
Pineapple	25
Raspberry	25
Strawberry	60
Tomato	10–40

Discoloration usually occurs within a few minutes after an apple is bruised or cut unless steps are taken to prevent it. Factors common to such discoloration include low ascorbic acid concentration and high phenolase activity. In order that browning can take place, three elements must be present, enzyme, substrate and oxygen. If one of these is absent or prevented from reacting, enzymic oxidation and browning will not take place. If the ascorbic acid has been oxidised, non-enzymic browning can still occur. Citrus fruits and berries are deficient in substrate and have higher ascorbic acid levels and so enzymic browning does not take place.

The enzyme, usually polyphenoloxidase, acts on a flavonoid or *ortho*-phenolic substrate in the fruit to cause, in the presence of oxygen, the production of *ortho*-quinone compounds. The initial products are reversibly-oxidised, coloured quinone structures but they are only transitory. Unless ascorbic acid or another hydrogen donor is present these quinone compounds will further oxidise in the presence of more oxygen and polymerise to form dark brown irreversibly oxidised compounds. As long as there is a sufficiency of ascorbic acid present, the *ortho*-quinone compounds are reduced back to *ortho*-phenolic compounds and so browning does not occur. As this oxidation and subsequent reduction continues, the enzyme activity decreases and eventually ceases, provided that enough ascorbic acid is present, otherwise the effect will be transitory and browning will only be delayed. It is necessary to add the ascorbic acid in such a way that the enzyme is inactivated rapidly before any appreciable oxidation can occur and then to maintain as much as possible of the natural colour and flavour. If oxygen remains in the head-space of the container when the ascorbic acid is exhausted, the flavonoids will be oxidised and polymerised. Removal of the oxygen will enable the added ascorbic acid to create an anaerobic environment.

7.1.2.1 *Fruit.* Frozen, canned and processed fresh fruits with low ascorbic acid levels (such as apples, apricots, peaches, bananas, pears and plums) benefit most from the addition of ascorbic acid. Intermediate level fruits (e.g. cherries, pineapple and some strawberry varieties) also benefit. Ascorbic acid is added either in the syrup or, where dry packs are involved, mixed with the sugar. Normally, an amount of ascorbic acid in the range of 275–550 mg/kg is sufficient. Where the syrup method is used it is important that the fruit should be kept below the surface and that freezing should take place as soon as possible after adding the syrup.

It is important to realise that the ascorbic acid treatment will not upgrade the original fruit quality nor will it cover up poor processing techniques.

Apples are difficult to treat due to their very porous nature. The problem of getting the ascorbic acid solution into sufficiently intimate contact with the fruit, which is important if the quality of the fruit is to be maintained, is solved by applying and releasing a vacuum so drawing the solution into the fruit. A level of 660 mg/kg of fruit is usually effective in apple halves.

Peaches and apricots benefit considerably in maintenance of flavour and colour by the addition of ascorbic acid in the region of 440 mg/kg.

The delicate flavour of properly ripened bananas may be preserved by spraying with a 1.5% solution of ascorbic acid in order to achieve a level of 440 mg/kg.

7.1.2.2 *Vegetables.* Much of what has been said about protecting colour and flavour of fruits in frozen and canned products also applies to similarly processed vegetables. Some examples of the ascorbic acid content of common vegetables are given in Table 7.3.

In processing vegetables, and fruits for that matter, blanching, or scalding, usually takes place immediately after cleaning and cutting and before canning, freezing or dehydration. Blanching is usually achieved by passing the food through a hot water bath or steam tunnel. The 'holding time' will depend on the nature and size of the material being processed but usually is only a few minutes at temperatures of 85–100 °C.

Table 7.3 Ascorbic acid content of some common vegetables

Vegetable	Ascorbic acid (mg/100 g edible portion)
Brussels sprouts (boiled)	40
Cabbage (raw)	55
Cabbage (boiled)	20
Cauliflower (cooked)	20
Lettuce	15
Onion	10
Parsnips (cooked)	10
Peas (frozen, boiled)	12
Peppers, green	100
Potatoes (raw)	8–19
Potatoes (cooked, various)	5–14
Spinach (boiled)	25
Sweet potato	25
Turnip (cooked)	17
Watercress	60

The purpose of blanching is to inactivate the enzymes present, to preshrink the material before further processing, to expel air, to reduce any initial infection and to aid in removing undesirable flavours and aromas. Sulphur dioxide is commonly used in the blanching process to prevent browning in peeled potatoes, apples and celery but it has disadvantages. A combination of ascorbic acid with sulphite has the advantage of replacing some of the sulphite with a substance natural to the food being processed. Adding a small amount of citric acid to keep the pH below 7 will also help by reducing any residual enzyme activity.

Preprepared potatoes for french fries are usually dipped in a sulphite solution to prevent discoloration. The taste of the sulphite can remain on the potato. Lower levels of sulphite can be used if the sulphite is partially replaced with ascorbic acid. Special mixtures based on ascorbic acid are available which allow the complete replacement of the sulphite in the treatment of potatoes.

Ascorbic acid is useful in the processing of other vegetable products. An example is canned mushrooms where, again due to oxidation, an undesirable brown colour develops. Ascorbic acid in an amount of 1.5–2.0 g/kg of mushrooms dissolved in the brine will prevent this discoloration. The ascorbic acid should not be added to the brine unless it is to be used fairly quickly. Ascorbic acid is also useful in maintaining the colour and taste of mushroom products such as mushroom soup.

7.1.2.3 *Fruit juices.* In some countries, ascorbic acid, and other substances, may not be added to fruit juices. Where it is allowed the addition is useful in preventing deterioration of flavour and colour. The solubility of oxygen in fruit juices is in the region of 7–9 mg/litre and that is equivalent to 75–100 mg ascorbic acid. Deaeration is not 100% efficient and conventional aseptic plastic/paper laminate packs do allow the ingress of some oxygen. Thus the natural ascorbic acid (already low in apple juice) is subject to loss and eventually the organoleptic quality will suffer.

7.1.2 Soft drinks

In common with fruit juices, soft drinks are subject to flavour deterioration due to oxidative changes brought about by dissolved oxygen and the oxygen in the head-space of the container. The amount of ascorbic acid contributed by the fruit juice content (if any) is not usually enough to neutralise the oxygen in the dissolved air. The oxygen in 1 cm^3 of air may be removed by 3.3 mg ascorbic acid. Thus, for example, if a one litre bottle has a head-space of 30 cm^3 of air, about 100 mg of ascorbic acid will be needed to remove the total oxygen which will otherwise be available to cause undesirable flavour spoilage.

The addition of ascorbic acid to some types of soft drinks, particularly those coloured with certain of the synthetic dyes, may cause colour loss when the product is exposed to sunlight. The manufacturer then has to choose between the alternatives of omitting the ascorbic acid, with consequent flavour deterioration or colour loss with good flavour retention, or reformulation of the colour component. The first is not usually chosen because of the demonstrable benefit of ascorbic acid.

7.1.4 Beer

Clarity, colour and flavour are important quality criteria for beer and the oxygen content plays an important role in maintaining them at their optimum. Turbidity may occur due to the oxidation of polyphenols, polymerisation and association with proteins. Heavy metals catalyse the process. High concentrations of dissolved oxygen have a serious effect on the flavour of the beer. Several compounds in the beer can readily be oxidised to give undesirable flavour compounds with very low taste thresholds. Another detrimental effect of oxygen is darkening of the clear beer colour. Ideally, a low oxygen content in the beer is highly desirable. The best way of protecting the beer from these deleterious changes is removal of the oxygen. A number of methods are known but the most usual is by the addition of ascorbic acid. In practice a dosage of 20–30 mg/litre should be sufficient. There is no harm in also taking all possible precautions to avoid any unnecessary pick-up of oxygen during processing. In an ideal plant where exposure to air is virtually eliminated, the amount of ascorbic acid addition may be reduced provided that a residual level after bottling of 10 mg/litre is maintained.

7.1.5 Wine

The addition of ascorbic acid to wine is gaining acceptance in practice. Apart from its antioxidant action, the ascorbic acid allows a substantial reduction to be made in the quantity of sulphite which is routinely used in wine making. In general, for white and sparkling wines, the addition of approximately 100 mg/litre should be enough.

Ascorbic acid added at 100 mg/litre can also help to prevent the turbidity caused by iron compounds, as ascorbic acid prevents the oxidation of ferrous iron compounds to the ferric form. Ascorbic acid also reduces the ferric iron compounds already in the wine. Thus, there is the possibility of removing turbidity which has already developed.

7.1.6 Flour and bread

Over a period of about a year, bread made at intervals from a white flour stored at ambient temperature shows steady improvement. The beneficial changes lie in greater loaf volume, finer and softer crumb cell structure and whiter crumb colour. After about a year, the loaf volume will begin to fall, indicating a decline in the baking quality of the flour. Unfortunately, such lengthy quality improvement is not feasible in commercial practice and usually the optimum is considered to have been reached in 4–6 weeks.

The period of maturation varies from flour to flour and with other factors, including a need for reduction in the moisture content.

A number of chemicals, mostly oxidising agents, are capable of effecting the same improvements when added to the flour at low levels. Such flour improvers have an added advantage in that the addition level may be adjusted to match the flour properties in such a way that the baker is able to rely on regular and predictable flour properties in standardised baking procedures. Among the twenty or so substances known to be effective are ammonium and potassium persulphates, potassium bromate and potassium iodide. Of these, the one most widely used has been potassium bromate. The potassium bromate was usually added to the flour at the mill at levels between 5 and 20 mg/kg flour and that became common practice in the USA and many European countries in the 1920s and 1930s.

Some twenty years after the discovery of the value of those chemical compounds as flour improvers, ascorbic acid was found to be a potent agent comparable in effectiveness with potassium bromate. This was an interesting finding bearing in mind the known reducing properties of ascorbic acid and the need for an oxidising agent.

For many years the price of ascorbic acid was too high to permit its wide scale use for flour improvement and the use of potassium bromate became the norm. However, as the commercial production of ascorbic acid increased and the price fell, the situation changed. Now, in a number of countries, only ascorbic acid is permitted as a flour improver.

At the mill the addition level is usually in the range of 5–30 mg/kg, careful precautions being taken to ensure complete and even distribution of the improver. Even 10 mg/kg is usually enough to halve the maturation time. The actual amount needed will vary according to whether the flour is freshly milled from newly harvested grain, the extraction rate and the gluten content. A rough guide to the addition level in mg/kg for a particular flour may be gained by taking two to four times the percentage ash content of the flour.

In the small bakery, ascorbic acid is useful as an improver. In the production of French bread (baguette), for example, by the conventional method, the addition of 40 mg/kg flour makes a significant increase in loaf volume. Production of the same bread on an industrial scale needs a slightly lower addition of ascorbic acid (30 mg/kg).

The Chorleywood Bread Process, introduced in 1961, was based on the concept of developing the dough by a specific and controlled quantity of intense mechanical energy during a dough development time of about three minutes. The overall process time from raw materials to finished bread was reduced from about five hours to about two hours. Classical chemical kinetics required that the concentration of some of the dough components should be increased in order to achieve the same end results in less time. Yeast concentration, for example, had to be approximately doubled. Higher concentrations of flour improvers, or more effective improvers were also needed. The obvious choice was potassium bromate, if only because it was already being added to the flour at about 5–15 mg/kg at the mill. Thus no change in the ingredient list would be needed. At the time there were reservations about the then unknown toxicological risk. Instead, the addition of ascorbic acid by the baker at 75 mg/kg was recommended. The use of 75 mg ascorbic acid per kg flour was later replaced by a mixture of 30 mg potassium bromate and 30 mg ascorbic acid per kg flour.

150　　　THE TECHNOLOGY OF VITAMINS IN FOOD

The added financial cost of the ascorbic acid was more than offset by the economies of the overall process and the Chorleywood Bread Process was rapidly adopted by the UK baking industry so that within a few years some 75% of bread in the UK was made by the new method. The proportion has remained high ever since.

Another way of rationalising time-consuming bread making is by the use of the activated dough development (ADD) process now quite widely used in many English speaking countries, especially Australia.

The basis of the method is the combination of reducing and oxidising agents, cysteine and ascorbic acid. The usual quantities are 10–40 mg cysteine hydrochloride together with at least the same amount of ascorbic acid per kg flour. The process shortens the mixing time by 20–50%.

7.1.7 Pasta

While pasta is associated especially with Italy, it is now an important item of diet in other European countries and Argentina. Consumption figures are given in Table 7.4 for several countries. The increasing popularity of pasta undoubtedly depends on its cheapness, ease of preparation and excellent storage properties.

Traditionally, pasta is prepared from durum wheat semolina and water. Other ingredients may be added. In some countries, where durum wheat is not available, pasta may be made from common flour. In such cases, the addition of ascorbic acid in the range of 200–400 mg/kg during processing improves colour, texture and taste and also reduces the loss of solid matter and protein during cooking.

7.1.8 Meat processing

Colour is one of the important factors influencing the choice of food and, perhaps, never more so than where meat is involved. Presentation is so important that it may not only determine choice of product but also whether a return visit to a particular retailer is likely to take place. These matters depend on the chemistry of the natural pigments of the meat.

Many of the reactions affecting the natural meat pigments involve oxidation or reduction. For example, ascorbic acid can be used to maintain the natural colour

Table 7.4 Pasta consumption in various countries

Country	Consumption (kg/caput/annum)
Italy	33
Italy (south only)	>60
Argentina	24
Switzerland	12
France	9
Greece	9
West Germany	8
Yugoslavia	5

of fresh meat. However, in many countries this is regarded as deception and consequently forbidden, even though the ascorbic acid *per se* cannot add colour and it has no preservative action. The established fact that ascorbic acid cannot be used to conceal poor quality raw materials or faulty processing makes it difficult to understand the charge of deception of the consumer.

For some years, a mixture of ascorbic acid and nicotinic acid was used for maintaining the fresh colour of unprocessed meat. The mixture was sprinkled on the cut surfaces of the meat at the point of sale. Repeated 'salting' of the meat in this way to ensure 'freshness' during whatever period elapsed between cutting and selling had unfortunate results. While the ascorbic acid was gradually destroyed, the concentration of the nicotinic acid (a very stable vitamin) increased. A small proportion of the population is sensitive to larger than normal intake of nicotinic acid and, inevitably, the treated meat caused such a reaction. The consequent media attention led to legislative action. The use of both vitamins on raw, unprocessed meat was forbidden. Had nicotinamide been used in place of nicotinic acid, the problem may perhaps not have arisen.

Ascorbic acid does have an important role in processed meat products, particularly those involving preservation with nitrite and nitrate. Cured meats depend for their stability on the presence of nitrite and salt. Consumer preference trends for lower salt levels together with legislative pressures against nitrite and nitrate make it necessary for the curing process to be at maximum efficiency.

Ascorbic acid is a powerful reducing agent which can accelerate the reduction of the nitrite chemically thus removing the dependence of the process on biological reactions. In practice, ascorbic acid is not used because it reacts with nitrite, even at chill temperatures. The result of the reactions which occur is the release of toxic nitrogen peroxide, a serious hazard to plant operators.

Ascorbic acid can be used with a system of injection with special double needles which ensure that the ascorbic acid solution does not come into contact with the nitrite solution until both are in the meat tissue. On the other hand, sodium ascorbate reacts fairly slowly with sodium nitrite and so may be added to the brine, although it is advisable that the mixed solutions be used soon after preparation in order that there should be no loss of either ascorbate or nitrite. The sodium salt is no less effective in meat curing.

The sensitivity of ascorbate to heavy metals, especially iron and copper, makes it essential that solutions must be prepared and, where necessary, stored in stainless steel, plastic or enamelled containers. Glass vessels would be ideal but not in a manufacturing environment. Solutions of ascorbate may be kept in a cool place, in suitable containers having a tightly fitting lid for three or four days without significant loss of strength. Similarly, should the solution be used for spraying, the spraying equipment must be such that the solution only comes into contact with suitable materials. The dry ascorbic acid and sodium ascorbate have very good stability provided that they are kept dry. On longer storage, particularly with sodium ascorbate, a faint yellow colour may develop. This coloration has no effect on the activity or usefulness of the products in any way.

152   THE TECHNOLOGY OF VITAMINS IN FOOD

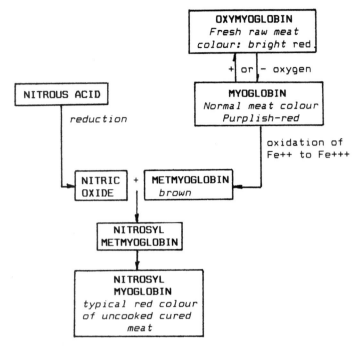

Figure 7.1 Colour changes accelerated by ascorbic acid during the curing of red meat.

In the curing process the ascorbate first accelerates the breakdown of the nitrite to nitric oxide and water. The nitric oxide combines with the muscle pigment, myoglobin to form nitrosylmetmyoglobin, and, almost instantaneously, the latter forms nitrosylmyoglobin, the typical red colour of cured, uncooked meat. The change is accelerated by ascorbic acid (Figure 7.1). Ascorbic acid, by its oxygen scavenging action, will prevent the oxygen from inhibiting colour development by oxidising the nitric oxide and causing the formation of brown discoloration by oxidation of the myoglobin to metmyoglobin.

The acceleration of the formation of nitric oxide by ascorbic acid increases the net yield, allowing a given degree of coloration to be achieved with an approximately one third saving in the amount of nitrite required. The residual nitrite is also lowered. The cured colour, once it has been formed, is stabilised by residual ascorbic acid.

In raw meat sausages, ascorbic acid (and here there is no chemical need to use the sodium salt although the greater solubility of the latter may be of some advantage) is widely used to stabilise the natural meat pigment, presumably by oxygen scavenging. The occasional formation of white spots in this type of sausage represents a very detrimental change in appearance. Adding more ascorbic acid confers no improvement. The effect is due to the onset of oxidation of the small, discrete fat particles in the sausage, especially just below the skin. The appearance of the white spots is accompanied by a change in the colour of the sausage

VITAMINS AS FOOD ADDITIVES

Table 7.5 Effect of sodium ascorbate and Ronoxan D20 on white spot formation in raw meat pork sausage

Addition	Days storage at 5°C						
	1	2	3	4	5	6	7
Special recipe: fresh back fat							
None	oo	ooo	oooo	oooo	oooo	oooo	oooo
Sodium ascorbate 0.025%		oo	ooo	oooo	oooo	oooo	oooo
Ronoxan D20 0.5%				o	o	o	o
Ronoxan D20 1.0%							
Special recipe: stored back fat							
None	oo	ooo	oooo	oooo	oooo	oooo	oooo
Sodium ascorbate 0.025%		oo	ooo	oooo	oooo	oooo	oooo
Ronoxan D20 0.5%				o	o	oo	oo
Ronoxan D20 1.0%						o	o
Commercial recipe: fresh and stored back fat							
None	o	oo	ooo	ooo	oooo	oooo	oooo
Sodium ascorbate 0.05%		o	oo	ooo	ooo	oooo	oooo
Sodium ascorbate 0.1%			o	oo	ooo	ooo	oooo
Ronoxan D20 0.05%				o	o	oo	ooo
Ronoxan D20 0.1%					o	o	oo
Ronoxan D20 0.25%							o
Ronoxan D20 0.5%							o

Ronoxan D20 concentrations are expressed on fat content basis.
Key: o - few white spots
    oo - intermediate number of white spots
    ooo - abundant white spots
    oooo - excessive coalescence of white spots
(Reproduced by kind permission of Roche Products Limited).

corresponding to deoxygenation of the oxymyoglobin. The problem is not common when the recipe involves fresh fat. However, when stored fat, especially frozen fat, is used, the spots can develop rapidly, often in a matter of only a few hours. The change in the appearance of the sausage makes the product unsaleable.

The fat soluble derivative of ascorbic acid, ascorbyl palmitate, may be used to prevent the development of these white spots. The ascorbyl palmitate is usually added as a proprietary synergistic mixture with *dl*-α-tocopherol and other ingredients on a water soluble base. The effect of this mixture in comparison with sodium ascorbate is shown in Table 7.5. The concentrations of the mixture, known commercially as Ronoxan D20^{TM}, are expressed on a fat content basis. The special recipe used in the first two parts of the table was one known to be particularly sensitive to the formation of white spot.

The observation that the formation of white spot in sausages coincided with a deleterious effect on the colour of the product led to the widespread use of the mixture for that reason, particularly as it offered a worthwhile economic advantage.

The commercial mixture has also found application in pork pies where it serves to protect the meat from discoloration during distribution.

7.2 Carotenoids (provitamins A)

A great deal of the beauty of nature is due to the colours of the carotenoids and their complexes with proteins. The importance of this group of compounds is not often obvious, for example, in vegetation the presence of chlorophyll masks the beauty of the carotenoids until autumn. In the animal world, the carotenoids and their complexes with proteins are major contributors to, for example, the rich plumage of birds and the colours of fish and the invertebrates, especially the crustacea. The most abundant carotenoid (fucoxanthin) is found in many marine algae. It has been estimated that the annual production of carotenoids by nature amounts to about 100 000 000 tonnes. Commercial production is insignificant in weight comparison terms but far outshines nature's effort in the commercial world of food additives.

The more important sources of the carotenoids in the diet are illustrated by Table 7.6. Only those foods with a content greater than 1 mg/kg have been listed. There are a great many foods with contents below that level.

Table 7.6 Carotene contents of some common fruits and vegetables

Food	Carotene (mg/kg)
Apricots, raw	6–15
Apricots, canned	7.4
Cherries, sour, canned	5
Blackcurrants, raw	2
Gooseberries, raw	1.8
Mandarins	1.3–2.8
Mangoes, raw	20.7
Melons, yellow, raw	11–20
Nectarines, raw	5
Olives, green, canned	1.5
Peaches, raw	3.4–5
Plums, raw	1.6–2.2
Asparagus, boiled	5
Beans, French, boiled	1.4
Broccoli, boiled	2.5–3.6
Brussels sprouts, boiled	4
Cabbage, raw or boiled	3
Carrots, raw or boiled	29–70
Endive, raw	20
Kale, boiled	50
Lettuce, raw	2–10
Mint, raw	110
Parsley, raw	50–70
Peas, boiled	3
Peppers, raw or boiled	2
Pumpkin, boiled and mashed	11–60
Spinach, boiled	60
Sweetcorn, on cob	2.4
Tomato juice, canned	5
Turnip greens, boiled	60
Watercress, raw	30

Carotenoids, in the form of extracts of plant materials, have been used as food colours for centuries. Extracts of carrots were at one time used for colouring cheese and red palm oil was used to colour margarine. Other carotenoid-rich extracts including annatto, saffron, paprika and tomato are still widely used but, as they are not vitamin active, must lie outside the scope of this chapter.

Over recent years 'consumer demand' has resulted in a reawakening of interest in natural extracts, notably the attempts to produce carotene from algae such as *Dunaliella*. While pure crystalline material has not yet been produced on a commercial scale, carotene rich extracts have been marketed. In its 1990 report the Joint FAO/WHO Expert Committee on Food Additives (WHO, 1990) recommended that there was insufficient data to evaluate any of the materials submitted for evaluation for the purpose of establishing an acceptable daily intake. It was also considered that carotene isolated from algal sources would be acceptable for food additive use if it was of sufficient purity to meet the already established specifications for synthetic β-carotene. Three years earlier the same committee had considered that toxicological data would be needed before carotene extracted from carrots with hexane and a carotene by-product from the manufacture of chlorophyll from alfalfa/grass meal by a process of solvent extraction and fractionation could be evaluated (WHO, 1987).

In addition to β-carotene, two other vitamin active carotenoids are produced commercially by chemical synthesis: β-apo-8'-carotenaldehyde (C30) and the ethyl ester of β-apo-8'-carotenoic acid (C30). These are more commonly known as apocarotenal and apocarotenoic ester, respectively. While β-carotene is regarded as having an activity of one sixth that of retinol, which is fairly universal, the other two carotenoids are taken to have one half of that activity (i.e. one twelfth of the activity of retinol) (WHO, 1967).

The commercial materials made by synthesis have very high purities and, as they are the principal sources of these materials for food additive use, consideration of the application of these food colours will be confined to them.

7.2.1 Properties

Carotenoids are, in general, not water soluble, have a low solubility in oils and fats and a slow rate of dissolution. At first sight these are not properties designed for ideal application as food additives. Add the high sensitivity to oxidation and it may be wondered how they have achieved any commercial significance. The oxidation is accelerated by light and by metallic catalysts, particularly copper, iron and manganese. Fatty acid hydroperoxides can initiate decomposition and thus poor quality fats are not suitable for coloration with carotenoids. Gamma radiation, through secondary reactions, can destroy carotenoids.

Because of the stability problems and the difficulty of handling the pure crystalline carotenoids, these colouring agents are very rarely, if ever, used in that form. However, for reference purposes and to aid understanding of the methods of application, the main physical properties are given in Table 7.7.

Table 7.7 Principal physico-chemical properties of crystalline, synthesised carotenoids

	β-carotene	Apo-carotenal	Apo-carotenoic ester
Colour of oily solution (depends on concentration)	yellow to orange	orange to orange-red	yellow to orange
Melting point (°C)	176–182	136–140	134–138
Solubility (g/100 ml 20°C)			
Water	ns	ns	ns
Ethanol	<0.01	~0.1	~30
Glycerol	ns	ns	ns
Chloroform	~3	~20	~30
Cyclohexane	~0.1	~0.8	~2
Fats, oils	0.5–0.80	0.7–1.5	~0.7
Orange oil	0.2–1.0	~3.1	3.4–4.4
Spectrophotometric data*			
max (in cyclohexane)	455–456 nm	460–462 nm	448–450 nm
A_{max} (absorbance)	456 nm >2400	461 nm >2530	449 nm >2440

*These values are for the *trans* isomer. In solution during storage or heating, isomerisation takes place: the predominant isomer at equilibrium will be the *trans* isomer.
ns: not soluble.

The obvious potential of these physiologically acceptable colouring agents has resulted in an interesting field of technology. Through the use of different physical forms (e.g. suspensions of micronised crystals in highly purified vegetable oil, emulsions and dry gelatin protected beadlets), a wide range of stable and easily used products has been developed (Kläui *et al.*, 1970). These various commercial products may be conveniently divided into those suitable for colouring oils and fats, and where coloration is best achieved through their oil content, and those suitable for coloration via the aqueous phase.

In the oils and fats application area the important products are micronised crystals of the pure carotenoids in a vegetable oil of very low peroxide value. Often an antioxidant such as *dl*-α-tocopherol is added for additional protection against oxidation. For further protection the suspension will be packed under nitrogen in lightproof containers (usually aluminium bottles) and stored under cool conditions.

The main aqueous-phase applications are served by three principal forms: dry beadlets in which the carotenoid is dispersed as very fine droplets in a solid matrix of gelatin and sugar, dry powders, in which the protective colloids are cold water soluble, and liquid emulsions. Each of these has its advantages for different applications.

7.2.2 Fat based foods

The solubilities of the carotenoids in oils and fats (see Table 7.7) are not high enough to allow them to be distributed commercially in solution. The preferred sales form is a suspension of 30 g of micronised crystals in 100 g of a suitable vegetable oil, particularly one with a very low peroxide value. The finely ground crystals will dissolve relatively quickly in warm liquid oils and fats. The usual procedure is to make a concentrated solution in a small part of the oil to be coloured and use it as a stock solution for colouring larger quantities. Care has to be taken, however, because the carotenoid will slowly recrystallise forming larger crystals which are much more difficult to redissolve. Fortunately, such recrystallisation is sufficiently slow to permit a stock solution to be prepared in advance for a full production day.

In a typical procedure for preparing a stock solution, a suitable oil would be warmed to 45–50 °C and the carotenoid suspension (not more than 2.5 g of the 30% suspension per litre of the oil) added with stirring. At first the mixture will be opaque in appearance but as the crystals dissolve the liquid will become dark and clear so that it will be easy to see when it is ready for use. More concentrated stock solutions may be made by raising the temperature of the oil but doing so will cause partial isomerism with some loss of colouring power and biological activity.

7.2.2.1 Margarine.
As the principle natural colour of butter is β-carotene it is clearly the preferred colouring agent for margarine where the closest similarity between the two products is desired. β-Carotene has a double advantage in this application: it also has a recognised vitamin activity, although there is often confusion over the equivalence. With these advantages it is not surprising that successful colouring trials were carried out on butter and margarine within the first year of the commercial introduction of synthesised β-carotene.

In international unit terms β-carotene and vitamin A are related by a factor of two. Thus, 1 IU of vitamin A is equivalent to 0.0003 mg retinol or 0.0006 mg β-carotene. In the USA, β-carotene is still commonly sold in terms of its value in international units, whereas in Europe it is offered in weight units. Quite frequently, for example, for labelling purposes, regulatory authorities lay down a conversion of 6 to 1. It is here that the confusion arises.

The difficulty can be resolved by considering both sides of the problem. On the one hand there is the question of the degree to which β-carotene is actually available for conversion into vitamin A, and on the other hand that of whether it is converted.

Only the β-carotene which has been absorbed into the system can become a vitamin A source. In order that absorption can occur, the food containing it must be broken down to release the β-carotene. On average, only about one third of the consumed β-carotene is absorbed into the system (WHO, 1967). Some foods, such as raw carrots, do not give up their β-carotene very readily while others, for example, where the carotenoid is in oil solution, have a better extraction and absorption.

It is clear, therefore, that a different factor would be required for all foods, possibly varying according to the method of preparation and cooking. The task of regulatory authorities in setting and controlling appropriate conversions for such different foods would be very difficult, if not impossible. Thus a sensible compromise has to be accepted. The UK authorities use a 3:1 factor for this aspect, as recommended by WHO (WHO, 1967).

However, that is not the whole story. Successful extraction and absorption does not necessarily mean that the carotenoids will be converted to vitamin A. The vitamin A status of the body is important. Conversion will only occur if the body needs the vitamin. Nevertheless, the factor accepted is 2:1. Overall, the equivalence is taken for dietary calculation and labelling purposes as 6:1.

In most countries where β-carotene may be used for margarine coloration, its equivalence to vitamin A may be counted towards the total vitamin potency of the end product. Thus it is quite common for stock solutions of the carotenoid, prepared as described above, to have the necessary balance of vitamin A and, where appropriate, vitamins D and E, added to it. Such a procedure simplifies manufacturing and can make quality control less expensive.

The amount of β-carotene used will depend on the end result required, some countries have a preference for low colour levels while others expect darker shades. Usually, concentrations in the region of 3 to 5 g of pure β-carotene will be enough to cover average tastes in one tonne of margarine.

The other commercially available carotenoids, apocarotenal and apocarotenoic ester, may also be used for adjusting the colour of margarine. By this means a wide range of colour shades from light yellow through golden shades towards rich dark orange can be obtained.

Low fat 'margarines' with a higher water content may need a higher concentration of β-carotene in the oil phase to counter the apparent lightening effect of the emulsification.

7.2.2.2 *Butter.* Not only is there variation in national preferences for the colour of butter but the product itself varies with the time of year. The variations reflect the differences between winter feeding cattle and their exposure to summer pastures. The month-by-month variation in β-carotene level is closely mirrored by the total vitamin A potency of the butter. The variations may be reduced by adding β-carotene to the butter, subject to the usual caveat regarding local regulations. Adding it via the animal feed is possible but quite uneconomic in practice.

Adding the micronised crystals directly to the churn is not a satisfactory way of achieving the desired result. The immediate colouring is apparently too low because, at the chill temperature, the carotenoid dissolves at best very slowly. After storage for a few hours the end product begins to develop intensely coloured spots where the crystals are slowly dissolving. The difficulty may be simply overcome by preparing a stock solution of the carotenoid before adding it to the churning butter.

Within the European Community various subsidies are available for surplus agricultural produce. Butter oil is one such item. The oil is available for manufacturing purposes. To ensure that it is not possible for it to be recycled, and thus for the subsidy to be claimed on more than one occasion, the oil is treated with a marker. One of the markers permitted is apocarotenoic ester. The addition level is quite high, making it unsuitable for use in products rich in butter but low in colour level.

7.2.2.3 *Other oils and fats.*
Seasonal variations and consumer colour preferences in liquid and solid oils, fats and shortenings can be adjusted by the addition of the carotenoids via the stock solution method recommended for margarine. Either by this means or by direct addition to the batter, the effect of egg-yolk colour variation on cakes, pastry and biscuits can be eliminated.

7.2.2.4 *Processed cheese.*
Compared with normal cheese a darker colour is characteristic of most processed cheeses. In many products, particularly in the USA, the product is almost orange in colour. Apocarotenal is eminently suitable for achieving the desired shades and, in the USA particularly, special colour blends are manufactured for this purpose. The colour preparation may be added directly to the hot melted cheese provided that the mixing is adequate. Alternatively, the colour can be predissolved in some of the melted cheese or in the butter oil.

7.2.2.5 *Other fat based foods.*
The question of whether to use the oil or water phase of a product as the vehicle for added colour depends very much on the product. Emulsions, if oil-in-water, are best treated as aqueous systems while the water-in-oil systems usually benefit from oil soluble coloration. Thus, salad creams, mayonnaise and similar products may effectively be coloured by adding the carotenoid to the oil phase.

At first sight adding the colour to the oil used for frying could represent a convenient method of improving the appearance of fried potato products or breaded fish, meat and poultry portions. There is no doubt that this works but the high temperatures involved, followed by the very large surface area of the end product results in stability problems on subsequent storage. The method is thus more suited for countries where relatively high levels of antioxidant protection may be used. Although such high temperatures are not employed, keeping the oil hot makes antioxidant precautions necessary in applying coloured oil to popcorn by spraying the freshly popped product. The shelf life of the end product, however, is usually short and the problem not quite so serious at that stage.

Using the oil phase as the application medium for the colour may be an advantage in some products even though the oil phase may be the minor part of the product. Water soluble colours will often migrate to the vegetable tissue when the latter is canned in a coloured sauce but oil soluble colours do not show this disadvantage. There are also occasions when the coloration of soft drinks is better when approached by this apparently anomalous route. This latter point will be considered together with the more conventional techniques under the appropriate heading below.

7.2.3 *Water based foods*

Although the properties of the carotenoids are far from ideal there are powerful incentives to overcoming the problems. Their inherent safety, coupled with their absolute identity with the naturally occurring substances, make them ideal food additives. Reinforcement comes from the heavy pressures of 'consumer demand' and more recently the discovery of the dietary benefits of β-carotene (Slater and Block, 1991).

In nature the carotenoids are not only found in lipid systems. Everyday examples of their occurrence as very fine dispersions in aqueous media are orange juice, tomato juice and carrot juice. The stability in these materials is well recognised and may, perhaps be due to the formation of complexes with proteins or lipoproteins. Then again, the presence of natural antioxidant systems or even the submicroscopic structure may enhance the stability.

The conversion of the pure crystalline carotenoids into forms suitable for colouring foods with an essentially aqueous nature includes the formation of emulsions, colloidal suspensions, dispersions in suitable colloids and, at least in theory, the use of emulsifying agents. (The last is not much used because of reluctance to use artificial substances to promote the application of natural or nature-identical food ingredients). Of the various methods employed commercially, dispersion into water soluble colloids and emulsification are most important.

Dispersions of carotenoids in suitable colloids are, in effect, reversible dried emulsions. The principal colloid used is gelatin. The carotenoid is dissolved in the smallest amount of a suitable vegetable oil and the solution emulsified into a solution of gelatin containing sugar as a plasticiser or humectant. The two solutions are mixed, emulsified and homogenised in a high speed homogeniser until the diameter of the oil droplets averages less than five microns. The size of the oil droplets is important for ensuring a satisfactory physical strength of the final dried beadlet as well as good emulsion stability when reconstituted.

After homogenisation, the emulsion is sprayed and dried (as distinct from spray-drying) by spraying into a cold atmosphere and catching the small jelly particles in oil, or sometimes in starch. Drying follows, resulting in the production of small, hard, spherical beadlets containing a fine dispersion of droplets inside. Usually, the beadlets contain 10% by weight of the carotenoid. A very thin film of starch on the surface helps to reduce the risk of the beadlets adhering to one another should adverse storage conditions be encountered.

As long as they are kept dry, the beadlets have excellent stability. In comparison with the artificial dyes, they are rather more clumsy in use but nevertheless are capable of application in a great many foods. Usually the beadlets are dissolved by adding the required amount to 10 to 15 times their weight of water at about 50°C with stirring. The beadlets will dissolve quickly to give a solution which should be used within the day, not because of carotenoid instability but due to the nutrient qualities of the gelatin–sugar solution and the consequent risk of microbial contamination.

The colour of emulsions prepared in this manner will be rather darker than would be expected from a true solution because of the physical properties of the coloured material. Lighter colours in shade as well as intensity can be obtained by emulsification of more dilute solutions into more readily soluble colloid bases. Such products have a greater range and convenience of application compared with the gelatin beadlets. Whereas the gelatin protected products are ideal for adding colour to a product in the course of manufacture by predissolving, the cold-water soluble powders are ideally suited to dry mixtures where the final preparation takes place in the domestic kitchen.

A relatively new product to appear on the market is a liquid emulsion of β-carotene. Containing 5% of the carotenoid, this preparation has immediate advantages in, for example, soft drinks manufacture. The need for prepreparation of the colour solution and certain other parts of the process are no longer necessary.

7.2.3.1 Confectionery.

7.2.3.1.1 Hard candy.

Two principal processes are used for producing hard candy. The traditional method involves boiling the sweet mass, usually sugar, glucose and water to about 135°C, depositing it on an oiled slab and kneading in the acids, flavours and colours while the mass cools. At the correct temperature the mass is then subdivided, moulded and wrapped. In the newer process all ingredients are mixed in the boiling pan and deposited in the liquid state at about 150°C into Teflon coated moulds from which, after cooling, the finished candies are removed mechanically and wrapped. The process is continuous and gives attractive, smooth-surfaced products with a high surface gloss.

It is clear that the traditional method involves less thermal stress than the continuous depositing process. Nevertheless, the carotenoids withstand the conditions in both processes without deleterious effect to give a range of most attractive colours. Table 7.8 gives suggested levels of the carotenoids for most of the popular flavours, subject to the limitations of the colour range. While the concentrations given are for individual carotenoids there is no reason why mixtures, or mixtures with other compatible colours cannot be used. Suitable grades of chlorophyll (usually copper chlorophyll) can, for example, give the lower levels of β-carotene an even more lemon colour.

7.2.3.1.2 Soft chewy candy.

In a typical process, the product is made by combining gelatin, sugar and fat to give a soft, chewy candy. The gelatin is dissolved with sugar and a whipping agent in water at room temperature and whipped to form a stable foam. The bulk of the sugar, together with glucose syrup

Table 7.8 Coloration of hard candy with carotenoids

Flavour	Colour preparation		Addition level (mg pure substance/kg candy)
Lemon	β-Carotene	1% CWS	2.5
Pineapple	β-Carotene	1% CWS	7.5
Orange	β-Carotene	10% WS	7.5
Orange	Apocarotenal	1% CWS	10.0
Mandarin	β-Carotene	10% WS	30.0
Strawberry	Apocarotenal	10% WS	30.0

and water is boiled to 128°C and added slowly with continuous stirring to the gelatin foam. The fat is warmed with an emulsifier to just above the fat melting point and slowly added with stirring to the mixed foam and syrup. Finally, appropriate amounts of citric acid and flavour are added.

The oil soluble carotenoids may be added by dissolving them in the melted fat. A good lemon colour is obtained with about 20 mg β-carotene (added as 66 mg of 30% suspension) per kg of finished candy. Apocarotenal 20% suspension at a concentration of 100 mg/kg (equivalent to 20 mg pure substance per kg) will provide an excellent apricot colour.

7.2.3.1.3 *Pectin jellies.* Flavour and colour (predissolved in water) are usually added to the product after boiling to the correct point (78% solids content, measured by refractometer). After thorough mixing, the mass is deposited into preformed starch moulds and set aside for one hour to gel. After removal from the starch bed, the jellies are brushed, steamed and coated with sugar.

The water soluble carotenoid preparations are eminently suited to colouring pectin jellies. Table 7.9 gives suggested concentrations for a number of popular flavours.

7.2.3.1.4 *Coloured crystallised fruits.* Cut pieces of pumpkin are sequentially soaked in sugar solutions of increasing concentration over a period of ten days. At stage one, the pieces are soaked in 30% sugar syrup containing the predissolved colour for 24 hours. Stages 2, 3 and 4 follow, being, respectively, two days in 30%, three days in 50% and four days in 60% syrups.

Good lemon and orange colours are obtained with β-carotene 1% cold water soluble (CWS) and 10% water soluble (WS), respectively. The colours are used at levels of 25 mg pure substance per litre of the first sugar syrup.

7.2.3.1.5 *Sugar coated dragees.* Under this heading are various confections including nuts, dried fruits, caramels, flavoured compressed sugar, chewing gum and, probably most well known, chocolate drops, all sugar coated. The principle is essentially the same for all, after coating the product with a white undercoating, the colour is applied by thin layers of coloured sugar syrup.

Table 7.9 Coloration of pectin jellies with carotenoids

Flavour	Colour preparation		Addition level (mg pure substance/kg product)
Lemon	β-Carotene	1% CWS	3.85
Pineapple	β-Carotene	1% CWS	7.7
Apple	β-Carotene	1% CWS	15.4
Orange	β-Carotene	10% WS	7.5
Orange	Apocarotenal	10% WS	7.5
Mandarin	β-Carotene	10% WS	30
Papaya	Apocarotenal	1% CWS	10
Mandarin	Apocarotenal	10% CWS	25.0

Table 7.10 Coloration of sugar coated dragees with carotenoids

Colour	Colour preparation	Application level (mg pure substance/kg dragees)
Yellow	Riboflavin–phosphate	0.15
Yellow	β-Carotene 1% CWS	6.0
Dark yellow	β-Carotene 10% WS	12.0
Orange	β-Carotene 10% WS	48.0
Orange	Apocarotenal 1% CWS	6.8

The water soluble carotenoid preparations are dissolved in water at twice the usually recommended concentration (one part of the colouring agent to four parts of water. This is necessary to ensure that when added to the sugar syrup (made with 40 parts water to 100 parts sugar at 106°C) it will not be seriously diluted.

Small amounts at a time of the coloured sugar syrup are added to the undercoated tablets in a revolving coating drum, drying between each addition. For most confectionery products some six to eight coatings will be enough to achieve the desired end product. More coatings with, perhaps, lower amounts of syrup at a time will give better quality. In the pharmaceutical industry the number of layers added in this way will be in the region of sixty. Finally, the application of a little polish (e.g. carnauba wax) will give the product its full glossy attractiveness. So great is the variation in the size and shape of the products under this heading that a firm guide to quantities is not possible. For regular tablet shapes (i.e. double convex circular tablets) of 9 mm diameter, the amount of pure colouring substance per tablet will be in the region of 0.005 to 0.05 mg depending, of course, on the end colour desired.

Table 7.10 gives an indication of the levels of carotenoids and riboflavin-5'-phosphate sodium salt needed to achieve attractive colours between lemon yellow and dark orange. Higher concentrations of apocarotenal can give shades very close to red. The quantities suggested are suitable for 1 kg (3000) cores.

7.2.3.1.6 Marzipan. Marzipan is usually made by warming ground almonds, sugar and a little water to make a sticky mass, icing sugar is then added with more water to produce the typical marzipan texture. The carotenoids may be added, as solutions of the water dispersible forms, just before the icing sugar is added. Shades of yellow and orange may be achieved using the amounts suggested in Table 7.11.

Table 7.11 Coloration of marzipan

Colour	Colour preparation	Addition level (mg pure substance/kg end product)
Yellow	Riboflavin phosphate	50.0
Yellow	β-Carotene 1% CWS	6.0
Light orange	β-Carotene 10% WS	20.0
Orange	β-Carotene 10% WS	30.0
Orange	Apocarotenal 10% WS	20.0

Table 7.12 Coloration of chewing gum

Colour	Colour preparation		Addition level (mg pure substance/kg end product)
Lemon	β-Carotene	1% CWS	6.0
Orange	β-Carotene	10% WS	15.0
Orange	β-Carotene	10% WS	30.0
Mandarin	Apocarotenal	10% WS	25.0

Marzipan shapes may also be coloured on the surface with carotenoid solutions, usually in the range of 0.1 to 1.0%. After the colour has dried, a protective varnish is customarily applied.

7.2.3.1.7 *Chewing gum.* Chewing gum may be coloured in either, or less often in both, of two ways. The thin sticks may be coloured in the mass while the sugar coated types may be coloured in the same way with or without colour in the sugar coating. The process for sugar coating is, in essence, the same as that for other sugar coated products described above.

In the standard method for preparing chewing gum the water soluble preparations may be added in solution together with the glucose syrup and gum base at the first kneading stage. Typical addition levels are given in Table 7.12.

7.2.3.1.8 *Fondant.* Aqueous solutions of the water dispersible carotenoid products will give attractive colour shades to fondant. The colour solution is added while the mass is warm and still reasonably fluid. Concentrations of 3 to 50 mg/kg in the end product produce pleasant pastel shades.

7.2.3.2 *Bakery products.*

7.2.3.2.1 *Cakes.* β-Carotene is a natural component of eggs, butter, milk and wheat flour, the essential ingredients for cake-making. The addition of β-carotene is thus logical for standardising the colour of cakes when variations occur due to season, or, in the case of the eggs, to the hen's feed. Adding more of the colour produces interesting shades. An addition in the region of 20 mg pure substance per kg will give an attractive yellow colour which would be suitable for a cake with added lemon flavour. Apocarotenal can be used to produce darker shades.

In cake mixes the stability of the carotenoids is good. The water soluble beadlet types are not normally used in dry mixes because the large particle size compared with that of the other ingredients could cause separation problems in mixing and filling. The more complete cake mixes, containing shortening, may not suffer from such separation. The fine particle size cold water soluble preparations are much more satisfactory, not only because the potential problems of separation are overcome but because their more ready dissolution in the batter gives rise to less risk of visible spots and streaks in the batter and, perhaps, even in the finished cake.

Ideal distribution would also result from adding the colour in the oil soluble form to the shortening component.

7.2.3.2.2 Biscuits. The carotenoids may be used to colour biscuit shells and wafer biscuits. The colour is predissolved and added to the batter.

7.2.3.2.3 Rusk (dried bread crumbs). Considerable quantities of coloured and uncoloured bread crumbs are manufactured commercially. Their uses include application to the surface of fish and meat portions, large amounts of which are distributed through the frozen food trade. Another important application is as an ingredient in other foods such as an extender in some meat products and as an ingredient in dessert mixes.

The stability of the carotenoids in these rusk or crumb products is excellent and the colour is stable during precooking, freezing, distribution and subsequent domestic cooking.

7.2.3.2.4 Icings and fillings for biscuits and cakes. It is common practice to decorate biscuits and cakes after baking. Cream fillings for biscuits are usually made of fat of a suitable plasticity, and sugar. A strong solution of the carotenoid is added to the mass towards the end of the mixing. Care is needed to ensure adequate colour distribution. The use of an antioxidant such as tocopherol and ascorbyl palmitate is helpful where the edges of the filling may be exposed, because of transparent or semi-transparent packaging, to fluorescent lighting, for example in a supermarket. Levels in the range of 2 to 30 mg pure substance per kg will give pleasant colours in the lemon yellow to orange range.

The carotenoids are also useful in colouring the fondant icing for cakes and in the harder icings applied to biscuits.

7.2.3.3 Dairy products. The carotenoids are valuable colours for milk drinks, milk based puddings, yoghurt, ice creams and sorbets (although the last are not, strictly, dairy products).

7.2.3.3.1 UHT milk drinks. The dry ingredients sugar, stabiliser, flavour and colour, are stirred into the milk which is then heated to 75°C, homogenised and flash sterilised at 139°C. The product is cooled immediately in process and

Table 7.13 Coloration of UHT milk drinks

Flavour	Colour preparation	Addition level (mg pure substance/litre end product)
Pineapple	Riboflavin phosphate	20.0
Banana	β-Carotene 1% CWS	3.0
Vanilla	β-Carotene 1% CWS	7.0
Apricot	β-Carotene 10% WS	15.0
Passion fruit	Apocarotenal 10% WS	15.0

Table 7.14 Coloration of instant pudding powders

Flavour	Colour preparation		Addition level (mg pure substance/kg powder)
Lemon	Riboflavin phosphate		270.0
Vanilla	β-Carotene	1% CWS	7.0
Orange	Apocarotenal	1% CWS +	80.0
	β-Carotene	1% CWS	40.0
Mango	Apocarotenal	1% CWS	41.0

aseptically filled into cartons. Suggested colour additions are given in Table 7.13. The product shows excellent stability on storage.

7.2.3.3.2 *Milk puddings.* In deciding which type of carotenoid preparation to use, consideration must be given to the end product and its use instructions. For example, an instant cold milk pudding will need a cold water soluble colour preparation. Suggested levels for various flavours are given in Table 7.14. For those puddings requiring cooking there is no need to use the cold water soluble colours, provided that the presence of dark spots in the powder will not be objectionable and that there will not be separation problems in mixing and filling.

7.2.3.3.3 *Yoghurt.* Although it is suggested that the carotenoids should be added after pasteurisation, it is not a practice which would be welcome in the modern dairy. The stability of the colours to the process temperature is such that they could better be added to the milk at the beginning. Addition levels are given in Table 7.15. The colours will be predissolved in water before addition to the milk.

7.2.3.3.4 *Ice cream and sorbet.* Processing conditions are mild and the carotenoids give good stable colours in these products. Addition levels suggested for a number of flavours are given in Table 7.16.

7.2.3.3.5 *Hard cheese.* Many hard cheeses, particularly cheddar, are coloured. The original colouring agent was carrot extract but this gave way to a water soluble, saponified annatto. Over more recent years the cheese by-product whey, has, after drying, assumed potentially greater commercial value as an ingredient in

Table 7.15 Coloration of yoghurt

Flavour preparation	Colour preparation		Addition level (mg pure substance/kg end product)
Lemon	β-Carotene	1% CWS	3.0
Pineapple	β-Carotene	1% CWS	7.0
Vanilla	β-Carotene	1% CWS	15.0
Apricot	β-Carotene	10% WS	15.0
Orange	Apocarotenal	10% WS	15.0
Tangerine	Apocarotenal	1% CWS	7.0

Table 7.16 Coloration of ice cream and sorbet

Flavour	Colour preparation	Addition level (mg pure substance/kg end product)	
		Ice cream	Sorbet
Vanilla	Riboflavin phosphate	50.0	
Vanilla	β-Carotene 1% CWS	10.0	
Lemon	Riboflavin phosphate		20.0
Apple	β-Carotene 1% CWS		3.0
Apricot	β-Carotene 1% CWS		15.0
Orange	β-Carotene 10% WS	30.0	30.0
Melon	Apocarotenal 1% CWS	20.0	

a considerable number of foods. A significant proportion of the norbixin in the annatto preparation remains in the whey reducing its value as a food ingredient.

A new emulsion form of β-carotene has now been developed which colours only the curd, leaving the whey unaffected. While the cost of the new preparation is rather higher in use than water soluble annatto, the cost is likely to be offset by the elimination of the decolorisation process now necessary, particularly where the whey powder is to be an ingredient in baby foods. The amounts of β-carotene needed are in the range 20–40 mg pure substance per kg of final cheese and give excellent results free from pinking, fading or effect on the flavour.

7.2.3.4 *Soft drinks.* Once again the close relationship between the commercial materials and those occurring in citrus fruits make them a natural choice for colouring citrus flavoured soft drinks. The range of soft drinks successfully coloured with the carotenoids includes carbonates, syrups, fruit based drinks and instant powders. The principal problem is that the carotenoids, being necessarily in oil solution, tend to float on the surface of the drink. Weighting agents such as brominated vegetable oils are often subject to restriction by regulatory authorities which exacerbates the problem. However, in spite of the difficulties, a number of techniques are available for successfully incorporating the carotenoids into soft drinks.

The carotenoids may be dissolved in the essential oil of the citrus drink although great care is needed to avoid over heating the sensitive oil. The method is more suited to the apocarotenoids whose solubility in these oils is greater.

Then, the water soluble beadlets may be used, after dissolving, to colour the fruit pulp, or fruit concentrate or a mixture of both with sugar and acid. After thorough homogenisation the coloured premix is diluted into the remainder of the beverage. Homogenisation greatly enhances the appearance and stability and may even contribute to a reduction in the amount of colour needed for a given colour intensity. Ascorbic acid considerably reduces the risk of oxidative loss of colour and flavour. The dissolved oxygen and that in the head space may be neutralised in that way. The stability of such products is excellent.

In canned drinks the carotenoids are stable and do not show the colour loss sometimes experienced with azo dyes in the presence of ascorbic acid.

β-Carotene is much more soluble in vegetable oil when heated. Use is made of this in another approach to soft drink coloration. Oil soluble β-carotene is heated strongly (e.g. to 130–140°C) in a small amount of a suitable oil (e.g. groundnut oil) under nitrogen until completely dissolved. Care must be taken not to heat for too long in order to avoid excessive isomerisation. The solution is then cooled to about 100°C and added with stirring to the orange concentrate, orange oil, if used, and the other ingredients. As soon as the addition is complete the mix is emulsified into aqueous gum acacia solution and homogenised in a two stage, high pressure homogeniser ($250–300$ kg/cm^2) The resultant base is then added to sugar syrup and finally diluted with carbonated water to make the finished product. Normal amounts of β-carotene in the finished product would lie in the range of 2–10 mg pure substance per litre.

Soft drink instant powders are made with the cold water soluble (CWS) carotenoid powders. β-Carotene, used as the 1% CWS product will give a good lemon colour at 4 mg pure substance per kg and a good orange at 70 mg/kg. Apocarotenal, also as the 1% CWS product, will give an alternative orange at 4 mg/kg and an apricot colour at 3 mg/kg. In all cases the dilution rate would be in the region of 120–130 g drink powder per litre of finished beverage.

7.2.3.5 *Other products.*

Apocarotenal is useful for standardising the colour of tomato soup or, in the case of some powdered products, contributing all of the colour. A small amount of β-carotene is often used in canned cream of chicken soup to make the colour a little more attractive. Similar applications include canned milk puddings, jams and the jam-like products used as a vehicle for incorporating the fruit into fruit yoghurt. β-Carotene and apocarotenal also have application in colouring salad dressings.

7.3 Riboflavin (vitamin B$_2$)

At first sight, riboflavin is not a particularly useful food additive. It is characterised by a very low solubility in water, approximately 0.007% at room temperature, and it is very sensitive to light. The phosphate ester sodium salt has a much higher solubility, approximately 3.0% at room temperature, but it is still light sensitive. In spite of the apparent difficulties, its striking, intense yellow colour is of value in some applications. Both compounds have strong bitter tastes although the parent compound is more pronounced in this respect. Applications must be limited to those where the product to be coloured does not have a particularly bland taste while at the same time needing a high colour level. In practice these problems are not likely to arise because of the relatively high price of the colouring material. The key to its successful use as a food colour seems to lie in the water activity of the product.

Thus, it is capable of giving a particularly attractive yellow coloration to boiled sweets by both the depositing and slab methods. It is also suited to colouring sugar coated products and, with considerable care, fondant and icings. Its use as a colour for salad cream is well established although difficult to explain. The vitamin is useful, although consideration of its light stability might seem to preclude it, in ice cream, sorbet and powdered drinks. Ice cream and sorbet are normally stored in the dark and thus are stable up to the point of consumption, the rate of consumption by the average child is considerably greater than the rate of destruction by light. Much the same applies to the soft drink powder where it is not subject to the effect of light until made up ready to drink. Everything then depends on the speed with which the beverage is consumed.

7.4 Niacin

It is now generally accepted that the name of this vitamin shall be niacin. Other names, which have led to some confusion are nicotinic acid, nicotinamide and niacinamide. The two chemicals are nicotinic acid and its amide, nicotinamide. In general, the acid is not recommended for use as a food additive simply because of the 'flushing reaction'. A transient tingling or flushing sensation in the skin after relatively large doses of nicotinic acid is a rather common phenomenon. Nicotinamide rarely causes that reaction. In the amounts which might reach the consumer in a food to which nicotinic acid had been applied as an additive it is not likely that the problem would arise. However, the possibility exists that a factory operative handling the material might be particularly sensitive.

Nicotinic acid was for some time listed in the British soft drinks regulations as an acidifying agent. Bearing in mind the nature of the flushing reaction, it is difficult to understand why.

The principal use of niacin in the food industry as an additive is in meat processing. In this context it is normally used in conjunction with ascorbic acid and therefore the two substances are considered together under that heading (p.150).

7.5 *dl*-α-Tocopherol (vitamin E)

dl-α-Tocopherol is more commonly known as vitamin E. The defined international unit of vitamin E is the biological activity of 1 mg *dl*-α-tocopheryl acetate. 1 mg *dl*-α-tocopherol is equivalent to 1.1 mg of the acetate. A number of tocopherols are known but, at least at the present, only the alpha isomer is manufactured commercially; thus for convenience in the present context, tocopherol is taken to mean *dl*-α-tocopherol.

Tocopherol is a clear, yellow, practically odourless, viscous oil. It is soluble in ethyl alcohol, insoluble in water and miscible with ethyl ether, chloroform, acetone and vegetable oils. In the absence of oxygen it is resistant to temperatures up to 200°C and to changes of pH. In nature, tocopherols occur in food fats and oils and fatty foods. Tocopherol is used commercially as a food antioxidant, often with synergists such as ascorbyl palmitate and lecithin.

7.5.1 Oils and fats

Ever since man began to store food, spoilage has been a problem. It is still a decisive factor in ensuring the supply of satisfying, acceptable, nutritious foods for the world's increasing population. The storage life of most food ingredients is limited. Food oils and fats in particular are very susceptible to deterioration. 'Rancidity' is a generally used term to describe the unpleasant tastes and odours which result from fat deterioration.

Fats may deteriorate by hydrolysis resulting in the formation of free fatty acids, polymerisation resulting from exposure of the fat to high temperatures, and autoxidation where mainly the unsaturated fatty acids are oxidised in the presence of oxygen.

By far the most important of these is autoxidation, an irreversible process of fat deterioration which can be retarded but not completely prevented. Autoxidation is especially noticable and significant when the oxidised products develop an unpleasant odour or taste. For example, the milk fats in milk and dairy products when oxidised may form aldehydes and other degradation products which can adversely affect the flavour in a concentration of only 1 mg/kg or less. The threshold values of some oxidation products of soya oil are as low as 1 µg/kg. The storage properties of some vegetables is influenced to a material degree by the stability of the low concentration of their relatively unstable lipids.

There are four main ways in which autoxidation can be retarded:

  i. reduction of energy input to the minimum
  ii. avoidance of contact with traces of heavy metals
  iii. exclusion of oxygen
  iv. removal of free radicals before they can react.

Reduction of energy input (i.e. light and heat) can be achieved by storing the fat in a cool dark place. These simple precautions are always the first preventive measures to be taken.

Traces of heavy metals, even in concentrations as low as 0.1–1 mg/kg can affect the stability of fats. Copper and iron equipment is not recommended for processing oils and fats. Traces may often be removed by using a complexing agent such as citric acid or lecithin or, where allowed, EDTA.

Contact with oxygen must be avoided. Any oxygen remaining or penetrating the packaging can be removed with oxygen scavengers such as ascorbic acid and its palmitic ester.

Radical scavengers can inactivate radicals and so interrupt the chain reaction. The radical scavenger first donates hydrogen to the free radicals and so combines directly with the free radicals to form inert products. Tocopherols are radical scavengers.

It is essential that antioxidants and synergists are added to fats and oils as early as possible in the manufacturing process and ideally to the peroxide free oil or fat.

A combination of tocopherol with ascorbyl palmitate shows marked synergism, increasing the effectiveness of the components and reducing the amount added. Vitamins E and C are natural components of many foods and thus such a combination is potentially more acceptable than some other antioxidant preparations. The low solubility and low rate of dissolution of ascorbyl palmitate is a problem which can be overcome by predissolving it in lecithin.

As a example of the strength of this synergism, the prolongation of the induction period of lard measured by the Rancimat method at 120 °C for a mixture of 500 mg/kg ascorbyl palmitate plus 1400 mg/kg lecithin is about 1.3 times that of the sum of the effects of the individual components. Under the same test conditions, 100 mg/kg tocopherol with 1400 mg/kg lecithin showed an enhancement of 1.8 times, while a mixture of all three substances at the same concentrations showed an enhancement of more than 2.5 times.

References

Kläui, H., Hausheer, W. and Huschke, G. (1970) In: *International Encyclopaedia of Food and Nutrition*, Volume 9, (ed. R.A. Morton). Pergamon Press, Oxford.
Slater, T.F. and Block, G. (1991) Antioxidant vitamins and beta-carotene in disease prevention. *Am. J. Clin. Nutr.* **53** (1), 189S–396S.
WHO (1967) Technical report series no. 362, *Requirements of Vitamin A, Thiamine, Riboflavin and Niacin*, FAO, Rome.
WHO (1987) *Evaluation of Certain Food Additives and Contaminants*. 31st Report of the Joint FAO/WHO Expert Committee on Food Additives, Technical Report Series 759, WHO Geneva.
WHO (1990) *Evaluation of Certain Food Additives and Contaminants*. 35th Report of the Joint FAO/WHO Expert Committee on Food Additives, Technical Report Series 789, WHO Geneva.

8 Vitamin analysis in foods
I. D. LUMLEY

8.1 Introduction

The determination of vitamins in foodstuffs poses significant problems for analysts, and results from recent interlaboratory studies and published papers indicate that this will be the case for many years to come. Biological assays, first used to determine many of the vitamins, were replaced by microbiological assays (MBAs) and some physicochemical methods, and subsequently gas liquid chromatography (GLC) and high performance liquid chromatography (HPLC) have been applied to the many and various problems of vitamin analysis.

The objective of this chapter is to provide an overview of methods which have recently been, or are currently applied to the determination of vitamins in foodstuffs. These methods, the majority of which involve HPLC and MBA, should produce useful data if used by experienced analysts. Many of the old physicochemical procedures, and some new applications (e.g. immunoassay) are not discussed in detail as the former cannot be considered to be sufficiently accurate (true and precise) and the latter currently have limited application and have not been fully validated.

It is not possible to describe methods in detail in a chapter of a book but an attempt has been made to provide some historical background where relevant and to provide an overview of methods or assay formats, drawing attention to problem areas. Selected references are included and some reference is made to methods used at the Laboratory of the Government Chemist of the UK (LGC), but analysts wishing to determine vitamins in foodstuffs must consult a wider literature base than is presented here.

The determination of vitamins in foodstuffs is like many areas of analysis: although it is relatively easy to produce results, it is very difficult to produce results which are accurate and valid.

8.1.1 Laboratory environment

The majority of the vitamins are sensitive to light and so a laboratory used for vitamin analysis should have windows covered with efficient blinds and artificial lighting should be provided by gold fluorescent tubes or lights which are covered in commercially available polystyrene film which filters out wavelengths below 500 nm.

8.2 Oil soluble vitamins

Vitamins A, D, E and the provitamin A active carotenoids are now usually determined by HPLC. The determination of vitamins A and E (individual tocopherols) is relatively straightforward for an experienced analyst, while the determination of the provitamin A carotenoids is somewhat more difficult because of the complexity of chromatograms obtained from many foodstuffs. The determination of vitamin D is the most challenging task, particularly if the analyst is to determine natural levels of this vitamin, however some elegant procedures and separations have been reported in the literature.

8.2.1 Vitamin D

8.2.1.1 Introduction. Dietary sources of vitamin D are ergocalciferol or vitamin D_2, and cholecalciferol or vitamin D_3. Vitamin D_2 is formed by ultraviolet irradiation of the provitamin ergosterol which is present in plants, yeasts and fungi. Vitamin D_3 occurs in animals and is formed by irradiation of the provitamin 7-dehydrocholesterol. Vitamin D is present in foods at very low levels, for example whole milk contains about 0.03 µg / 100 g, eggs about 1.6 µg / 100 g, butter 0.8 µg/100 g and margarine 8 µg / 100 g, thus the analyst faces a difficult challenge when asked to determine natural levels of vitamin D in foodstuffs.

In the 1950s and early 1960s vitamin D was determined in fortified products after saponification and column chromatographic separation from vitamin A, sterols and carotenoids, followed by measurement of the transient pink colour formed on reaction with antimony trichloride (Nield modification of the Carr-Price reagent, 1940). During the mid 1960s and early 1970s gas chromatographic procedures were developed for the determination of vitamin D and the method used at the LGC involved saponification of the sample, extraction of vitamin D from the unsaponifiable matter, removal of interferences such as cholesterol by precipitation, chromatography on dry alumina, Sephadex and Florisil, then conversion of the vitamin D to the isovitamin prior to formation of the trimethylsilyl derivative and gas chromatography (Bell and Christie, 1973, 1974). When determining vitamin D_3 in foodstuffs, vitamin D_2 was added as an internal standard prior to saponification of the sample (and vice versa for determination of vitamin D_2). GLC methods and the procedures used for sample extract 'clean-up' laid a solid foundation for development of HPLC methods for the determination of vitamin D in the mid 1970s, and HPLC methods have continued to evolve as equipment and column technology have improved. HPLC currently offers the analyst the most suitable means of determining vitamin D in a wide range of foodstuffs at fortified and natural levels.

8.2.1.2 Vitamin D extraction. Saponification of foodstuffs with ethanolic KOH is the usual procedure used to hydrolyse lipid and degrade some potential interfering compounds and release vitamin D from the sample matrix. Saponification is

Table 8.1 Saponification conditions for vitamin D

Reference	Conditions
Thompson et al. (1982)	15 ml milk + 1% ethanolic pyrogallol + 6 g KOH, 18 h in dark at ambient. Hexane extraction.
	15 ml 5% margarine in hexane + 15 ml 1% pyrogallol + 2 ml aqueous KOH (1:1 w / v), shake overnight ambient. Hexane extraction.
Reynolds and Judd (1984)	10 g skimmed milk powder + 1 g ascorbic acid + 40 ml ethanol + 10 ml 60% w / v KOH + 10 ml ethanol, nitrogen flush, reflux 30 min. Extraction with diethyl ether: petroleum ether (1:1).
Indyk and Woollard (1984)	10 g milk powder + 40 ml 1% (w / v) ethanolic pyrogallol + 10 ml of 50% (w/v) KOH, nitrogen flush, seal, ambient overnight with stirring. Extraction with petroleum ether: diethyl ether (90:10).
Sertl and Molitor (1985)	30 ml liquid feed, 4 g milk powder + 15 ml ethanol + 0.4 g ascorbic acid + 7.5 g KOH, 30 min at 75°C. Extraction with diethyl ether and petroleum ether.
Monard et al. (1986)	4 g cod liver oil + 60 ml ethanol + 10 ml 10% sodium ascorbate + 30 ml 50% (w / v) KOH. Reflux with stirring at 50°C for 12 min. Extraction with pentane.
Lumley and Lawrance (1990)	15–25 g foodstuff + 1 g pyrogallol + 150 ml 34% (w / v) ethanolic KOH, nitrogen flush, reflux 30 min. Extraction with diethyl ether: petroleum ether (1:1).
AOAC (1990a)	50 g milk powder + 100 ml ethanol + 25 ml 25% sodium ascorbate + 25 ml 50% (w / v) KOH, reflux 45 min on steam bath. Extract with ether then pentane.

performed at room temperature or by heating under reflux in the presence of antioxidants; nitrogen flushing is also recommended. Table 8.1 shows a selection of saponification conditions taken from the literature. The ratio of sample weight to amount of potassium hydroxide varies, but the fat content of the sample is the key factor. Additional KOH should always be added to the saponification mix if it is apparent that not all of the fat in the sample has been saponified. Several workers recommend the use of ambient overnight saponification but it is worth noting that Indyk and Woollard (1984) reported a 10–20% loss of vitamin D when nitrogen flushing of the flask prior to sealing was not performed.

Mixed ethers, i.e. diethyl ether and petroleum ether (40:60) have traditionally been used to extract vitamin D from the saponification mixture, but alternatives have been used. Indyk and Woollard (1984) used a 90:10 petroleum ether–diethyl ether mixture and reported that the diethyl ether improved the extraction of vitamin D but fatty acid soaps are soluble in diethyl ether and these have to be washed from the ether extract. This can be done with water or saline solution and washes continued until they are neutral to phenolphthalein. Hexane is also used to extract vitamin D and has the advantage that it does not extract fatty soaps to the same extent as diethyl ether. However, extraction efficiency is dependent upon the amount of fatty acid soaps in the digest and therefore care must be exercised if an internal standard is not being used. Thompson et al. (1982) extracted saponified milk, margarine and infant formula with hexane which was washed with 50% KOH to remove traces of polar constituents, including soaps, then with water, then with 55% aqueous ethanol to remove lipids which were more polar than vitamin D.

One potential problem associated with the saponification of samples at temperatures above ambient is the formation of previtamin D by thermal isomerisation which may not be accounted for in all analytical procedures. When GLC was used to determine vitamin D, previtamin D was converted to the corresponding isotachysterol along with vitamin D, therefore GLC of isotachysterols resulted in the determination of 'total' vitamin D. If the HPLC procedure used separates the previtamin from the vitamin D both must be quantified otherwise low results will be obtained. This is not always possible as previtamin D is not easily separated from other chromatographed unsaponifiable components of the sample matrix.

If an external calibration is performed, the sample vitamin D peak can be measured and a correction factor applied to account for the previtamin formed under the saponification conditions employed (deVries and Borsje, 1982). An alternative is to saponify the vitamin D standard in parallel to the sample and then use this standard for the external calibration (Koboyashi et al., 1982). If saponification at elevated temperatures is used (and this appears to be the most common approach now) the potential problem of the formation of previtamin D is most easily overcome if vitamin D_2 is used as an internal standard for the determination of vitamin D_3 (and vice versa) as any thermal isomerisation will be compensated for. If this is the approach taken, the analytical HPLC procedure used must be capable of separating vitamin D_2 from D_3.

8.2.1.3 HPLC of vitamin D.

Solvent extraction of a saponified food will obviously not only extract vitamin D but also other lipid components such as sterols and other oil soluble vitamins and sample degradation products. This can make direct analytical HPLC difficult because of the many potential interferences. Sample degradation products are difficult to quantify but if we consider a semi-skimmed milk with a vitamin D content of 0.01 µg / 100 g, then cholesterol and vitamin E, both potential interferences, are present at levels of 10^6 and 3×10^3 times that of the vitamin D, respectively.

An HPLC procedure used in the LGC in the late 1970s (Wiggins et al., 1982) involved isomerisation of the vitamins D (D2 added as internal standard) to their isotachysterols, removal of interferences on one or two dry alumina columns followed by reverse phase HPLC separation of the vitamin D_2 and D_3 isotachysterols from each other and sample matrix interferences. This procedure was applicable for the determination of vitamin D at fortified and natural levels in foodstuffs but suffered from the fact that alumina column chromatography was lengthy (up to four hours), column preparation was sometimes irreproducible and large volumes of chlorinated solvents were used. Since the early 1980s many papers have appeared in the literature describing methods for the determination of vitamin D and a selection of preparative chromatographic and analytical HPLC conditions reported are shown in Table 8.2.

Indyk and Woollard (1984) reported a procedure which involved the use of two reverse phase cartridges in series for the determination of vitamin D in fortified

Table 8.2 Selection of preparative chromatographic and analytical HPLC conditions for determination of vitamin D

Reference	Preparative chromatography/HPLC	Analytical HPLC
Thompson et al. (1982)	Supelcosil LC-S: 15cm × 4.6 mm. Propan-2-ol: cyclohexane-hexane (1 + 1)	Spherisorb 10 ODS or Radial Pak C18, 5 µm. Methanol: acetonitrile (1 + 9)
Reynold and Judd (1984)	Sep-pak C_{18} cartridge	Spherisorb C_{18}, 5 µm: 25 cm × 4.6 mm. Methanol–water (97.5 +2.5)
Indyk and Woollard (1984)	No chromatographic clean-up	2 Radial-Pak cartridges C18, 5 µm, in series. Methanol–THF (99:1)
Sertl and Molitor (1985)	Partisil 5 PAC: 25 cm × 4.6 mm. Hexane–amyl alcohol (99.2:0.8).	Apex 3 µm silica: 15 cm × 4.5mm. Hexane–amyl alcohol (1997 ml+ 3 ml).
Monard et al. (1986)	R-Sil CN 10 µM: 25 cm × 4.6 mm. Hexane–amyl alcohol 98.4:1.6 v/v).	Sperisorb ODS 5 µm: 15 cm × 4.6 mm. CH_3CN: H_2O Phosphoric acid–aqueous CH_3CN 96% (0.2:99.8 w/v).
Lumley and Lawrance (1990)	Partisil 5 PAC: 25 cm × 4.6 mm. Hexane–amyl alcohol (99:1) + Sep Pak C18 and silica clean-up.	Zorbax ODS: 25 cm × 4.6 mm. CH_3CN–methanol (90:10)
AOAC (1990a)	Sil-60D-10CN: 25 cm × 4.6 mm. Hexane containing 0.35% amyl alcohol.	Partisil, 5 µm: 25 cm × 4.6 mm. Hexane containing 0.35% amyl alcohol.
Wiggins et al. (1982)	µBondapack NH_2: 30 cm × 7.8 mm. Methylene chloride: iso-octane: propan-2-ol 600:400:1.	Zorbax ODS 25 cm × 4.6 mm. CH_3CN–methylene chloride (0.001% triethylamine) (70:30).
Koboyashi et al. (1982)	Nucleosil 5 C_{18}: 30 cm × 7.5 mm. CH_3CN–methanol (1:1 v/v).	Zorbak Sil: 25 cm × 4.6 mm. 0.4% propan-2-ol in hexane.

whole milk powder; no preliminary chromatographic removal of interferences was performed prior to analytical HPLC. The authors state that their aim was to produce a method for routine use which did not involve dual or tandem HPLC column techniques. The method was adequate for their purpose but it was unsuitable for analysis of milk powders with added vitamin E and/or vegetable oil as present in infant formulations.

Reynolds and Judd (1984) described a rapid procedure for the determination of vitamin D in fortified skimmed milk which involved addition of D_2 or D_3 as an internal standard and clean-up of the unsaponifiable material on a Sep-pak C18 cartridge prior to analytical HPLC.

Although these fairly simple approaches to the determination of vitamin D are ideal for specific fortified products, more complex procedures are generally needed if the analyst requires a method with wider applicability. The most common approach now taken is the use of HPLC in a semi-preparative mode for partial sample extract clean-up followed by analytical HPLC to determine the amount of vitamin D present. Thompson *et al.* (1982) used a normal phase column for semi-preparative chromatography, a band-cut was taken which contained the vitamin D with subsequent analytical HPLC on a reverse phase column. This procedure was used for fortified milks; for margarine and infant formulae an additional clean-up on a short column of alumina was required prior to HPLC. Table 8.2 shows a selection of combinations of semi-preparative and analytical HPLC columns used by other analysts.

The paper by Kobayashi *et al.* (1982) provides some useful background information on HPLC methods used for the determination of vitamin D and interesting chromatograms from their semi-preparative and analytical HPLC systems. These workers used reverse phase HPLC for the semi-preparative stage and normal phase for analytical HPLC which is not the most common assay format. The advantage of the use of a reverse phase analytical column is that vitamins D_2 and D_3 can be separated and therefore one vitamer can be introduced prior to saponification as an internal standard (see earlier comments on previtamin D and thermal isomerisation). In a procedure involving saponification, solvent extraction, cartridge clean-up steps, semi-preparative HPLC and band collection prior to analytical HPLC, there are many points in the protocol where some of the vitamin D to be determined can be lost; total recoveries may be quite low, therefore it is desirable to use an internal standard. Although standards should be weighed for preparation, concentrations must be calculated by spectrometry using recognised molar extinction coefficients.

Villalobos *et al.* (1990) recently published results of a comparison of three HPLC columns for the chromatography of vitamin D_2 and D_3. One point that is clear if the literature is reviewed, and from communication with other vitamin analysts, is that non-aqueous reverse phase chromatography (NARP) is preferred because it offers the advantages of greater selectivity and prevents the prolonged retention of very hydrophobic compounds on the column packing.

For a method to be generally applicable to a wide range of foodstuffs, analysts must be prepared to use silica and/or reverse phase commercially available cartridges for preliminary 'clean-up' of the solvent extracted non-saponifiable material, followed by semi-preparative and analytical HPLC stages. UV detection is used, preferably with diode-array detection for assessment of peak purity.

A procedure used at the LGC (Lumley and Lawrance, 1990) on a wide range of sample types for the last eight years has involved addition of vitamin D_2 to the sample as an internal standard, addition of antioxidant, saponification with nitrogen flushing, solid phase clean-up on silica and / or C18 mini-cartridges, vacuum concentration, semi-preparative HPLC on an amino–cyano column followed by analytical separation of D_2 and D_3 on an ODS column and quantitation of D_3. Figure 8.1 shows semi-preparative and analytical chromatograms obtained from a sample of margarine using this analytical approach.

Many other HPLC procedures and associated clean-up techniques have been reported in the literature and the analyst who wishes to adopt a method for the determination of vitamin D should consult widely before embarking on practical work.

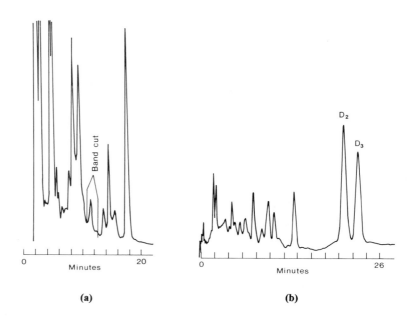

Figure 8.1 HPLC determination of vitamin D_3 in margarine. (a) Semi-preparative HPLC column: Partisil PAC, 25cm × 4.6mm. Mobile phase: hexane–amyl alcohol 99:1, 1 ml min^{-1}. UV detection 265 nm. (b) Analytical HPLC Column: Zorbax ODS 25cm × 4.6mm. Mobile phase: acetonitrile–methanol 90:10, v/v, 1.2 ml min^{-1}. Diode array detection.

The following general guidelines should be borne in mind if vitamin D is to be reliably determined:

1. Saponification should include the use of antioxidants and nitrogen flushing. Sufficient alkali must be used to ensure complete saponification of samples with a high fat content. Saponification under reflux conditions appears to be most commonly used.
2. Semi-preparative HPLC (usually bonded normal phase column) followed by analytical HPLC (usually reverse phase) is the most universal assay format. Additional sample extract clean-up using silica and C18 commercially available cartridges may be required after saponification and before analytical HPLC.
3. Vitamin D_2 should be used as an internal standard for vitamin D_3 (and vice versa) to compensate for losses during sample extract clean-up and thermal isomerisation.
4. Standard concentrations must be calculated after spectrometry using accepted molar extinction coefficients.

8.2.2 Vitamin A

8.2.2.1 Introduction. Vitamin A is used as a generic term to describe retinol and its isomers. Retinol is mainly found in products of animal origin, the main forms being all-*trans* retinol and 13-*cis*-retinol. 13-*cis*-Retinol has about 75% of the biological activity of the all-*trans* form. Vitamin A usually occurs as esters of long chain fatty acids, but it can occur in the unesterified 'free' form. Foodstuffs are fortified with esterified vitamin A, e.g. retinyl acetate or palmitate, as the vitamin is more stable in this form.

Vitamin A is usually extracted from foodstuffs by saponification and extraction of the non-saponifiables into solvent. During alkaline hydrolysis it is essential that the reaction mixture is protected from light, antioxidants are present and nitrogen flushing is used. Amber glassware should be used for all solutions containing vitamin A, and laboratory windows must be covered to exclude daylight.

Vitamin A used to be determined by colorimetry after reaction with antimony trichloride (Carr–Price procedure) after a range of lengthy open-column chromatographic procedures designed to remove interferences. UV spectrometry has similarly been used but these methods are very susceptible to interferences and sensitivity is limited. HPLC now offers the analyst a relatively easy way of determining the amount of all-*trans*- and 13-*cis*-retinol in foodstuffs.

8.2.2.2 Extraction of vitamin A. Vitamin A esters which are naturally present in a foodstuff or added as a nutritional supplement (e.g. retinyl palmitate) can be extracted by dissolution of the sample in solvent prior to HPLC. Thompson *et al.* (1980) dispersed milk samples (2 ml) in ethanol (5 ml) and hexane (5 ml), then water was added (3 ml); after centrifugation an aliquot of the hexane layer was injected directly onto a normal phase HPLC column. Margarine (5 g) was dissolved in hexane (100 ml), then a 5 ml aliquot of the hexane was washed with 60% ethanol and the mixture centrifuged; an aliquot of the hexane layer was again taken for HPLC and determination of retinyl palmitate. Woollard and Woollard (1981) dispersed samples of milk powder in a non-aqueous solvent mixture of dimethyl sulphoxide/dimethyl formamide/chloroform with subsequent partition of retinyl palmitate or acetate into hexane. Again normal phase HPLC was used for the determination of the vitamin A esters. Woollard and Indyk (1989) later used the two extraction procedures mentioned (Thompson *et al.*, 1980; Woollard and Woollard, 1981) to study the esterification patterns of retinol in milks using normal and reverse-phase chromatography.

As previously stated the most common extraction procedure for the determination of all-*trans* retinol and 13-*cis*-retinol is saponification; any retinyl esters present naturally or by addition in the sample will be hydrolysed to the alcohol.

Stancher and Zonta (1982, 1984) used a mixture of water (25 ml), potassium hydroxide (25 ml, 100 g / 100 ml H_2O), ethanol (50 ml) and sample (5–10 g) with an addition of 0.5–0.7 g of ascorbic acid for the saponification of cheese and fish samples prior to the determination of retinols. The flask and contents were flushed with nitrogen prior to overnight saponification at ambient temperature. Retinol was extracted into diethyl ether, the ether washed with portions of phosphate buffer (pH 7.4), and aliquots of the dried hexane (passage through anhydrous sodium sulphate) used for HPLC.

As a consequence of COST 91 Brubacher et al. (1985) recommended a procedure for the determination of vitamin A (retinol and retinyl esters) by HPLC. The sample is weighed into a flask and ethanol (3–4 times the weighed portion in grams, but at least 30 ml), potassium hydroxide solution (50% w / v, volume as weighed portion) and hydroquinone (100 mg) added; the mixture is refluxed with constant stirring and flushing with nitrogen at 80°C for 30 min. The proposers of the procedure state that overnight saponification at ambient results in considerable loss of retinol and therefore reflux for a shorter time is recommended. Retinol is extracted from the saponification mix using 1:1 diethyl ether–petroleum ether (40 / 60); the combined solvent extracts are washed with water until colourless to phenolphthalein.

Thompson and Duval (1989) have reported an overnight saponification procedure for milk and infant formulae which includes the use of ethanolic pyrogallol and they report that vitamin A was stable in these digests for several days. They add that the procedure was not as suitable for cheese and meats. This difference in evidence means that analysts must satisfy themselves that the saponification procedure they are using is liberating all the vitamin A from samples but not leading to its subsequent degradation.

The procedure used at the LGC for all sample types is very similar to the COST 91 method and is detailed by Kirk and Sawyer (1991b).

Once retinol has been extracted from the saponification mixture it must be protected from light and oxidation. Before rotary evaporation of the mixed ether extracts to reduce volume prior to HPLC, an antioxidant (e.g. butylated hydroxytoluene, BHT) and a small amount of a high-boiling protective solvent (e.g. hexadecane) should be added to protect the retinol.

8.2.2.3 HPLC of vitamin A. Normal phase HPLC has been used to determine retinyl esters and various isomers which may be found in some sample extracts. Thompson et al. (1980) used a microparticulate silica column to determine retinyl palmitate in milks and margarine, while Woollard and Woollard (1981) used a similar chromatographic procedure to determine tocopheryl acetate and palmitate in milk powders. Woollard and Indyk (1989) later used a silica packed cartridge to determine total vitamin A in solvent extracts of milk and to collect the all-*trans*-retinyl fraction prior to subsequent chromatographic separation of the various esters by reverse phase HPLC.

Stancher and Zonta (1984) separated 13-*cis*-retinol, 11-*cis*-retinol, 13-*cis*-dehydroretinol, 9,13-di-*cis*-retinol, 11-*cis*-dehydroretinol, 9-*cis*-retinol, all-*trans*-retinol and all-*trans*-dehydroretinol from fish livers on a silica column.

Table 8.3 Typical columns, mobile phase and detection combinations HPLC of Vitamin A

Reference	Column	Mobile phase	Detection
Thompson et al. (1980)	LiChrosorb Si-60	Hexane–diethyl ether (98:2 v/v)	UV 325 nm
Woollard and Woollard (1981)	μ-Porasil	Hexane–CHCl$_3$ (92:8 v/v)	UV 313 nm or 340 nm or fluorescence
Woollard and Indyk (1989)	z-Module, silica cartridge	Hexane–propan-2-ol (99.93:0.07 v/v)	UV 325 nm
Stancher and Zonta (1984)	LiChrosorb Si-60	Hexane–propan-2-ol (99.6:0.4 v/v)	UV 340 nm
Brubacher et al. (1985)	C$_8$ or C$_{18}$	Methanol–water (e.g. 85: 15–98:2 v/v)	UV 325 nm
Kirk and Sawyer (1991)	Partisil 10 ODS 2	Methanol–water (90:10 v/v)	UV 325 nm
Egberg et al. (1977)	Vydac TP201C18	CH$_3$CN–water	UV 328 nm
Bui-Nguyen and Blanc (1980)	Nucleosil C$_{18}$	CH$_3$CN–water (95:5)	UV 328 nm

As stated previously, food analysts will most probably want to determine total retinol in a foodstuff after saponification and reverse phase HPLC is the procedure commonly used. UV detection provides sufficient sensitivity and selectivity for most applications but fluorescence detection can be used. Table 8.3 shows some typical columns, mobile phase and detection combinations reported by analysts. The COST 91 protocol (Brubacher et al., 1985) recommends a C8 or C18 column packing with a mobile phase of methanol–water in ratios to be determined by experiment (suggests 85:15, 90:10, 95:5). In the LGC procedure (Kirk and Sawyer, 1991) the system suitability test involves the HPLC separation of a mixed standard of all-*trans*-retinol and 13-*cis* retinol; the system is suitable for use if the resolution factor is equal to or greater than 0.6. Mobile phase composition can be adjusted if this condition is not achieved. A typical chromatogram obtained from the LGC procedure is shown in Figure 8.2.

Standard all-*trans* retinol for calibration must be prepared on the day of use and the concentration checked by UV spectrometry (e.g. 10 μg ml^{-1} all-*trans* retinol standard in methanol; true concentration to be calculated from its absorbance at 325 nm in a 10 mm cell using $E_{1cm}^{1\%} = 1832$). 13-*cis*-Retinol can be determined by reference to the all-*trans* retinol peak after multiplying by the ratio of their $E_{1cm}^{1\%}$ at 325 nm to account for the difference in molar absorptivity if UV detection is being used at this wavelength. COST 91 recommend that the all-*trans*-retinol standard be taken through the saponification and extraction procedure prior to HPLC to account for possible vitamin losses during extraction.

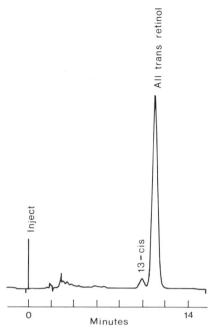

Figure 8.2 HPLC determination of vitamin A in a milk powder. Column: Partisil 10 ODS 25 cm × 4.6 mm. Mobile phase: methanol–water 90:10 v/v, flow rate 1 ml min^{-1}. Detection: UV 265 nm.

The analyst can report the amount of all-*trans* retinol and 13-*cis* retinol in a product by weight of each vitamer but it is also common for results to be reported as all-*trans* retinol equivalents which is the sum of the all-*trans* retinol plus the contribution from the 13-*cis*-retinol after correction to account for its lower biological activity (75% of all-*trans*-retinol). It is important that the units used for reporting are made clear.

Summary. If vitamin A is to be reliably determined in foodstuffs by HPLC the following points should be borne in mind:

1. Sample extracts and standard must be protected from light and oxidising conditions.
2. Extraction media should contain antioxidants.
3. Saponification under reflux (30 min) with antioxidants and nitrogen flushing is preferable to overnight ambient saponification.
4. Antioxidant must be added to solvent extracts prior to reduction of volume by rotary evaporation (temperatures below 40°C).
5. Adequate chromatographic resolution of vitamers of interest must be ensured by correct choice of column type and mobile phase.
6. Standards must be prepared on the day of use and concentrations calculated by UV spectrometry.
7. Units used for reporting results must be clearly defined.

Table 8.4 Selection of saponification conditions reported in the literature

Reference	Reported saponification conditions
Khachik et al. (1986)	Solvent extraction of carotenoids, petroleum ether (40/60)–acetone/diethyl ether. Extracts treated with methanolic KOH, 3 h, ambient, nitrogen atmosphere Samples: green vegetables
Quackenbush and Smallidge (1986)	Sample extract + ethanolic KOH, nitrogen flush, 55–60°C, 30 min Samples: orange juice, other foods
Fisher and Rouseff (1986)	Methanol extract + methanolic KOH, 1 h, 23°C Sample: orange juice
Indyk (1987)	Sample + ascorbic acid + ethanol + KOH solution. 10 min at 70°C
Wills et al. (1988)	Acetone–diethyl ether extract + methanolic KOH, ambient overnight Samples: fruit and vegetables
Lumley and Lawrance (1990)	Sample + ethanolic KOH + pyrogallol + nitrogen flush, reflux, 30 min Samples: various
Speek et al. (1986)	Sample + ethanolic KOH + ascorbate + sodium sulphide, reflux 30 min Samples: vegetables

8.2.3 Pro-vitamin A carotenoids

8.2.3.1 Introduction. Carotenoids are pigments which are found in a wide variety of plants and vegetables as well as in some animals. About 600 carotenoids have now been identified but few have vitamin A activity. The provitamin A active carotenoids contain a β-ionone (cyclohexyl) ring at either one or both ends of the polyene chain and about 50 are predicted to have provitamin A activity. Of these the most important is β-carotene, but α-carotene and α- and β-cryptoxanthin are found in many foodstuffs; they have approximately 50% of the provitamin A activity of β-carotene. It is assumed that 6 µg of β-carotene has the biological equivalence of 1 µg of retinol; this equivalence takes into account the conversion losses and the lower absorption of β-carotene from dietary sources. The food analyst who has an interest in the determination of provitamin A active carotenoids must obviously be able to separate them from the numerous non-active carotenoids and associated compounds in food extracts. Traditionally large scale open column chromatography has been used in an attempt to separate carotenes from interferences prior to spectrometric determination, but this approach is time consuming and carotenes may degrade during analysis, and it is non-specific, and can lead to an overestimate of provitamin A activity if carotenes other than β-carotene are present. HPLC offers the analyst the means of separating, identifying and quantifying the carotenoids of interest in foodstuffs.

8.2.3.2 Extraction of carotenoids from foodstuffs. Carotenoids are extracted from foodstuffs using procedures similar to those used for vitamin A. Carotenoids are extremely sensitive to light, oxidation and catalysts such as iron and copper and therefore measures must be taken to limit degradation. Light must be excluded from solutions of carotenoids, antioxidants should be added as appropriate, and nitrogen flushing of solutions and containers is necessary.

Carotenoid esters can be extracted using organic solvents but saponification is usually the procedure of choice either at ambient or at various temperatures up to reflux conditions. The presence of antioxidants and nitrogen flushing during saponification is highly recommended and sodium sulphide can be added to reduce the oxidative effects of metal ions. There are so many saponification conditions reported for various types of sample matrix that the analyst should consult the literature to

Table 8.5 Selected HPLC conditions reported in the literature

Reference	Column and mobile phase	Detection
Bureau and Bushway (1986)	Partisil ODS 25 cm × 4.6 mm CH_3: THF: water (85:12.5:2.5) 2 ml min^{-1}	470 nm
Khachik et al. (1986)	Michrosorb C_{18} 25 cm × 4.6 mm	450 mm
	A. methanol. CH_3CN: methylene chloride: hexane (22:55:11.5:11.5) isocratic	
	B. methanol: CH_3CN: methylene chloride: hexane (15:75:5:5 to 15:40:22.5:22.5) isocratic for 12 min then gradient to 27 min	
Quackenbush and Smallidge (1986)	Vydac 201 TP, C_{18} 25 cm × 4.6 mm methanol: chloroform (90:10) 1 ml min^{-1}	475 nm
Fisher and Rouseff (1986)	Zorbax ODS 25 cm × 4.6 mm CH_3CN: methylene chloride: methanol (65:25:10, 70:20:10) CH_3CN: THF: methanol (70:20:10) CH_3CN: ethyl acetate: methanol (70:20:10) 1.5 ml min^{-1}	450 nm
Indyk (1987)	C_{18} 25 cm × 4.6 mm Alltech CH_3CN: dichloromethane: methanol: 2-octanol (70:20:10:0.1) 2 ml min^{-1}	436 nm
Wills et al. (1988)	Novapak C_{18} or Radialpak C_{18} × 2 various mobile phases, mainly methanol: CH_3CN (25:75)	440 nm
Lumley and Lawrance (1990)	Zorbax ODS 25 cm × 4 mm CH_3CN: dichloromethane: methanol (70:20:10) 1.3 ml min^{-1}	450 nm Diode array 200–600 nm
Khachik et al. (1991)	Microsorb C_{18}, 25 cm × 4.6 mm	Diode array
	methanol: CH_3CN: dichloromethane: hexane (1:1) 10:85:5, isocratic 0–10 min, final composition at 40 min (10:45:45) 0.7 ml min^{-1}	
Speek et al. (1986)	Hypersil ODS 3 μm, 25 cm × 4.6 mm	445 nm
	methanol: CH_3CN: $CHCl_3$: water (200:250:90:11)	

identify appropriate conditions. Table 8.4 shows a selection of saponification conditions reported in the literature. Analysts should consult the original papers to obtain further relevant information rather than risk using any of the extraction conditions out of the intended context.

8.2.3.3 HPLC of carotenoids. A large number of papers have been published in the last ten years on the HPLC separation of a whole host of carotenoids and xanthophylls in many sample types and it is difficult to select HPLC conditions which will satisfy all applications. Table 8.5 shows some selected HPLC conditions reported in the literature.

The HPLC conditions reported by Lumley and Lawrance (1990) were found at the LGC to be suitable for the separation and determination of α-, β- and γ-carotene and β-cryptoxanthin in a wide range of vegetables and fruits (see Figure 8.3). Slight changes in mobile phase can be used to expand or compress various parts of the chromatogram to match the analyst's requirements. Single wavelength monitoring is

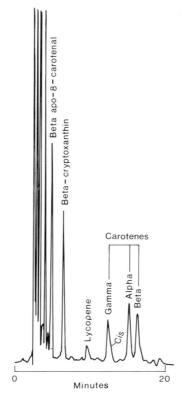

Figure 8.3 HPLC separation of carotenoids. Column: Zorbax ODS 25cm × 4.6mm. Mobile phase: acetonitrile–dichloromethane–methanol (70:20:10) v/v/v, 1.3 ml min^{-1}. Detection: diode array, chromatogram at 450 nm.

satisfactory but diode-array detection provides invaluable information on peak identity and purity.

Other workers have presented rather elegant separations of carotenoids using chromatographic conditions outlined in Table 8.5 and Khachick *et al.* (1991) have published a comprehensive review of carotenoids in fruit and vegetables, with details of their HPLC procedure used to compile the database with a host of relevant references. Again before selecting HPLC conditions analysts should consult the literature to assess the potential applicability of the separations reported.

Concentrated carotenoid standards can be kept at freezer temperature in the dark for perhaps a week or two without significant degradation occurring but working standards must be prepared on the day of use and the actual concentration calculated by spectrometry using recognised molar extinction coefficients. It is also advisable to scan and record the UV–visible spectrum of new standard solutions to ensure conformity with an authentic spectrum of the carotenoid. Commercial carotenoid standards are not always as pure as one would like.

8.2.3.4 Summary. It should be possible for the competent analyst to determine reliably provitamin A active carotenoids if all the precautions summarised for vitamin A are adhered to in addition to having an appreciation of the relevant literature.

8.2.4 Vitamin E

8.2.4.1 Introduction. The term vitamin E is used as a collective term for α-tocopherol, β-tocopherol, γ-tocopherol and δ-tocopherol and their corresponding tocotrienols. The tocopherols and tocotrienols occur naturally in vegetable oils and animal lipid and processed foods may be supplemented with α-tocopheryl acetate or palmitate for nutritional purposes or to act as an antioxidant. The tocopherols and tocotrienols have different biological activity and so it is important that they be individually quantified if biological vitamin E activity is to be determined.

Vitamin E was traditionally determined in foodstuffs by colorimetry after reaction with ferric chloride in the presence of 1,1-dipyridyl or 4,7-diphenyl-1,10-phenanthroline (Tsen modification of the Emmerie–Engel reaction); samples were saponified to remove lipids, major interferences removed by chromatography and tocopherol isomers separated prior to colorimetry. This procedure is susceptible to many interferences and is labour intensive. Gas chromatography was used reasonably successfully for several years (Bell and Christie, 1973) but again lengthy sample preparation was necessary and β and γ tocopherols could not be easily separated. HPLC now offers a rapid and reliable means of separating and determining tocopherols and tocotrienols in foodstuffs.

8.2.4.2 Extraction of vitamin E. Normal phase HPLC is usually used for separation of tocopherols and tocotrienols with fluorescence detection. Reverse phase procedures have been used but β and γ-tocopherols cannot be separated. UV detection has also been used but lack of sensitivity and specificity are problems and the poor results obtained in collaborative studies when UV and fluorescence methods have been compared have highlighted the desirability of fluorescence detection.

Many oils and fats can be dissolved in hexane, or a hexane-based HPLC mobile phase, and aliquots injected onto the silica analytical column without further treatment. Some analysts have extracted vitamin E from foodstuffs using various solvent combinations prior to HPLC. Thompson and Hatina (1979) extracted vitamin E by placing samples in boiling propan-2-ol for ten minutes while Hakansson et al. (1987) extracted tocopherols, tocotrienols and α-tocopheryl acetate from wheat products by Soxhlet extraction with hexane for four hours in the presence of BHT; this procedure was not tested on other foodstuffs. Woollard and Blott (1986) used a non-aqueous solvent mixture of dimethyl sulphoxide, dimethylformamide and chloroform (2:2:1) containing ascorbic acid to extract vitamin E acetate from milk powder formulations; the vitamin was subsequently extracted into hexane. Probably the most commonly used procedure for the extraction of vitamin E from foodstuffs involves alkaline saponification of samples followed by solvent extraction of the nonsaponifiable material which contains the vitamin E. If saponification is used any α-tocopheryl ester which has been added as a supplement will be determined as α-tocopherol. Tocopherols and tocotrienols are extremely sensitive to oxidation, particularly in alkaline conditions, therefore all extractions must be performed in the presence of antioxidants (e.g. ascorbic acid, pyrogallol, BHT), and saponification should be performed with nitrogen flushing of the reaction mixture. Thompson and Hatina (1979) used solvent extraction and saponification to extract vitamin E from lipids and foods respectively. Pyrogallol was used as an antioxidant and the ethanolic sample suspension was refluxed for three minutes prior to addition of the ethanolic potassium hydroxide to exclude air from the saponification system. Weber (1984) used nitrogen flushing during saponification at 90°C, while Piironen et al. (1984) used alkaline saponification at ambient temperature overnight. The sample homogenate was mixed with aqueous ethanol, ascorbic acid was added, then sodium hydroxide solution; the flasks were purged with nitrogen before overnight saponification with gentle shaking.

8.2.4.3 HPLC of tocopherols and tocotrienols. Typical HPLC analytical columns and mobile phases used to separate tocopherols and tocotrienols are shown in Table 8.6. With correct match of column and mobile phase it should be relatively easy to separate all vitamers and obtain baseline separation of β- and γ-tocopherols. A procedure for the determination of tocopherols and tocotrienols in oils and fats which was developed and validated by interlaboratory study for the IUPAC has been published by Pocklington and Dieffenbacher (1988). This paper provides useful information on the validation of the procedure over four interlaboratory trials and between up to 19 laboratories. Samples for the collaborative studies included a blended soya/maize oil and wheat germ oil to provide a range of tocopherol levels and palm oil to provide a source of tocotrienols. The last study included margarine samples to test the applicability of the procedure for samples containing tocopheryl esters. The procedure recommends the use of an HPLC analytical column (25 cm × 4 mm) packed with microparticulate silica, e.g. 5 μm LiChrosorb Si 60 or Spherisorb

Table 8.6 HPLC conditions reported for vitamin E

Reference	Analytical column and mobile phase	Detection
Thompson and Hatina (1979)	LiChrosorb Si 60, 25 cm × 3.2 mm 5% diethyl ether in moist hexane 2 ml min^{-1}	Ex 290 nm Em 330 nm
Weber (1984)	Ultrasphere-Si 25 cm × 4.6 mm 1.2%–1.3% isopropanol in hexane 1.25–1.45 ml min^{-1}	Ex 205 nm Em 330 nm
Hakansson et al. (1987)	LiChrosorb si 60 25 cm × 4 mm 7% diisopropyl ether in hexane 2.1 or 2.3 ml min^{-1}	Ex 292 nm Em 324 nm
Pocklington and Dieffenbacher (1986)	LiChrosorb Si 60 or Spherisorb S5 25 cm × 4 mm 0.5% propan-2-ol in hexane 0.7 to 1.5 ml min^{-1}	Ex 290 nm Em 330 nm
Hakansson et al. (1987)	LiChrosorb Si-60-5 25 cm × 4.6 mm 7% diisopropyl ether in hexane 2.5 ml min^{-1}	Ex 290 nm Em 330 nm
Woollard et al. (1987)	Rad Pak Silica column, 5 μm, Waters 0.07% propan-2-ol in hexane	Ex 280 nm Em 335 nm

S5, and a mobile phase of propan-2-ol in hexane (0.5 / 99.5 v / v) with fluorescence or UV detection. The interlaboratory studies showed that fluorescence detection was ten times more sensitive than UV and that UV detection was prone to interferences, making fluorescence the detection mode of choice. Standards are prepared in hexane; an aliquot is taken, the hexane removed by rotary evaporation, a measured amount of methanol added and the absorbance measured at a specified wavelength appropriate for each tocopherol for calculation of standard concentration.

The working HPLC parameters are optimised by injecting mixed standards onto the column and adjusting the propan-2-ol content of the mobile phase to achieve an α-tocopherol retention time of not less than five minutes and a resolution factor for the separation of β- and γ-tocopherol of not less than 1.0. Mobile phase flow rates of between 0.7 and 1.5 ml min^{-1} were recommended and a column efficiency of 10 000 plates per metre calculated on the δ-tocopherol peak was quoted as being achievable. Oils and fats were dissolved in hexane prior to HPLC and a cold saponification suggested for samples which contain tocopheryl esters. Figure 8.4 shows a typical separation of a saponified oil blend obtained at the LGC.

Some laboratories reported low recoveries of α-tocopherol from a margarine sample spiked with α-tocopheryl acetate which was surprising as the cold saponification protocol had been adopted from a successful collaborative study carried out in Germany. The laboratory organising the studies (LGC, UK) determined the recovery of α-tocopherol from margarine spiked with α-tocopheryl acetate and obtained a mean recovery of 92% and showed that time of saponification (5, 10, 20 min) has little effect on free tocopherols but can significantly affect the recovery of α-tocopheryl acetate. It is suggested that analysts satisfy themselves that any saponification used is validated for a particular application by analysing matrix matched spiked samples in parallel with sample submitted for analysis.

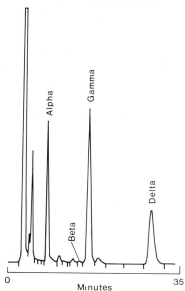

Figure 8.4 HPLC determination of tocopherols in a vegetable oil blend. Column: LiChrosorb Si60 25 cm × 4 mm. Mobile phase: Hexane–propan-2-ol (99.5:0.5) v/v 2 ml min^{-1} Detection: fluorescence, x 290 nm Em 330 nm.

Rammell and Hoogenboom (1985) reported the use of an amino-cyano polar phase HPLC column (Partisil PAC) to obtain some good separations of tocopherols and tocotrienols and one laboratory used this type of column to good effect in the IUPAC study (Pocklington and Dieffenbacher, 1988). This polar phase may have advantages over silica columns which are worth investigating.

8.2.4.4 HPLC of α-tocopheryl acetate. Hakansson et al. (1987) separated α-tocopheryl acetate from tocopherols and tocotrienols (the acetate eluted before α-tocopherols) as did Woollard et al. (1987). Despite the low fluorescent activity of α-tocopheryl acetate (9% of that of α-tocopherol) the described procedures were sufficiently sensitive to allow the determination of the acetate in supplemented foodstuffs in addition to naturally occurring tocopherols and tocotrienols.

Although unsaponified sample extracts can be injected directly onto the HPLC analytical column if fluorescence detection is used without introducing visible interferences on the chromatogram, the analyst must take care when identifying tocopherol and tocotrienol peaks. The lipid thus injected onto the silica column may cause local chromatographic conditions on the silica surface to be partially reverse-phase which may result in slightly different retention times in comparison to standards injected in solvent only.

8.2.4.5 Reporting results. As stated earlier the various tocopherols and tocotrienols have differing vitamin E activity. The results of analyses can be reported as mg/100 g of sample (or $\mu g\ g^{-1}$) or a value for total vitamin E activity can be reported as α-tocopherol equivalents if the following factors are used:

α-tocopherol × 1.00
β-tocopherol × 0.40
γ-tocopherol × 0.10
δ-tocopherol × 0.01
α-tocotrienol × 0.30
β-tocotrienol × 0.05
γ-tocotrienol × 0.01

It is obviously important that analysts make it quite clear how results are reported and clearly define the units used.

8.2.4.6 Summary. To ensure that the vitamin E content of foodstuffs is determined reliably the analyst must ensure that:

1. The extraction procedure used involves the use of antioxidants and nitrogen flushing to avoid distruction of the vitamers. Low actinic amber glassware should be used for all vitamin solutions.
2. Fluorescence detection is used in conjunction with an analytical column and mobile phase which is capable of separating tocopherols and tocotrienols from each other and from interferences.
3. Standards are freshly prepared and their concentration has been determined by UV spectrometry using recognised molar absorptivities.
4. Results are reported in an unambiguous manner either as individual tocopherols and tocotrienols or as vitamin E equivalents.

8.3 The B-group vitamins

The B-group vitamins have traditionally been determined by MBA but many applications of HPLC have appeared in the literature over the last twenty years. MBAs have many disadvantages in that they require dedicated laboratory facilities, microbiologically trained staff, and are lengthy and prone to known and unknown interferences. Despite these drawbacks, MBA often provides the only procedure capable of determining the amount of a particular vitamin in a wide range of foodstuffs with reasonable accuracy. HPLC offers the potential advantages of selectivity, so that we can determine individual vitamers of interest, as opposed to a total growth response from MBA, and sometimes it provides rapid results, but it does suffer from disadvantages related to sensitivity, extraction of the vitamers from foodstuffs and often the need for pre-HPLC removal of interferences. Neither MBA or HPLC offer the analyst the solution to all problems associated with the determination of the B-group vitamins, they both have advantages and disadvantages. For many studies they should be regarded as complimentary techniques, the results from both providing the analyst with essential information.

In this section HPLC applications are discussed for each vitamin. The assay format for MBA is common to all B-group vitamins and this is described below. MBA detail such as test organism and assay range is detailed under each vitamin heading.

8.3.1 Microbiological assays

Microbiological assays for the determination of vitamins are dependent upon the specific growth requirements of selected microorganisms, usually lactic acid bacteria. The first MBA for riboflavin was performed by Snell in 1939 using lactic acid bacteria because their growth requirements had been studied by dairy bacteriologists, and as a group their requirements were complex but specific. Lactobacilli have the additional advantage that their growth can be easily followed by turbidometric or optical density measurement or by titration of the lactic acid produced during growth. Their choice as assay organisms is amply justified as they are still used today for the determination of B-group vitamins.

The basis of a MBA is addition of a dilution series of the sample extract to a basal medium which contains all the growth requirements for the test organism except the vitamin to be determined; the mixture is then inoculated with the test organism and incubated. The test organism will grow in proportion to the vitamin content of the sample extract and quantitation is achieved by inclusion of a range of vitamin standards in the MBA. The growth of the test organism is measured as mentioned above.

The basic procedure for a MBA is the same for all the vitamins and can be broken down into a series of stages as follows:

1. Preparation of media for maintaining stock cultures of the test organisms.
2. Preparation of a basal medium deficient in the vitamin to be determined.
3. Preparation of the inoculum medium and inoculum culture.
4. Extraction of the vitamin from the samples.
5. Setting up the assay.
6. Sterilisation of the assay tubes and media.
7. Inoculation of assay tubes with the test organism.
8. Incubation (18–24 h).
9. Measurement of the growth response of the test organism and calculation of results.

Stages 1–3 require media which are capable of supporting the growth of the test organism to be used and for lactic acid bacteria they must contain amino acids, vitamins, purine and pyrimidine bases, fermentable carbohydrate, mineral salts and buffers. These media are available commercially which eliminates the possible variation in composition experienced when media are prepared from basic ingredients in the laboratory. Test organisms must be obtained from a recognised national culture collection and stock cultures have traditionally been maintained as agar stabs. Preparation of the basal assay medium is conveniently performed by rehydrating commercially available dried media according to suppliers instructions. The preparation of the inoculum has shown variation over time, and to some extent is dependent upon the method to be used at the end of the incubation period to measure bacterial growth. One classic procedure involves taking a stab-culture from the stock culture into a sterile lactobacilli broth and incubating overnight before required. The broth is centrifuged, the supernatent discarded, the bacteria washed several times with sterile saline and then suspended in sterile saline and used for inoculation of the assay tubes. This technique is not very satisfactory for 18–24 h turbidometric assays (as opposed to 72 h titrimetric assays) as the bacteria are in the lag phase of growth and need time to pass through the acceleration phase and into the desired exponential phase which is required for the assay. A technique used by Bell (1984) and still employed at the LGC involves growth overnight in lactobacilli broth and the following morning one drop of the broth is subcultured into the appropriate assay medium which contains a controlled amount of the vitamin to be determined. After six hours growth the inoculum is diluted with sterile assay medium and used for inoculation of the assay. This ensures that the test organism is approaching or in the exponential phase of growth at the start of the assay. Another way of preparing inocula is the use of glycerol cryoprotected lactobacilli (Wilson and Horne, 1982) which offers the advantage of standardised inocula strength and is becoming increasingly popular. A large volume of overnight inoculum is prepared and incubated overnight, glycerol added, and then stored in 2 ml vials at -70 °C. A thawed vial is then used for each assay. Extraction of the vitamins from samples is discussed elsewhere in this chapter.

Setting up the assay (stage 5) is almost the same for all vitamins. Test tubes are most commonly used in racks of perhaps 90 tubes. 0–5 ml of sample extracts and standards are pipetted into the tubes, the volume made up to 5 ml as necessary and 5 ml of assay medium added to each tube. An alternative procedure (Bell, 1984) is to add 0–400 µl of sample extracts and standards to assay tubes followed by the addition of 10 ml of assay medium. This has the advantage that the pH of sample extracts is less critical and single strength assay media can be used, and smaller volumes are handled, making semi-automation of the system easier. After addition of the media, tubes are capped and sterilised by autoclaving.

Inoculation (stage 7) involves adding the same amount of inoculum to each assay tube (e.g. 1 drop or 100 µl) then the whole rack of tubes is placed in a water bath at the required temperature for incubation. Air incubators are not suitable because of the temperature differential experienced between assay tubes. After incubation for 18 to 24 h bacterial growth in standards and sample assay tubes should be visible and the most convenient way to measure growth is by the use of a nephelometer or by measurement of optical density. It is essential that the measurement of growth is determined after carefully defined times if results are to be valid, and these times must be determined within the laboratory using a defined assay protocol. Growth measurements from standards are plotted against vitamin concentration and results for samples obtained by interpolation from this calibration line. Setting up the assay and measurement of the growth response can be labour intensive operations but these operations can be semi-automated (Lumley and Lawrance, 1990).

MBAs have been scaled down to microtitre plate assay formats and this development shows much promise (Newman and Tsai, 1986).

The AOAC manual of official methods (AOAC, 1990b) provides MBA procedures for vitamins B_{12}, folic acid, niacin, pantothenic acid and riboflavin.

MBAs are simple in theory but often very difficult to perform. The laboratory must be dedicated to the routine performance of MBAs; they cannot be used on an *ad-hoc* basis in the way many chemical methods of analysis are used. Staff must be well trained in the application of standard microbiological procedures and laboratory facilities and glassware and equipment must be scrupulously clean. Cleaning regimes for glassware, etc. must be committed to a defined protocol and all sources of reagents and water must be of defined quality. Glassware for individual vitamin assays must be segregated. For very sensitive assays, e.g. B_{12} and folate, glassware will probably have to be subjected to a cleaning regimen similar to the one used in this laboratory which involves boiling in a water bath containing detergent (the type is important) for three hours prior to thorough washing in an automated washing machine with deionised water rinses, followed by soaking in dilute acid, further deionised water rinses and baking in an oven at 100 °C overnight before use. If cleaning or reagent quality is compromised, assays will fail; a result which is extremely expensive in staff time.

There are many other factors which must be considered but analysts should refer to specialist texts on the subject; this chapter cannot provide all the information necessary in a few pages. Each section which follows provides some basic information on appropriate assay organisms for each vitamin discussed, but again more detailed texts need to be consulted in order to embark on the MBA of vitamins.

8.3.2 *Thiamin – vitamin B_1*

8.3.2.1 Introduction. Vitamin B_1 exists in tissue as thiamin, thiamin monophosphate, thiamin diphosphate (pyrophosphate, cocarboxylase), thiamin triphosphate and in protein bound forms. The most common way to determine the vitamin B_1 content of foodstuffs has been to release and extract thiamin and its phosphate esters using acid hydrolysis followed by enzymatic dephosphorylation of the esters and subsequent determination of thiamin. The determination of thiamin has traditionally been achieved by fluorimetry after oxidation of thiamin to the fluorescent compound thiochrome, by using microbiological assay (MBA) and more recently by high performance liquid chromatography (HPLC). MBA and HPLC are discussed in this section.

8.3.2.2 Extraction of thiamin. The first critical step in the determination of thiamin is the efficient extraction of thiamin and its phosphate esters from foodstuffs and the subsequent dephosphorylation of the esters. Table 8.7 shows selected extraction procedures used for thiamin and Table 8.8 shows procedures reported for the extraction of thiamin with other B-group vitamins.

Table 8.7 Extraction procedures reported for thiamin

Sample	Extraction procedure	Reference
All foodstuffs	121°C for 30 min with 0.1 M HCl; incubate with Takadiastase at 37°C overnight	Bell (1984)
All foodstuffs	95–100°C or 121°C for 30 min; diastatic/phosphorolytic enzyme treatment 45–50°C for 3 h, pH 4–4.5	AOAC (1990c)
Fortified breakfast cereal	(a) reflux with 1:1 MeOH–0.2 M HCl for 15 min; (b) 120°C for 30 min with 0.13 M H_2SO_4; incubate with Takadiastase for 45 min, ppt protein with TCA (c) 95°C for 60 min with 0.13 M H_2SO_4; incubate with pancreatin, lipase and Takadiastase at 60°C for 45 min (d) 120°C for 15 min with 0.1 M H_2SO_4; incubate with Claradiastase at 45°C for 2 h (e) 121°C for 1 h with 0.03 M H_2SO_4; incubate with Takadiastase at 45°C for 3 h (f) 100°C for 30 min with 0.1 M HCl; incubate with α-amylase at 45°C for 2 h (g) 90°C for 45 min with 0.1 M HCl; incubate with α-amylase at 65°C for 45 min	Nicolson et al. (1984)
Total diets	Homogenised samples + 0.1 M H_2SO_4 at 45°C for 1 h; incubation with Claradiastase for 2 h at 45°C	Botticher and Botticher (1986)
Chicken	121°C for 20 min with 0.1 M HCl; incubate with Takadiastase and papain at 45°C, pH 4.5 for 3 h	Bertelsen et al. (1988)
Legumes, milk powder, pork	121°C for 15 min with 0.1 M HCl; incubate with Takadiastase at 48°C, pH 4.45 for 3 h	Vidal-Valverde and Reche (1990b)

Table 8.8 Extraction procedures reported for thiamin and other B-group vitamins

Sample	Extraction procedure	Reference
Rice products	121°C for 30 min with 0.05 M H_2SO_4; incubate with Takadiastase and papain at 35°C, pH 4.5 overnight. For B_1, B_2 and niacin	Toma and Tabekhia (1979)
Meats	121°C for 30 min with 0.1 M HCl; incubate with Takadiastase and papain at 42–45°C, pH 4–4.5 for 2.5–3 h. For B_1 and B_2	Ang and Moseley (1980)
Potato	Reflux for 30 min with 0.1 M HCl; incubate with Takadiastase at 45–50°C for 2 h. For B_1 and B_2	Finglas and Faulks (1984a)
General foodstuffs	121°C for 30 min with 0.1 M HCl incubate with Claradiastase at 45–50°C for 3 h, pH 4–4.5	Wills et al. (1985)
Meats	As Lumley and Lawrance (1990). For B_1 and B_2 and niacin	Dawson et al. (1988)
Dietetic foods	100°C for 30 min with 0.1 M HCl; incubate with α-amylase and Takadiastase at 37°C, pH 4.5 overnight. For B_1 and B_2	Hasselmann et al. (1989)
Soyabeans and tofu	Autoclaved (20 psi) for 20 min in water adjusted to pH 2 with 5 M HCl. No enzyme treatment. For B_1 and B_2	Fernando and Murphy (1990)
General foods	121°C for 15 min with 0.1 M HCl; incubate at pH 4–4.5 with Takadiastase at 48°C for 3 h	Fellman et al. (1982)

Some of the extraction procedures shown are more rigorous than others; several are not to be recommended. Sulphuric acid has been used in some references for the acid hydrolysis step, but McRoberts (1960) has reported that this may affect the thiamin content and Fellman et al. (1982) reported that hydrochloric acid is preferable for the hydrolysis stage.

Once thiamin and its phosphate esters are released from the food matrix by acid hydrolysis, enzymatic dephosphorylation is performed using commercial enzyme preparations such as Takadiastase or Clarase which are mixtures of amylases which contain phosphatases as 'impurities', i.e. we make use of the adventitious presence of phosphatase activity. The analyst who wishes to determine thiamin in a foodstuff must ask which of the extraction procedures in Tables 8.7 and 8.8 will release all of the thiamin from the food matrix and also its phosphate esters and why do some workers use an overnight incubation with enzyme while others only use 45 minutes and some do not employ an enzyme treatment at all?

The extraction procedure reported by laboratory (a) in Nicolson et al. (1984) (Table 8.7), may be suitable for extracting free thiamin from a fortified product but a total figure for thiamin content will not be obtained and this procedure is of very limited use. Fernando and Murphy (see Table 8.8) state that the use of Takadiastase failed to increase the yield of thiochrome when analysing soybeans and tofu; this may be true but their extraction procedure is of very limited use for other foodstuffs.

Laboratories (f) and (g) referred to by Nicolson et al. (1984) (Table 8.7) used α-amylase for the enzymatic hydrolysis; is this enzyme satisfactory for general application? α-Amylase preparations as purchased are reasonably pure, therefore their level of phosphatase activity is low and one would not expect them to be very efficient for dephosphorylation of thiamin. This was shown in a study by Hasselmann et al. (see Table 8.8) who reported that unsatisfactory results were submitted by laboratories who used β-amylases because of the large variations in phosphatase activity depending on the source of the enzyme; results were also on average lower than those obtained using Takadiastase. The highest yield of thiamin from foods was achieved using a mixture of Takadiastase and α-amylase. The use of acid phosphatase only resulted in about 25% release of thiamin indicating that the presence of other types of enzymes in the Takadiastase and β-amylase mixture are necessary for efficient release of thiamin.

Bell and Christie (1973) prescribed an overnight incubation with Takadiastase because this was convenient timing within the analytical protocol and efficient hydrolysis of thiamin phosphates was achieved if batches of Takadiastase varied in phosphatase activity. Are short incubation times, e.g. 45 min, satisfactory? The answer is definitely no unless the foodstuff being examined contains a low level of thiamin phosphate and the enzyme preparation used has a high phosphatase activity and these factors are not always known.

Sources of Takadiastase, Clarase and similar preparations vary in their phosphatase activity and vary between batches. The AOAC (1990c) recommends that enzyme preparations should be checked for their ability to convert thiamin mono-, di- and tri-phosphate to thiamin before use. A conversion of greater than 85%

should be achieved; few references are made in published papers to this criterion for enzyme use. Defibaugh (1987) evaluated a range of commercially available enzyme preparations for potential use in the AOAC procedure for the determination of thiamin. Substrate thiamin ester, time/temperature and pH at two levels of enzyme addition were studied and results expressed in terms of mean percentage conversion (MPC). The AOAC procedure (1990c) defines a three hour incubation period and not all of the enzymes were satisfactory for this application after statistical evaluation of results. Defibaugh and the current AOAC procedure recommend three commercial enzyme preparations for the dephosphorylation of thiamin phosphates.

A three hour incubation with Takadiastase may be sufficient for total hydrolysis of thiamin phosphate esters if the enzyme preparation has a high phosphatase activity but this time should be considered as a minimum. In addition, the activity of each enzyme preparation obtained should be measured so that efficiency can be recorded and appreciated by the analyst. Factors such as incubation temperature, sample to enzyme ratio, and pH of the reaction mixture need to be optimised in the laboratory; these conditions need to be evaluated in conjunction with the enzyme activity studies.

8.3.2.3 Determination of thiamin. The determination of thiamin by fluorimetry involves extraction of thiamin as previously discussed, ion-exchange column chromatography of the sample extract to separate thiamin from interferences, oxidation of thiamin to thiochrome and measurement of fluorescence against thiamin standards taken through the oxidation procedure. Interfering compounds can coelute with thiamin and interfere with the oxidation procedure and also quench fluorescence. It is not, therefore, the method of choice for general application to foodstuffs. This method perhaps has its applications if similar sample types are always to be analysed (e.g. for quality control purposes) and the analyst is aware of the possible interferences and their consequent effect on results (by comparison of results with those obtained by other methods). Despite the many criticisms of this procedure in the literature it is still prescribed for use in the AOAC (1990c) for the determination of thiamin in foods.

Microbiological assay (MBA) offers the advantage of sensitivity and general applicability to a wide range of foodstuffs and feeds without appearing to suffer significantly from the interfering effects of the sample matrix. The MBA for thiamin is one of the most 'well behaved' assays in the portfolio of vitamin MBAs. The experienced laboratory should have few problems with this assay, but as indicated in the introduction to this chapter, MBAs cannot be used intermittently as a chemical method of analysis would be used. The laboratory must be performing assays regularly or continuously.

Many papers have been published describing the determination of thiamin by HPLC and this procedure has considerable potential application if the analyst selects the appropriate methodology.

8.3.2.4 High performance liquid chromatography.

Reverse-phase HPLC is most commonly used to attempt to separate thiamin from the many other components of a food sample digest. UV or fluorescence detection can be used, the latter being preferable. UV detection offers limited sensitivity and poor selectivity; the chromatographic separation of thiamin from the many UV absorbing coextractants found in a food digest is very difficult so that complex chromatograms are often obtained and the potential for coeluting interferences is high. Oxidation of thiamin to its fluorescent derivative thiochrome either before chromatography or by post-column oxidation, with subsequent fluorimetric detection offers the only HPLC procedure which will be applicable to a wide range of sample types. Post-column oxidation procedures have several disadvantages which are:

(i) Post-column plumbing and mixing of reagents can be difficult to maintain.
(ii) Chromatographic efficiency is usually wasted because of increased 'dead volume' in tubing and eluent dilution with oxidising reagent.
(iii) Oxidising reagents can affect the fluorescent intensity of thiochrome.

Pre column oxidation of thiamin and reverse-phase HPLC of thiochrome is the preferred option.

Four of the six participating laboratories in the study by Nicolson *et al.* (1984) used UV detection for thiamin and this probably explains in part the poor agreement between laboratories; chromatograms were not provided in the publication and so further comment is not possible.

Vidal-Valverde and Reche (1990b) describe an HPLC procedure which includes a fairly complex purification stage which concentrates the thiamin and removes some interferences so that UV detection can be used. After extraction of the thiamin (see Table 8.7) and filtration, an aliquot of the extract was passed through an Amberlite CG-50 resin column, the column was washed with water and thiamin eluted with HCl. The HCl eluent was evaporated to dryness, the residue dissolved in aqueous buffer and then passed through a C_{18} Sep Pak cartridge (conditioned with ion-pair reagents). Interferences were removed by washing the Sep Pak and thiamin was finally eluted. The eluent was chromatographed on a reverse-phase column with UV detection at 254 nm. A measure of the difficulty of using UV detection can be understood by the fact that column temperatures (between 30 and 50°C) and mobile phase compositions (based on methanol, water, acetic acid, ion pair reagents) had to be selected according to the type of sample being analysed. Chromatograms provided by the authors are more complex than one would obtain if a fluorescence procedure had been used.

Table 8.9 shows selected literature references with brief details of the pre-column oxidation conditions used and column type and mobile phase employed for the determination of thiamin as thiochrome.

After oxidation some workers extract thiochrome into butan-2-ol and inject this solution onto the HPLC column, whilst others inject aliquots of the neutralised aqueous oxidation mixture. Hasselmann *et al.* (1989) passed the oxidation mixture

Table 8.9 Determination of thiamin as thiochrome–pre-column oxidation conditions, column type and mobile phase.

Reference	Oxidation procedure	HPLC column	Mobile phase	Detection
Botticher and Botticher (1986)	2 ml sample extract + 3.75 butan-2-ol, 0.375 ml 50% w/v NaOH, 0.1 ml 5% m/v potassium hexacyanoferrate- III	Nucleosil–NH$_2$ 5 μm 25 cm × 4.6 mm	K$_2$HPO$_4$ buffer pH 4.4/CH$_3$CN (25:75)	Fluorescence Ex 370 nm Em 425 nm
Bertelsen et al. (1988) Finglas and Faulks (1984a)	5 ml sample extract + 3 ml 0.03 M Potassium ferricyanide in 3.75 M	μ-Bondapak C$_{18}$ 25 cm × 4.6 mm	Water/MeOH (70:30)	Ex 365 nm Em 435 nm
Dawson et al. (1988)	10 ml sample extract + 5 ml 1% potassium ferricyanide in 15% NaOH. Extract into 10 ml butan-2-ol	Alltech C$_{10}$ μm	0.02M phosphate buffer pH 7/MeOH (70:30)	Ex 378 nm Em 430 nm
Hasselmann et al. (1989)	1 ml sample extract + 3 ml alkaline potassium ferricyanide (1 ml 1% Potassium. ferricyanide +24 ml 15% sodium hydroxide). Thiochrome extracted onto C$_{18}$ Sep Pak.	μ-Bondapak C$_{18}$ acetate buffer	0.05 M sodium Em 435 nm pH 4.5/MeOH, (40:60)	Ex 366 nm
Fernando and Murphy (1990)	2.5 ml sample extract + 2.5 ml 1% Potassium ferricyanide in 15% NaOH neutralised after 45 sec	Ultrasphere C$_{18}$ 5 μm 15 cm × 4.6 mm	CH$_3$CN/0.01 M acetate buffer pH 5.5 (13:87)	Ex 364 nm Em
Fellman et al. (1982)	10 ml sample extract + 5 ml 1% potassium ferricyanide aq. in 15%NaOH. Thiochrome extracted onto C$_{18}$ Sep Pak	C$_8$ Radial Pak 10 cm × 18 mm	MeOH/0.01M phosphate buffer pH 7 (37:63)	Ex 360 nm Em 415 nm

through a C_{18} Sep Pak cartridge which was washed with 0.05 M sodium acetate buffer and the thiochrome then eluted with methanol–water (60:40 v / v) prior to chromatography. Injecting aliquots of the oxidation mixture directly onto the HPLC column can shorten column life and thus lead to decrease in column efficiency, although this approach does save some preparation time.

At the LGC a procedure is used that involves oxidising 5 ml of sample extract with 3 ml of alkaline ferricyanide solution for 30 min, addition of a measured amount of dilute phosphoric acid to stop any further reaction (to pH 5.5), and addition of phosphate buffer to 25 ml. Aliquots of this buffered thiochrome solution are injected directly onto Spherisorb ODS columns (25 cm × 4.6 mm) using a mobile phase of methanol–0.01 M KH_2PO_4 buffer (65:35) with fluorescence detection. Figure 8.5 shows a chromatogram obtained from a sample of dried pork which contains approximately 2.5 mg / 100 g of thiamin. Analytical columns have been used for 6 months without signs of drastic loss of chromatographic efficiency.

8.3.2.5 Summary. If thiamin is to be determined reliably using HPLC the following guidelines should be borne in mind:

1. Use hydrochloric acid for the hydrolysis stage; reflux or autoclave conditions.
2. Use Takadiastase or a similar enzyme preparation only after investigating

Figure 8.5 HPLC determination of thiamin in a dried meat product. HPLC of thiochrome derivative. Column: Spherisorb ODS 25cm × 4.6mm. Mobile phase: methanol–0.01 M phosphate buffer 65:35 v/v, 1 ml min^{-1}. Detection: fluorescence Ex 365 nm Em 435 nm

its ability to hydrolyse thiamin phosphate esters to thiamin. It may be beneficial to also use α-amylase in conjunction with Takadiastase. Papain may assist in the digestion of samples with a high protein content, e.g. meats.
3. Use a significantly long incubation time with the enzyme preparation to ensure complete conversion to thiamin.
4. Ensure that the oxidation conditions used for sample extracts and standards are identical and reproducible and that reaction times are optimised.
5. Use a reverse-phase column and buffered mobile phase capable of separating thiochrome from sample matrix interferences.
6. A calibration line should be prepared for each assay as opposed to single point calibration.
7. A reagent blank should be taken through the whole assay system.

This list is not exhaustive but includes the major critical points in the HPLC determinations of thiamin.

8.3.2.5 MBA of thiamin. *Lactobacillus fermenti* was proposed for the determination of thiamin in 1944 but this organism was shown to respond to pentoses, reducing agents, fructose, maltose and heat degradation products of glucose. Attempts to compensate for inhibitory and stimulatory effects were not practicable and this organism was superseded by the use of *Lactobacillus viridescens* (ATCC 12706, NCIMB 8965) for which commercial assay media can be obtained. Thiamin is extracted from foodstuffs usually by autoclaving with acid followed by treatment with an enzyme preparation and dilution to a suitable volume. The assay range using *L. viridescens* is typically 0–25 ng per assay tube, and the assay should be incubated at 30 °C for 18–20 hours for turbidometric measurement. Thiamin assays take about three days to produce results because of the overnight enzyme incubations during extraction (Bell, 1984).

8.3.3 Riboflavin – vitamin B_2

8.3.3.1 Introduction. Natural riboflavin occurs in foods as free riboflavin or as the protein bound riboflavin-5′-phosphate (FMN, flavin mononucleotide) and flavin adenine dinucleotide (FAD). Extraction of these bound forms of the vitamin is most commonly achieved by hydrolysis with a dilute mineral acid (e.g. 0.1 M HCl); this stage in the extraction also hydrolyses most FAD to FMN. If we are to determine 'free' riboflavin, the FMN has to be further hydrolysed enzymatically using a commercial enzyme preparation such as Takadiastase or Clarase. As stated previously, these enzyme preparations contain phosphates (as 'impurities') in addition to amylases, therefore, FMN can be dephosphorylated to yield riboflavin; the amylase has the advantage of hydrolysing starch which aids sample digestion when carbohydrate rich foods are being examined.

Riboflavin can be determined fluorimetrically and the AOAC prescribe a manual and an automated procedure for foodstuffs as final actions (AOAC, 1990d).

The suitability of this method has been questioned for many years because it is time consuming and subject to many potential interferences. The procedure has been adapted to a flow injection procedure which shortens analysis time but Russell and Vanderslice (1992) report that the method is not universally suitable for all sample types.

Microbiological assay, despite its drawbacks, is still routinely used for the determination of riboflavin, and HPLC methods have been published and show promising results. These two procedures will be discussed.

8.3.3.2 Extraction of riboflavin from foodstuffs. The procedures used to extract riboflavin from foodstuffs are similar to those used for thiamin; in fact both vitamins can be extracted from foodstuffs using a common procedure. The problem of enzyme activity can affect the quantitative release of riboflavin in a similar way to the problems outlined for thiamin. Table 8.10 shows some reported extraction procedures used for riboflavin; in addition, extraction procedures given in Table 8.8 for 'thiamin and other B-group vitamins' can be consulted.

Lumley and Wiggins (1981) reported that samples of Takadiastases varied considerably in their phosphatase activity and acid hydrolysates of samples with a

Table 8.10 Extraction procedures reported for riboflavin

Sample type	Extraction procedure	Reference
Various foods	121°C for 30 min with 0.1 M HCl; incubate with (10% w/v) Takadiastase at 40°C for 2–4 h	Lumley and Wiggins (1981)
Milk, yogurt, cheese	Pass filtered liquid milk or buffer/sample homogenate (filtered) through a suitably conditioned C_{18} Sep-Pak cartridge	Ashoor *et al.* (1985)
Fruits and vegetables	99°C for 30 min with 0.1 M HCl; incubate with (5% m/v) amylase at 38°C overnight	Watada and Tran (1985)
Various foods	121°C for 20 min with 0.2 M H_2SO_4; incubate with Takadiastase at pH 4.5 and 45°C overnight	Ollilainen *et al.* (1990)
Legumes, milk	121°C for 15 min with 0.1 M HCl; incubate with Takadiastase (6%) at pH 4–4.5 at 48°C for 3 h. Sep-Pak extraction	Vidal-Valverde and Reche (1990a)
Potatoes	Reflux for 30 min with 0.1 M HCl; incubate with Takadiastase (1.25 w/v) at pH 4.6 at 50°C for 2 h	Stancher and Zonta (1986)
Various foods	121°C for 30 min with 0.1 M HCl; incubate with Clarase (5 ml 10% suspension) at 45–50°C for 3 h	Wills *et al.* (1985)
Dietectic foods	100°C for 30 min with 0.1 M HCl; incubate with Takadiastase 500 mg and β-amylase, 50 mg, at pH 4.5 at 37°C overnight	Hasselmann *et al.* (1989)

high fat content required longer incubation periods to convert all of the FMN to riboflavin. Enzyme preparations from several suppliers were evaluated and Takadiastase from one source was recommended, but because of batch-to-batch variation it is essential that analysts check the efficiency of each new enzyme preparation before use. Ollilainen *et al.* (1990) investigated the use of Claradiastase and two sources of Takadiastase; they found that one source/batch of Takadiastase was not sufficiently efficient to hydrolyse all FMN to riboflavin under the conditions used. Conclusions from experiments in this laboratory and conclusions reached by Ollilainen *et al.* indicate that incubation with a commercial enzyme preparation, e.g. Takadiastase, is best carried out overnight; this usually ensures complete hydrolysis of FMN to riboflavin and is convenient in the laboratory routine. Shorter incubation times can be used as long as the analyst is sure that quantitative conversion of FMN is taking place if only riboflavin is being measured on the chromatogram. Some HPLC procedures separate, and allow the determination of FMN and riboflavin (Lumley and Wiggins, 1981; Reyes *et al.*, 1988), so that total conversion of FMN to riboflavin is not essential. However, this approach is not to be recommended since FMN is not always completely separated from impurities and fluorescent components which derive from Takadiastase. A chromatographic system which allows identification of FMN is, however, useful as enzyme efficiency can more easily be monitored.

Ashoor *et al.* (1985) and Stancher and Zonta (1986) extracted free riboflavin from dairy products without the need for acid hydrolysis or enzyme hydrolysis. Ashoor *et al.* (1985) reported that autoclaving acidified milk at 121°C for 30 min did not increase the yield of riboflavin from whole milk, non-fat dry milk, yogurt, processed and cheddar cheese when compared with their reported solid phase extraction. The solid phase extraction consisted of passing liquid milk, reconstituted milk or an acetate buffer extract of cheese, through a conditioned (5 ml MeOH, 5 ml H_2O) C_{18} Sep-Pak cartridge, washing the column with water (2×10 ml) and eluting the riboflavin with acetate buffer and methanol (0.02 M acetate buffer pH 4: methanol, 1:1, 5 ml). The eluent was subsequently loaded onto the HPLC column. Up to 25 ml of sample extract was loaded onto a cartridge and eluted with 5 ml of buffer–methanol, thus incorporating a significant concentration step in addition to removing potential interferences. Stancher and Zonta (1986) used a more complex extraction procedure for Italian cheese and did not employ solid phase extraction.

The approach of Ashoor *et al.* (1985) is rather attractive for dairy products but obviously caution must be exercised for other types of products which may contain protein bound forms of riboflavin. Riboflavin is strongly bound to reverse-phase materials when in an aqueous environment and this fact was used by Lumley and Wiggins (1981) for HPLC on-column sample concentration.

Watada and Tran (1985) employed acid hydrolysis for the digestion of fruit and vegetables and concluded that the temperature of hydrolysis must be near 99°C because lower riboflavin results were obtained when lower temperatures were used, and the enzyme hydrolysis was required for complete extraction of riboflavin.

Reyes et al. (1988) used acid hydrolysis with no subsequent enzyme hydrolysis to compare an HPLC procedure against the AOAC fluorimetric and microbiological methods for riboflavin determination on a range of foodstuffs. The chromatograms provided clearly show the considerable amount of FMN in many of their sample extracts. These workers obtained the total riboflavin contributions, however, as stated earlier this approach has the potential for greater error. The analytical aim should be to use extraction conditions which convert all bound FAD and FMN to free riboflavin which can then be determined by measurement of one peak on a chromatogram.

Riboflavin is extremely light sensitive and so all sample extracts and standards must be protected from sunlight and strong artificial light at all times.

8.3.3.3 High performance liquid chromatography. Reverse-phase HPLC is almost universally used to separate riboflavin from other components of sample digests and fluorescence detection offers the only generally acceptable detection method.

Ashoor et al. (1985) and Stancher and Zonta (1986) used UV detection for the determination of riboflavin in dairy products (see Table 8.11 for chromatographic conditions) with apparent success. As discussed earlier, Ashoor et al. used a C_{18} Sep-Pak sample extract clean-up and concentration step and produced evidence of riboflavin peaks which appear to be well resolved from interferences. Detection at 270 nm was used. Stancher and Zonta extracted riboflavin from cheese but did not employ any 'clean-up' of the extract. The chromatogram obtained using UV detection at 267 nm shows the riboflavin peak amongst a mass of interfering peaks; to overcome this detection at 446 nm, the secondary riboflavin UV maxima, was used to produce clear chromatograms of riboflavin. This approach is satisfactory for the application reported but sensitivity may not be sufficient for samples with riboflavin contents lower than is found in cheese, and interferences may be a problem with sample extracts from pigmented foods (natural or artificial). The chromatographic conditions used are also perhaps limiting because the retention time of riboflavin is only five minutes; longer retention may result in better selectivity and resolution of riboflavin from potential interferences.

Vidal-Valverde and Reche (1990a) also report on the use of UV detection for the determination of riboflavin in chick peas, beans and milk powder (see Table 8.11 for chromatographic conditions). Sample hydrolysates were loaded onto a Florisil column, riboflavin selectivity eluted, the eluent evaporated to dryness, dissolved in methanol / water, loaded onto a Sep-Pak C_{18} cartridge and the riboflavin selectively eluted with a methanol–water–ion pair reagent. To resolve riboflavin from interferences by HPLC two mobile phases were used and differing column temperatures investigated, with mobile phase flow rates of between 0.5 and 2.0 ml min^{-1}. This somewhat complicated procedure highlights the difficulties of using UV detection for the determination of riboflavin. The authors state that a disadvantage of the use of fluorescence detection is that fluorescent compounds which derive from Takadiastase or similar preparations interfere with the analysis

Table 8.11 HPLC conditions reported for the determination of riboflavin

Column	Mobile phase	Detection	Reference
Bio-sil ODS-5S C_{18} 25 cm × 4 mm	H_2O: MeOH: acetic acid (65:35: 0.1) m/v 1 ml min^{-1}	UV at 270 nm	Ashoor et al. (1985)
Altex ultrasphere ODS 5 μm 25 cm × 4 mm	MeOH: H_2O (40:60) v/v + 0.005 M heptane sulphonic acid, pH 4.5 0.8 ml min^{-1}	Fluorescence, Ex 450 nm Em 530 nm	Watada and Tran (1985)
LiChrosorb RP 18 5 μm 25 cm × 4 mm	CH_3CN: H_2O (20:80) v/v 1 ml min^{-1}	UV at 446 nm	Stancher and Zonta (1986)
Sherisorb S50DS 2 5 μm, 25 cm × 4.6 mm	Me OH: H_2O (35: 65) v/v 1 ml min^{-1}	Fluorescence Ex 445 nm Em 525 nm	Ollilainen et al. (1990)
μ Bondapack C_{18} 10 μm	(a) MeOH: acetic acid: H_2O (32:1:67) + sodium hexanesulphonate, 5 mM (b) MeOH: acetic acid: H_2O (31: 0.5:68.5) + sodium heptanesulphonate 5 mM/ hexanesulphate, 5 mM 0.5–2.0 ml min^{-1}	UV at 245 nm	Vidal-Valverde and Reche (1990a)
Hibar LiChrosorb RP-8, 25 cm × 4 mm	H_2O: MeOH (60:40) + 0.005 M 1-hexane sulphonic acid 1 ml min^{-1}	Fluorescence Ex 440 nm Em 565 nm	Reyes et al. (1988)
μ Bondapak C_{18} 25 cm × 4.6 mm	H_2O: MeOH (70: 30) v/v 2 ml min^{-1}	Fluorescence Ex 450 nm Em 410 nm	Finglas and Faulks (1984a)
μ Bondapak C_{18} radial-pak	MeOH: H_2O + 5 mM PIC B_6 (40:60) v/v 1.5 ml min^{-1}	Fluorescence Ex 360 nm Em 500 nm	Wills et al. (1985)
μ Bondapak C_{18}	MeOH:0.05 M sodium acetate, pH 4.5 (60:40) v/v 1 ml min^{-1}	Fluorescence Ex 422 nm Em 522 nm	Hasselmann et al. (1989)

of riboflavin but the experience of this author shows that if chromatography is efficient there should be no problems separating riboflavin from these interferences, although separation of FMN from them can sometimes be difficult. The fluorescence methods outlined in Table 8.11 are all similar. Methanol–water mobile phases, some with buffer or ion-pairing reagent used in conjunction with a C_{18} column (Finglas and Faulks, 1984a; Wills et al., 1985; Hasselmann et al., 1989) have been described for the simultaneous or sequential determination of riboflavin and thiamin.

The fluorescent intensity of riboflavin is pH dependent and therefore it is best to use a mobile phase which contains buffer, preferably at the pH of the riboflavin fluorescence maxima. If buffering is not done, local pH differences in the column eluent may affect riboflavin peak height.

The HPLC procedure used at the LGC (based on Lumley and Wiggins, 1981)

VITAMIN ANALYSIS IN FOODS

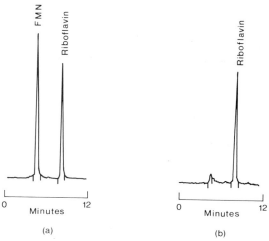

Figure 8.6 (a) HPLC of FMN and riboflavin standards. (b) HPLC determination of riboflavin in a total diet sample. Column: Spherisorb ODS 25cm × 4.6mm. Mobile phase: methanol –phosphate buffer, 0.01 M, 0.7 ml min^{-1} Determination: fluorescence, Ex 450 nm Em 520 nm.

involves autoclaving the sample with 0.1M HCl for 30 min followed by overnight incubation with Takadiastase, filtration and injection onto a Spherisorb ODS column (25 cm × 4.6 mm) with a mobile phase of methanol 0.01M phosphate buffer (40 / 60 v / v) and fluorescence detection. Figure 8.6 shows the chromatograms obtained from a total diet sample found to contain 0.18 mg / 100 g riboflavin. A small amount of FMN may be present in this extract but peak identity was not confirmed.

The paper by Ollilainen *et al.* (1990) provides some interesting information on their investigations into the chromatography of riboflavin. Using the conditions outlined in Table 8.11 they determined riboflavin in 21 food items and did not experience problems from interferences.

Finglas and Faulks (1984a), Wills *et al.* (1985) and Hasselmann *et al.* (1989) describe the simultaneous or sequential determination of riboflavin and thiamin. This can be achieved by:

- pre-column oxidation and passing the HPLC column eluent through two fluorescence detectors in series set at appropriate excitation and emission wavelengths for riboflavin and thiochrome;
- by loading two injections onto the HPLC system with a change of excitation and emission wavelengths between injections to allow detection of riboflavin and then thiochrome;
- by passing the eluent through a fluorescence detector set for riboflavin, then into a reaction coil where thiamin is oxidised to thiochrome and then through a second fluorimeter set for thiochrome.

The simultaneous or sequential determination of both vitamins is an attractive option but the analyst must firstly ensure that they are getting the right result for each vitamin independently; getting the correct result must be the prime concern.

8.3.3.4 Summary. If riboflavin is to be determined reliably by HPLC the following guidelines should be borne in mind:

1. Ensure that the vitamers are efficiently released from the sample matrix, e.g. by hydrolysis with dilute hydrochloric acid at 100 or 121°C for 30 min.
2. Use Takadiastase or a similar enzyme preparation only after investigating its ability to hydrolyse FMN to riboflavin. Use a sufficiently long incubation time to ensure complete conversion of FMN to riboflavin; overnight incubation may be necessary.
3. Use a reverse-phase HPLC column (usually C_{18}) with a buffered methanol–water mobile phase capable of separating riboflavin from interferences. Short retention times for riboflavin, e.g. 5 min are best avoided; mobile phase compositions which provide riboflavin retention times of 10–15 minutes should provide the chromatographic selectivity required.
4. Use fluorescence detection for selectivity and sensitivity. UV detection is prone to too many interferences.
5. A calibration line should be prepared for each assay; this is preferable to a single point calibration.
6. A reagent blank must be run with each assay as some Takadiastases and similar preparations contain riboflavin. Allowance must be made for this.
7. C_{18} cartridges may be used to concentrate and 'clean-up' sample extracts prior to HPLC.
8. Protect all sample extracts and standards from light.
9. Inject FMN standard onto the HPLC column and identify its retention time relative to riboflavin. This may be useful for identification of residual FMN in sample extracts if enzyme hydrolysis is not complete.

8.3.3.5 MBA of riboflavin. The riboflavin assay developed by Snell in 1939 used the organism *Lactobacillus casei* and this organism is still used today and is recommended for use by the AOAC (1990d), but it can be stimulated by starch and both inhibited and stimulated by fatty acids and lipids. The assay range is typically 0–200 ng per assay tube. The test organism *Enterococcus faecalis* (ATCC 10100, NCIMB 7432) has the advantage of being more than ten times more sensitive to riboflavin (assay range 0–10 ng per tube) and much less sensitive to stimulation or inhibition. Riboflavin is extracted from foodstuffs usually by autoclaving with dilute acid. *E. faecalis* responds equally to FMN and riboflavin and so an enzymatic hydrolysis during extraction is not required. Assays are usually incubated at 37°C for 22 to 24 h (Bell, 1984).

8.3.4 Niacin

8.3.4.1 Introduction. Niacin is the collective name for nicotinic acid and nicotinamide. Nicotinic acid is readily converted to the physiologically active form, nicotinamide which functions as a component of the coenzymes nicotinamide adenine dinucleotide (NAD) and nicotinamide adenine dinucleotide phosphate (NADP). Nicotinic acid, NAD and NADP occur in almost all foodstuffs and provide sources of the vitamin which are available to man, but the vitamin does occur bound as nicotinyl esters to polysaccharides, peptides and glycopeptides which are not metabolised by the human gut. If we are to determine 'available' niacin then care has to be exercised over choice of extraction conditions used. Acid hydrolysis will release niacin from NAD and NADP but will not hydrolyse nicotinyl esters; alkali hydrolysis is required to achieve this. Thus if we wish to determine 'available' niacin, acid hydrolysis should be used, whereas alkali hydrolysis will yield 'total' niacin. Bound forms of niacin occur mainly in cereal products. Pellagra (niacin deficiency disease) has been common in areas where maize is a staple food because the niacin is present in a bound form and tryptophan is lacking in maize protein (tryptophan is converted to nicotinic acid in the body), but in Mexico where maize is usually eaten as tortillas Pellagra has not been a problem. This is because in the preparation of tortillas, maize is mixed with lime which hydrolyses the nicotinyl esters of the bound niacin and the vitamin becomes nutritionally available.

If sufficiently robust acid hydrolysis conditions are used for niacin extraction, nicotinamide is hydrolysed to nicotinic acid and so one compound can be determined as opposed to two.

Niacin has been determined by microbiological assay and by colorimetric procedures based on the Konig–Zincke reaction which involves the use of cyanogen bromide and sulphanilic acid. The colorimetric procedure suffers from the common disadvantages of all non-specific colour procedures and involves the use of cyanogen bromide which is a major disadvantage. The microbiological assay for niacin is one of the easiest MBAs to perform in a correctly equipped and experienced laboratory; a niacin assay should always work, unlike those for folate and B_{12}. MBA procedures are discussed later.

Papers describing the use of HPLC for the determination of niacin have appeared over the last thirteen or so years but this approach for foodstuffs cannot yet be considered as generally applicable or suitable for routine use for all sample types. As with all procedures it is likely that suitable HPLC assay protocols will be developed and some approaches taken to date are discussed.

As mentioned above, in human metabolism tryptophan is a precursor of niacin; the accepted relationship is that 60 mg of tryptophan will produce 1 mg of niacin so that determination of niacin will provide a figure for probably the majority, but not all the niacin activity of a foodstuff. The tryptophan content should also be taken into account.

8.3.4.2 Extraction of niacin from foodstuffs. As discussed in the introduction, the extraction procedure used will depend on whether the analyst wishes to determine nutritionally available niacin or 'total' niacin. Table 8.12 shows some extraction procedures which have been taken from the literature. In most instances the food analyst will probably be concerned with the production of data which

Table 8.12 Extraction procedures reported for niacin

Food	Extraction procedure	References
All foodstuffs	Sample extracted with 1 M HCl at 121°C for 15 min. Cool, adjust pH to 4.5, filter.	Bell (1984)
Cereal samples	Samples extracted with Ca(OH)$_2$ suspension, 100°C for 30 min autoclave at 121°C for 30 min. Cool overnight and centrifuge; use supernatent.	Tyler and Shrago (1980)
Potatoes	Samples extracted with 0.5 M H$_2$SO$_4$ at 121°C for 30 min, neutralised with CaCO$_3$, centrifuge; portion of extract evap. to dryness, residue taken up in small quantity of water.	Finglas and Faulks (1984b)
Various foods	(1) Samples extracted with 0.1 M H$_2$SO$_4$ at 121°C for 1 h; incubate with diastase enzyme for 3 h at 45°C.	van Niekerk et al. (1984)
	(2) Samples extracted with Ca(OH)$_2$, 30 min at 100°C then autoclave at 121°C for 30 min. Cool and centrifuge.	
Coffee	Instant coffee dissolved in hot water (80°C)	Trugo et al. (1985)
Legumes and meat	Samples extracted with 0.1 M HCl (30 ml) and 6 M HCl (1ml) at 121°C for 15 min. pH adjusted to 4–4.5, add 6% soln Takadiastase, incubate 48°C for 3h.	Vidal-Valverde and Reche (1991)
Various foods	Sample extracted with 40% sodium hydroxide and added water; on steam bath for 30 min. Cool, neutralised with HCl, methanol added, mixture filtered. Bulked to known volume and rotary evaporated to dryness. Residue taken up in water.	Hirayama and Maruyama (1991)
Foods	Samples extracted with 0.5 M H$_2$SO$_4$ (>10 × sample wt). Autoclave at 121°C 30 min.	AOAC (1990e)

represent nutritionally available niacin and therefore acid extraction will be used. Most acid hydrolysis procedures involve the use of 1 M HCl or 0.5M H$_2$SO$_4$ with autoclaving at 121°C for 15–30 min. This should ensure complete liberation of nutritionally available niacin as nicotinic acid. Although some workers have used acids of lower concentration, e.g. 0.1 M HCl, care should be taken as nicotinic acid and nicotinamide may be present in the sample, and thus if an HPLC procedure is used which only determines nicotinic acid, low results may be obtained. This will not be of concern if microbiological methods employing *Lactobacillus plantarum* are used as this organism utilises both nicotinic acid and nicotinomide, and the Konig–Zincke colorimetric method is common to all substituted pyridines, including bound niacin.

The extraction of niacin is similar to thiamin and riboflavin in that the analyst must consider the extraction procedure in conjunction with the analytical procedure to be used, and ensure maximum liberation of the vitamin in the form appropriate for measurement. Incorrect and inappropriate results will be obtained if these considerations are ignored. Van Niekerk et al. (1984) used acid and alkaline hydrolysis to determine niacin in a range of foodstuffs (see extraction conditions, Table 8.12) and compared HPLC and microbiological results. They reported that there was no significant difference between microbiological results obtained from the acid, and alkaline extractions (somewhat surprising as the majority of the samples were cereal based) but that HPLC results from the acid extracts were lower than those obtained from the alkali extracts. The authors state that this may indicate that part of the niacin in the acid extracts was not present as free niacin but existed in a form available to microorganisms. This could explain the low HPLC results but another possible explanation not considered by van Niekerk *et al.* is that they used 0.1 M H_2SO_4 and this dilute acid may not have resulted in complete hydrolysis of nicotinamide to nicotinic acid; the HPLC procedure was only designed to determine nicotinic acid not nicotinamide, whereas the MBA would have determined both vitamers.

Trugo *et al.* (1985) used hot water to extract nicotinic acid from instant coffee and this is satisfactory for 'free' nicotinic acid but not for foods containing NAD, NADP or bound niacin. Nicotinic acid in coffee mainly results from the demethylation of trigonelline during roasting and is easily extractable.

Vidal-Valverde and Reche 1991 autoclaved samples with 0.1 M HCl (30 ml) + 1 ml 6 M HCl for 15 min and determined nicotinic acid by HPLC but there is no comparative data supplied to indicate if these extraction conditions were sufficiently rigorous to produce a quantitative yield of nicotinic acid. Nicotinic acid is acid and heat stable and therefore it is advisable to use stronger acids (e.g. 0.5–1 M) to ensure quantitative release of nicotinic acid from samples.

8.3.4.3 HPLC of niacin. Reverse-phase and ion-exchange HPLC procedures have been used to determine the niacin content of foodstuffs but this application is more difficult than thiamin or riboflavin because of the lack of a selective detection procedure. Nicotinic acid and nicotinamide are not conveniently converted to fluorescent derivatives so that UV detection (at between 254–263 nm) has to be used and many sample coextractants in addition to the vitamers of interest absorb at these wavelengths. Efficient separation of nicotinic acid and nicotinamide is therefore essential, but in practice difficult. The majority of workers wish to determine nicotinic acid rather than nicotinamide. With this in mind the determination of this vitamer will be discussed in more detail here. Table 8.13 shows some reported HPLC conditions for nicotinic acid. Tyler and Shrago (1980) removed interferences from sample extracts using anion-exchange column chromatography, permanganate oxidation, and a C_{18} Sep Pak prior to analytical HPLC on a reverse-phase column with ion pairing. Chromatograms for semolina and bread show reasonable separation of 'niacin' from interferences; 'niacin' is the term used by the authors for the standard used. PIC A was used for HPLC (tetrabutyl ammonium phosphate) so that it is assumed they are referring to the determination of nicotinic acid. Trugo *et al.* (1985) and Takatsuki *et al.* (1987) similarly used reversed-phase HPLC with ion pairing (tetrabutyl ammonium ion) for coffee and beef extracts respectively without prior 'clean-up' of sample extracts. Vidal-

Table 8.13 HPLC conditions reported for the determination of nicotinic acid

Reference	HPLC conditions
Tyler and Shrago (1980)	Sample extract 'cleaned-up' on anion exchange column; niacin eluent oxidised with permanganate, passed through C_{18} Sep Pak and 0.45 µ filter. Eluent chromatographed on µ Bondapak C_{18} column, 9% or 5% methanol in water with PIC A as mobile phase, 2 ml min^{-1}.
Finglas and Faulks (1984b)	Concentrated sample extract chromatographed on Partisil SAX 10 µm column (25 cm × 4.6 mm); elution gradient of 10% aqueous 0.05 M KH_2PO_4 pH 4.5:90% water for 6 min to elute niacin then 100% phosphate buffer for 13 min.
Monard et al. (1986)	Sample extract chromatographed on Nucleosil 5 C_{18} (15 cm × 4.6 mm) using 0.1 M acetic acid pH 3; niacin band cut loaded onto Nucleosil 5 SB anion exchange column (15 cm × 4.6 mm) and chromatographed using 0.4 M acetic acid, pH 3. 0.1 M acetic acid/methanol (5:95) used to elute interferences from C_{18} column.
Tngo et al. (1985)	Sample extract chromatographed on Spherisorb ODS-2 (15 cm × 5 mm) using methanol 8% (v/v) containing 0.005 M tetrabutylammonium hydroxide, pH 7.0, 1.5 ml min^{-1}.
Takatsuki et al. (1987)	Sample extract chromatographed on µBondapak C_{18} (30 cm × 3.9 mm) using methanol–water (1:9) containing 5 mM PIC-A (tetrabutylammonium ion).
Hamono et al. (1987)	Sample extract chromatographed on Partisil SCX 10 µm (25 cm × 4.6 mm) using 50 mM phosphate buffer at pH 3.0.
Vidal-Valverde and Reche (1991)	Sample extracts 'cleaned-up' on ion-exchange column then chromatographed on µBondapak C_{18} (10 µm 30 cm × 3.9 mm) or Spherisorb ODS 2 (10 µm) with methanol :0.01 M sodium acetate buffer (pH 4.66) (1:9 v/v) containing 0.005 M tetrabutylammonium bromide.
Hirayama and Maruyama (1991)	Sample extracts 'cleaned-up' on anion then cation exchange then chromatographed on Asahipak NH2P (25 cm × 4.6 mm, 5 µm) with acetonitride–water (60: 40 v/v) containing 0.075 M sodium acetate.

Valverde and Reche (1991) removed interferences from extracts of meat and legumes by ion exchange chromatography (3 cm × 1 cm column) prior to analytical reverse-phase HPLC again using the tetrabutyl ammonium ion for pairing. Chromatographic conditions, i.e. rigorous control of mobile phase pH and column temperature, were selected for optimal resolution of nicotinic acid from interferences for each sample type. The authors state that omission of the ion-exchange purification step led to the consistent overlap of the nicotinic acid peak with interferences. C_{18} Sep Pak cartridges were tried as a purification step but the authors state that they only achieved a nicotinic acid recovery of 60%; a CN cartridge resulted in a 98% recovery of nicotinic acid but failed to remove interferences and permanganate oxidation was also ineffective for the sample extracts examined.

Finglas and Faulks (1984b) concentrated neutralised potato extracts by rotary evaporation and chromatographed the solubilised residues on a strong anion-exchange HPLC column. A mobile phase of an aqueous solution of 0.05 M KH_2PO_4, pH 4.5/water (10:90 v/v) was passed through the column to elute the niacin (6 min) and then 100% phosphate buffer (for 13 min) to remove other sample components from the column. No chromatograms were supplied by the authors and so it is difficult to assess the efficiency of this procedure, however the stated five minutes retention time for niacin (nicotinic acid) perhaps is not long enough to enable selective separation from interferences. The authors report that results were substantially lower than values reported in food tables.

Haimano *et al.* (1988) used a strong cation exchange HPLC column to determine nicotinic acid and nicotinamide in aqueous extracts of beef and pork. Nicotinic acid eluted close to interfering peaks (Rt 10 min), but this procedure was only used for free niacin; hydrolysed sample extracts would probably contain many more potential interferences.

Van Niekerk *et al.* (1984) have reported a dual HPLC column technique with column switching to determine niacin in samples of plant / cereal origin. Sample extracts were chromatographed on a reverse-phase column (mobile phase 0.1 M acetic acid, pH 3) and the eluent switched from this first column to an anion-exchange HPLC column when the niacin band eluted. The mobile phase used with this first reverse-phase column ensured that the niacin band-cut was retained at the top of the anion-exchange column. The niacin was subsequently chromatographed on the anion exchanger and eluted using another mobile phase (0.4 M acetic acid, pH 3). Only one chromatogram of a millet sample extract is provided by the authors but niacin (nicotinic acid) appears to be reasonably well resolved from interferences after about forty minutes. The authors state that this procedure would have an advantage over MBAs when a small number of samples are to be analysed but for a large number of samples the method could only compete with MBA if an autosampler and overnight operation was employed.

Hirayama and Maruyama (1991) reported an HPLC procedure designed to determine small amounts of niacin in foodstuffs. Alkali hydrolysates were concentrated by rotary evaporation and chromatographed on an anion-exchange column (laboratory prepared), the niacin eluted and concentrated by rotary evaporation. The solubilised residue was chromatographed on a commercial cation-exchange cartridge column and again the eluted niacin fraction was concentrated by rotary evaporation. Analytical HPLC was performed on this last concentrate using an amino bonded column. A chromatogram of a strawberry jam extract is provided by the authors (nicotinic acid content 0.07 mg / 100 g) which shows reasonable separation of nicotinic acid from interferences but the procedure is lengthy and not convenient for large sample numbers because of the intensive sample extract 'clean-up' prior to analytical HPLC.

At the LGC the use of C_{18} cartridges for 'clean-up' of acid hydrolysates of cereal and milk samples followed by analytical HPLC on a strong cation-exchange column with a phosphate buffer mobile phase has been investigated. Interferences were found to have been removed prior to analytical HPLC; this can be partially achieved using C_{18} cartridges but recovery of nicotinic acid is only about 80%. More efficient clean-up procedures need to be developed.

8.3.4.4 Summary. It is apparent from the literature that several approaches have been taken to the HPLC determination of nicotinic acid in foodstuffs but most published procedures only describe applications to a very limited number of sample types. 'Clean-up' of sample hydrolysates is often necessary and time consuming and analytical HPLC is not always capable of completely resolving nicotinic acid from interferences.

As stated in section 8.3.4.1, HPLC cannot yet be considered as generally applicable or suitable for routine use for the determination of nicotinic acid in all sample types. If HPLC is to be attempted for the determination of niacin the analyst must consider the following:

1. Is the extraction procedure quantitatively releasing niacin from the sample matrix in a form which will be measured by the HPLC procedure to be used?
2. What clean-up procedure(s) will remove interferences from the sample extract but allow acceptable recovery of niacin; we should consider ion-exchange, reverse-phase and perhaps polar bonded phases?
3. Will reverse-phase with ion pairing or ion-exchange provide the best analytical mode of HPLC to resolve niacin from interferences?

These questions involve significant analytical effort if meaningful answers are to be obtained and results for niacin in foodstuffs are to be valid.

8.3.4.5 MBA of niacin. Niacin and nicotinamide are extracted from foodstuffs using acid under autoclave conditions and after pH adjustment to 4.5, filtration and dilution an MBA is performed using the test organism *Lactobacillus plantarum* (ATCC 8014, NCIMB 6376) (Bell, 1984; AOAC, 1990e). The typical assay range is 0–125 ng niacin per assay tube, and incubation at 37 °C for 22–24 h is usual. Niacin MBAs are usually the the most reproducible and reliable of all the vitamin assays.

8.3.5 Vitamin B_6

8.3.5.1 Introduction. Vitamin B_6 is found in biological tissues in six free forms, namely pyridoxamine (PM), pyridoxine (PN), pyridoxal (PL), pyridoxamine phosphate (PMP), pyridoxine phosphate (PNP) and pyridoxal phosphate (PLP). The relative amounts of these vitamers found in foodstuffs can vary significantly depending upon the type of food, for example in cereals, vegetables and fruits pyridoxine (PN) predominates, whereas in meats and dairy products pyridoxamine (PM), pyridoxal (PL) and their phosphates (PMP and PLP) are the major contributors to vitamin B_6 activity.

The vitamers are unstable if exposed to UV light and PL and PLP are unstable in alkaline conditions, but all are stable under acid conditions. Interconversions between the vitamers can occur due to the presence of a range of enzymes in fresh foodstuffs, and conversion of PL and PLP to PM and PMP by transamination can occur during heating or cooking of foodstuffs which contain high levels of protein,

e.g. meats. The vitamers may be bound to proteins in some foodstuffs which may or may not make them biologically available to man, for example PLP is often bound to the ε-amino group of lysine whilst PN can be present as a glucoside in plant products. It has been reported that B_6 vitamers as 5'-o-β-D-glucosides and conjugated forms are biologically available to the rat but they appear to be generally unavailable to man.

These factors and more must be borne in mind when the analyst is considering extraction procedures for the vitamin B_6 vitamers which are compatible with the method of determination to be used and the information required. Microbiological methods for the determination of vitamin B_6 using the yeast *Saccharomyces carlsbergensis* have been used for many years and still provide analysts with a relatively robust means of assessing the vitamin B_6 content of a wide variety of foodstuffs and feeds and where large numbers of analyses are required. One disadvantage of the MBA procedure is that *S. carlsbergensis* does not respond to PM to the same extent as PN and PL and therefore care has to be exercised in calibration of the assay for samples which contain appreciable amounts of PM, i.e. meats and dairy products. Another disadvantage is that without using three different assay organisms with differing growth requirements (which is lengthy and difficult) MBAs using the test organism *S. carlsbergensis* provide only a total vitamin B_6 content; we cannot obtain information on the individual vitamers present. Over the last fifteen or so years, HPLC procedures designed to separate the naturally occurring forms of B_6 have been developed and published but it is only in the last five or six years that methods suitable for routine use and liable to provide reproducible results have become available. Despite these advances in HPLC methodology there remain many potential pitfalls and problems to confront the unsuspecting analyst.

8.3.5.2 Sample extraction. Vitamin B_6 has traditionally been extracted from foodstuffs prior to MBA using HCl or H_2SO_4, but the strength of the acid, hydrolysis time and temperature often differ depending upon the reference consulted. The AOAC (1990f) prescribes the use of 0.44 M HCl for 2 h at 121°C for plant products, and 0.055 M HCl for 5 h 121°C for animal products. Toepfer and Polansky (1970) prescribe the same conditions but with the use of 0.2 M HCl for food composites for 5 h at 121°C.

At the LGC we use 0.22 M H_2SO_4 for 5 h at 121°C as prescribed by Bell (1984) prior to the microbiological determination of vitamin B_6. Addo and Augustin (1988) have published information on the effect of acid strength and autoclave time on the extraction of vitamin B_6 from potatoes; 2 h using 0.44 M HCl appeared to show adequate extraction, but the analyst should be aware that these conditions may not be adequate for all sample types.

Sample preparation prior to vitamin B_6 extraction by acid hydrolysis and determination by MBA must be performed under conditions which will conserve total B_6 but problems of interconversion of vitamers is not a considerable problem because the analyst is measuring total vitamin content.

Acid hydrolysis will liberate B_6 vitamers from glycosides (the degree of liberation depending on hydrolysis conditions) and therefore we may overestimate the B_6 which is biologically available to man in some types of foodstuff. An alternative procedure is to use a buffer extraction or a more dilute acid for a shorter period of time followed by enzymatic hydrolysis but the analyst needs to be satisfied that this procedure is releasing the optimum amount of the vitamin. If the analyst is to use HPLC to determine individual B_6 vitamers, steps must be taken to ensure that residual enzyme activity of a foodstuff will not catalyse changes in vitamer profile and that the extraction used similarly does not induce changes. For example if a low pH is used during extraction, particularly for meat samples, PL and PLP will convert to PM and PMP (transamination) but not quantitatively, therefore this will cause problems with quantitation of individual vitamers by HPLC. If HPLC is to be used to separate and quantify PN, PM, PL, and their phosphate esters, conditions which will induce hydrolysis of the esters must be avoided, however if the HPLC procedure is designed to separate and determine PN, PM, and PL only, the analyst must use extraction conditions which quantitatively hydrolyse the esters without affecting the relative proportions of PN, PM, and PL.

A tentative HPLC procedure recommended by COST 91 (Brubacher *et al.*, 1985) involves autoclaving the sample (121°C) for 30 min with 0.1 M H_2SO_4 prior to the separation of PM, PL and PN, although the majority of recently published procedures prescribe extraction with perchloric (PA), trichloroacetic (TCA), sulfosalicylic (SSA) or metaphosphoric (MPA) acids. These acids release bound vitamin B_6 from foods but do not hydrolyse phosphate esters under the prescribed conditions of use. After pH adjustment of these acid extracts enzyme preparations can be used quantitatively to hydrolyse B_6 phosphate esters to allow HPLC determination of PM, PL and PN. The analyst must ensure that the enzyme preparations used are sufficiently active to achieve this quantitative hydrolysis; this potential problem has already been discussed for thiamin and riboflavin extraction.

8.3.5.3 HPLC of vitamin B_6. UV detection is not sufficiently sensitive or selective for the detection and determination of B_6 vitamers in foodstuffs (unless they are fortified) so fluorescence detection must be used. Vanderslice and Maire (1980) described an ion-exchange HPLC procedure with fluorescence detection for the determination of all B_6 vitamers but the procedure is complicated by a complex HPLC system which involves column switching and the changing of excitation and emission wavelengths during chromatography to account for the differing fluorescent characteristics and intensities of the vitamers. Various forms of derivatisation have been used to enhance the fluorescence of B_6 vitamers and Coburn and Mahuren (1983) reported the use of post-column reaction with bisulphate ion which allowed sensitive detection of all vitamers after ion-exchange separation. This post-column procedure has found favour with many analysts and has been applied with success to food extracts.

Table 8.14 Vitamin B_6 reported HPLC conditions

Reference	Extraction	Column and mobile phase	Detection
Gregory and Feldstein (1985)	SSA	3 μm C_{18} 3 cm × 4.6 mm	Post-column addition, of 1M KH_2PO_4 + 1 mg/ml sodium bisulphite pH 7.5, 0.2 ml min^{-1} into column eluent Ex 330 nm Em 400 nm
Gregory and Ink (1987)		Solvent A1: 0.033 M KH_2PO_4, 8 mM octanesulphonic acid, pH 2.2 Solvent A2: 0.033 M KH_2PO_4, 8 mM octanesulphonic acid, 2.5% propan-2-ol, pH 2.2 100% A1 to 100% A2 in 12 min, after 15 min switch to 100% B Total flow rate 1.8 ml min^{-1}	
Addo and Augustin (1988)	SSA	Radial compression C_{18} 10 cm × 8 mm 4 μm, Solvent A: 4 mM octane and heptane sulphonic acids in water: propan-2-ol (98.5: 2.5), pH 2.2. Solvent B: 0.33 M phosphoric acid in (82.5:17.5) water: propan-2-ol, pH 2.2. Solvent B: 1 h, 10 min before injection Solvent A, 3 min after injection solvent B.	Post column addition of 1 M KH_2PO_4 + 1 mg ml^{-1} sod. bisulphite 0.3 ml min^{-1} Ex 338 nm Em 425 nm
Ang et al. (1988)	MPA	C_{18} BioSil ODS-SS, 25 cm × 4 mm. Mobile phase 0.066 M ICH_2PO_4, pH 3, 1 ml min^{-1}	Fluorescence Ex 290 nm Em 395 nm
Bitsch and Moller (1989)	PA	LiChrospher RP-18, 5 μm, 12.5 cm × 4 mm Solvent A: methanol Solvent B: 0.03 M phosphate buffer, pH 2.7 + 4 mM octanesulphonic acid. 90% B at 2 min to 60% B at 12 min: 60% B and 40% A 12 to 17 min, 60% B 90% B 17 to 19 min. Flow 1.5 ml min^{-1}	Post-column addition of 0.5 M phosphate buffer, pH 7.5 + bisulphite 3.7 mg ml^{-1} 0.07 ml min^{-1} Ex 330 nm Em 400 nm
Sampson and O'Connor (1989)	MPA	Ultremex C_{18} 15 cm × 4.6 mm 3 μm Solvent A: 0.033 M phosphoric acid 0.008 M 1-octanesulphoric acid, pH 2.2 Solvent B: 0.033 M phosphoric acid, 10% (v/v) propan-2-ol pH 2.2 sample injection then linear gradient to 100% B in 10 min, 100%B for 15 min, linear to 100% A in 4.5 min, 5.5 min 100% A.	Post-column addition of 1 M potassium phosphate buffer pH 7.5 + 1 mg sodium bisulphite 0.12 ml min^{-1} Ex 330 nm Em 400 nm

Ion-exchange HPLC has the disadvantage of long retention times for the B_6 vitamers with concomitant effects on chromatographic efficiency and complex mobile phase gradients usually have to be used which are difficult to reproduce.

Gregory and Feldstein (1985) used reversed-phase HPLC with post-column bisulphite reaction to form the highly fluorescent hydroxy sulphonate derivative of PLP and PL and permitted the determination of PMP, PM and PN at the same excitation and emission wavelengths. Samples of plant and animal tissue were extracted with sulfosalicylic acid (SSA) and 4-deoxypyridoxine (4-dPN) was added as an internal standard. Sample extracts were purified on a preparative ion-exchange chromatographic system to remove SSA prior to injection onto the analytical HPLC column (see Table 8.14 for HPLC conditions). Elegant separations of PLP, PMP, PL, PM, PN and 4-dPN were achieved from various food extracts. The sample extract purification on the semi-preparative system is mainly to remove SSA which interferes with the fluorescence detection; the authors state that the use of trichloroacetic acid (TCA) or perchloric acid (PA) many allow direct HPLC analysis of sample extracts. In a subsequent paper Gregory and Ink (1987) extended the application of the HPLC procedure to include the determination of pyridoxine-β-glucoside which occurs in samples of plant origin. A β-glucosidase treatment was used to prove peak identity in addition to NMR spectrometry. Again elegant chromatograms were published.

Addo and Augustin (1988) examined the vitamin B_6 content of potatoes during storage and extracted total B_6 by acid hydrolysis (including B_6 from glucosides), by SSA extraction and by buffer extraction of unbound B_6. MBA was performed on buffer, acid and SSA extracts (SSA including hydrolysis with acid phosphatase and β-glucosidase), and HPLC was performed on SSA extracts; interesting comparisons of results were made. The SSA extracts were purified by ion-exchange chromatography prior to reverse-phase HPLC (similar to the work of Gregory and Feldstein, 1985 and Gregory and Ink, 1987) but a binary mobile phase was used to separate PLP, PM, PN-glucoside, PL, PN, 4-dPN and DM (see Table 8.14).

Ang *et al.* (1988) used metaphosphoric acid (MPA) for extraction of B_6 vitamers from raw and cooked chicken with direct injection onto a C_{18} analytical column using an isocratic phosphate buffer for the mobile phase and fluorescence detection with no post-column enhancement of fluorescence. Chromatographic separation of vitamers from interferences in sample extracts appears to be just adequate for purpose, but PLP and PL were not completely resolved from an unidentified peak and PMP and PM were not totally resolved. The procedure has the advantage of simplicity when compared to the extraction and chromatographic conditions used by Gregory *et al.* The described procedure may be just suitable for extracts of chicken but care would have to be exercised if applied to other foodstuffs because of the potential for coelution of interferences with B_6 vitamers.

Bitsch and Moller (1989) extracted foodstuffs with cold perchloric acid and after pH adjustment dephosphorylated a portion of the extract with alkaline phosphatase. Chromatography of all vitamers or PL, PM and PN after enzyme treatment was achieved by injection onto a C_{18} column using a binary gradient system with ion pairing. Post-column reaction with bisulphite was employed and 4-deoxypyridoxine hydrochloride (4-DPN) used as internal standard. All vitamers were separated, including 4-pyridoxic acid (product of B_6 metabolism), with the exception of PNP, but this occurs only in small amounts in foodstuffs. Vitamers were chromatographed within fifteen minutes and the authors state that the lowest detectable amount of substances separated corresponds to 0.4–0.7 pmol. The separation of PLP, 4-PA and PMP from interferences in food extracts may cause problems and so peak identity was confirmed by enzymatic dephosphorylation and subsequent HPLC analysis.

Sampson and O'Connor (1989) extracted biological tissue with metaphosphoric acid and used a binary gradient system with a C_{18} column and 4-DPN as an internal standard and obtained good resolution of all B_6 vitamers of interest within thirty minutes. Resolution of vitamers appears sufficiently good to warrant use for food extracts although the authors did not report this.

8.3.5.4 Summary. As stated earlier, there are many potential problems associated with the determination of B_6 vitamers by HPLC and the analyst must take great care to select the most appropriate sample extraction and HPLC conditions for the sample types to be examined. Fairly complex mobile phase gradients are required if the analyst wishes to attempt to separate all vitamers from sample matrix interferences prior to quantitation, but quantitation of PL, PM and PN after enzymatic desphosphorylation of the sample extract provides less complex chromatograms and slightly relaxes the need for complex mobile phases. The choice of approach will obviously depend upon the reason for obtaining the B_6 content of the foodstuff and the degree of detail necessary.

If an analyst wishes to determine vitamin B_6 vitamers by HPLC the following points should be borne in mind:

1. The extraction procedure: MPA, PA or TCA are better than SSA because they do not interfere with fluorescent detection.
2. An internal standard should be used.
3. What vitamers are to be measured? If PL, PM and PN only are to be determined, enzymatic dephosphorylation of the sample extract may provide more simple chromatograms. If this approach is taken the efficiency of enzyme dephosphorylation must be monitored.
4. Reverse-phase HPLC with post-column reaction with bisulphite provides the most reproducible and sensitive method of analysis.
5. Analysts should be aware of the possible presence of pyridoxine-β-glucoside in extracts of plant origin.

8.3.5.5 MBA of vitamin B_6. As mentioned earlier, the test organism usually used to determine 'total' vitamin B_6 in foodstuffs is the yeast *Saccharomyces carlsbergensis*. The organism is maintained on agar slopes and when required a subculture is taken and the inoculum grown overnight at 26 °C in a shaking water bath in the assay medium containing a controlled amount of vitamin B_6. Sample extraction is usually by acid under autoclave conditions and after pH adjustment to 4.5, filtering and dilution, the assay is set up covering the range 0–10 ng B_6 per assay tube. Incubation is at 27 °C for 20 h in a shaking water bath. A glass bead is added to each assay tube to assist the agitation (Bell, 1984).

One disadvantage is that *S. carlsbergensis* does not respond to PM to the same extent as to PN and PL. If samples contain appreciable amounts of PM (e.g. meats) then this can be used to prepare the calibration line instead of PN.

Differential assays using other organisms can be performed but they are labour intensive. *Saccharomyces faecalis* responds to PL and PM, *Lactobacillus casei* responds to PL and neither respond to PN, thus by using results obtained from these two organisms and *S. carlsbergensis* it is possible to calculate the amount of each vitamer present in a sample. Other workers have used *S. carlsbergensis* after chromatographic separation of the B_6 vitamers.

8.3.6. Folates

8.3.6.1 Introduction. This term 'folate' refers to a group of compounds based on pteroylglutamic acid or folic acid. The biologically active forms are derivatives of folic acid reduced at the 5, 6, 7 and 8 positions (5, 6, 7, 8-tetrahydrofolate) and these compounds occur conjugated with one or more (usually 5–7) γ-glutamyl residues. The total folate content of the majority of foodstuffs is very low, (e.g. about 6 µg / 100 g in milk, 15 µg / 100 g in beef and 60 µg / 100 g in lettuce), therefore determination of total folate is very difficult and determination of individual folates more difficult still. Microbiological assay is the most common way to determine the total folate content of foodstuffs because of the sensitivity of the technique. HPLC procedures have been published but the methods are not yet widely applicable because of the very low detection levels required and the associated problems of interferences from the sample matrix. Assays using folate binding protein (FBP) have been used but the performance of these assays on wider ranges of samples needs to be further assessed before performance can be fully characterised and validated.

Folates are very labile and therefore care has to be taken when extracting the vitamins from foodstuffs, also enzyme deconjugation of the folates is usually required so that the analyst must be sure that the enzyme preparation used is suitable for the assay to be performed. A large number of papers have been published giving details of the determination of folates in food and this review can only discuss some of the more recent or significant developments. It is worth stressing again that the reliable determination of folates is extremely difficult and the analyst must be familiar with all the relevant literature if this determination is to be performed.

8.3.6.2 Extraction of folates from foodstuffs. Folates in plant and animal tissues are usually tetrahydrofolic acid (THF), 5-methyl tetrahydrofolic acid (5-Me-THF), 10-methyl-tetrahydrofolic acid (10-Me-THF) and 5-formyl tetrahydrofolic acid (5-CHO-THF). Folic acid is probably not significant in biological tissue but may be present in sample extracts as a result of the oxidation of THF. These folates will be present as polyglutamates as mentioned earlier. Folates are usually extracted from foodstuffs using buffers at neutral or slightly acid pH at 100°C and 121°C for 3 through to 60 min. An antioxidant must be present during extraction and other critical steps in the analytical protocol to prevent degradation of the labile folates. Ascorbic acid is the most common antioxidant but Wilson and Horne (1983) reported some interconversion and degradation of folates when using a neutral buffer extraction (pH 7.4) at 100°C for 10 minutes and attributed this degradation to ascorbate. These workers later reported the use of a buffer containing ascorbate and 2-mercaptoethanol (pH 7.85) which overcame these problems (Wilson and Horne, 1984); extraction was at 100°C for 10 min. In this buffer 10-CHO-THF was preserved. Schulz et al. (1985) reported that they did not detect any degradation of 5-Me-THF when using sodium ascorbate in Tris–HCl buffer (pH 8) when heating for 60 min at 100°C and so the analyst should bear these reports in mind. Ascorbate is more stable at acid pH and this may account for the greater stability observed for folates when extracting solutions with a pH of less than 7 are used. Gregory et al. (1984) prescribed the use of a 0.05 M sodium acetate buffer at pH 4.9 containing 1% of ascorbic acid with nitrogen flushing and 100°C for 60 min for extraction of folates prior to HPLC; in a recent paper Gregory et al. (1990) compared this extraction with the Wilson and Horne (1984) extraction and concluded the pH 7.85 buffer was superior for MBA of total folate in beef, liver and peas.

Prior to MBA the AOAC (1990,) prescribes blending samples with water (> 10 times dry weight of sample) and 2 ml of ammonium hydroxide and autoclaving for 15 min at 121°C, whilst Bell (1984) prescribes phosphate–ascorbate buffer (0.2 M) pH 6.1 at 121°C for 15 min.

There are many extraction procedures prescribed in the literature and analysts must satisfy themselves, probably by experiment, that the procedure that they are using is suitable for the samples to be analysed and the method of analysis to be used. Gregory et al. (1984) selected their extraction conditions so that all of the 10-CHO-THF was converted to 5-CHO-THF which simplified HPLC chromatograms and allowed measurement of total CHO-THF as one peak which was not possible with shorter extraction times or at other pHs. The pH 7.85 extraction medium of Wilson and Horne (1984) preserves 10-CHO-THF, thus the analyst must select the extraction conditions carefully. Methods should not just be abstracted from the literature and used without careful consideration of many factors.

As stated previously, folates occur as polyglutamates, and most analytical procedures require that these are hydrolysed to mono-or di-glutamates prior to measurement. *Lactobacillus casei*, the most commonly used microorganism for MBA of folates, has been reported to respond equally to folates having mono-, di-or tri-glutamate residues, but much less so to higher conjugates (Tamura et al., 1972) and the number of glutamic acid residues drastically affects the retention of folates on HPLC columns and perhaps avidity for folate binding proteins, making deconjugation necessary prior to assay.

Chicken pancreas and hog kidney are the most common sources of folate deconjugase enzymes (γ-carboxypeptidase) and enzymes can be isolated from fresh tissue or from acetone powders of the tissues which are commercially available. It is generally accepted that chicken pancreas produces diglutamates whilst hog kidney deconjugase produces monoglutamates. Kirsch and Chen (1984) compared the use of fresh hog kidney and chicken pancreas with each other and dried chicken pancreas prior to MBA and optimised conditions of buffer type, pH, incubation time and other variables. No significant difference in results was observed between the three sources of deconjugase enzyme. Pedersen (1988) performed a similar study and also concluded that results from hog kidney and chicken pancreas deconjugases gave similar results on a range of foodstuffs. These results are not in agreement with those reported by Phillips and Wright (1983) who reported that pig kidney deconjugase would result in higher results than those obtained by chicken pancreas. Martin *et al.* (1990) investigated the use of chicken pancreas deconjugase, α-amylase and pronase mixtures prior to MBA and reported significant increases in folate yield for some sample types (mean increase 19%) while Engelhardt and Gregory (1990) reported on the possible inhibition of hog kidney deconjugase by food extracts, and on the conditions of incubation time and enzyme concentration to overcome these effects.

Goli and Vanderslice (1992) recently reported on the use of human plasma and chicken pancreas deconjugases and the effects of folate glutamates on the growth response of *L. casei*. The authors reported that in their experiments the relative *L. casei* growth response to pteroyl di- and tri-glutamates were 88% and 59%, respectively of the response to the monoglutamate in the upper half of their calibration line, which are not in agreement with the results of Tamura *et al.* (1972). This evidence, plus the fact that we require folates in the monoglutamate form for reliable HPLC, highlights the analyst's need to appreciate the products of the deconjugase treatment used. The authors report that plasma deconjugase treatment at pH 4.5 results in substantial degradation of folate but at pH 6–7 degradation is not so marked; the end product is a monoglutamate at pH 4.5 and 6.0, while a small amount of diglutamate is present at pH 7. Chicken pancreas yielded mono-and di-glutamates at pH 6.0 and 7.0. The authors conclude that human plasma, chicken pancreas and hog kidney deconjugase treatments each have their own inherent problems and that no entirely satisfactory procedure can yet be developed for deconjugation of folates prior to MBA or HPLC analysis of total folates.

8.3.6.3 HPLC of folates. HPLC separation of various folates has been reported since the mid 1970s using anion-exchange, ion-pair and reverse-phase mechanisms, but methods suitable for routine application to foodstuffs are rare. A selection of HPLC conditions reported in the literature are shown in Table 8.15.

Day and Gregory (1981) used C_{18} and phenyl columns in series with an isocratic mobile phase of 0.033 M phosphate buffer (pH 2.3)–acetonitrile (90.5:9.5) and UV detection to determine THF, 5-Me-THF and 10- and 5-CHO-THF in beef liver, but interferences were experienced when cabbage was examined. Gregory *et al.* (1984) later used a phenyl analytical column and a linear 15 min gradient of 7.2 to 11.3% acetonitrile in a mobile phase of 0.033 M phosphate buffer (pH 2.3) with fluorescence detection (Ex 295 nm, Em 356). Sample extracts from rat liver, cabbage,

Table 8.15 Conditions reported for the HPLC of folates

Reference	Column and mobile phase	Detection
Day and Gregory (1981)	μ Bondapak phenyl (3.9 mm × 30 cm) 0.033 M phosphate buffer, pH 2.3, linear gradient of CH_3CN from 7.2 to 11.3%	Fluorescence Ex 295 nm Em 356 nm
Holt et al. (1988)	Rainin C_{18} microsorb short-one 10 cm × 4.6 mm, 3 μm Phosphate buffer, pH 6.8/methanol + tetrabutylammonium ion (84: 16) or 0–20% methanol gradient	Fluorescence Ex 238 nm Em >340 nm Post-column pH adjustment
Gounelle et al. (1989)	Spherisorb ODS, 5 μm, 30 cm × 3.9 mm 30 mm phosphate buffer pH 2/CH_3CN (89.5:10.5)	Fluorescence Ex 300 nm Em 360 nm
White et al. (1991)	Waters Nova-Pak C_{18} 3.9 mm × 7.5 cm, 4 μm Zorbax ODS 4.6 mm × 25 cm 5 μm A: 10% methanol in phosphate/acetate buffer (pH 5) + 0.005 M TBAP B: as above but 30% methanol (TBAP = tetrabutylammonium phosphate) column switching	Electro-chemical

milk and orange juice were partially purified by chromatography on a small column of DEAE-Sephadex A-25 prior to analytical HPLC. The excitation and emission wavelengths had to be changed after elution of THF and 5-Me-THF and before elution of 5-CHO-THF because of the relatively weak fluorescence of the latter vitamer. Silica pre-columns were used to protect the analytical column from the acidic mobile phase.

Holt et al. (1988) used reverse-phase HPLC (C_{18}, 10 cm column, 3 μm) and a phosphate buffer (pH 6.8)–methanol mobile phase containing tetrabutylammonium dihydrogenphosphate as an ion-pair reagent to determine 5-Me-THF in milk. 16 % methanol was used in the mobile phase as only 5-Me-THF was found in milk but a methanol gradient was required (0–20 %) to separate other folates. As fluorescence of the folates is dependent upon pH post-column pre-detector addition of phosphoric acid to the column eluent to achieve a pH of less than three was necessary to achieve required sensitivity.

Gounell et al. (1989) examined the fluorescence intensity and reverse-phase chromatographic retention of folates at various pHs and selected a pH 2 mobile phase for optimum separation and detection of THF, 5-Me-THF and 5-CHO-THF in rat liver extracts. Folates were extracted from sample extracts using an anion-exchange cartridge prior to analytical HPLC. The authors state that the native fluorescence of 5-Me-THF is twice as strong as that of THF and ten times stronger than that of 5-CHO-THF.

White et al. (1991) used column switching and electrochemical detection to determine 5-Me-THF in citrus juice. Folate was retained on the first C_{18} column using a buffer–methanol (90:10) mobile phase containing an ion-pair reagent and then eluted onto the analytical HPLC column using a mobile phase containing 30%

methanol. Several other workers have used electrochemical detection for folates but applications to food analyses are scant. However, the approach is worth further consideration because of potential sensitivity, although it may not be as selective as fluorescence detection.

The commercial folate standards used for quantitation are of variable purity and it is advisable to check the concentration of prepared standards using appropriate molar absorptivity values. Prepared standards must contain antioxidant, e.g. ascorbic acid, and should be nitrogen flushed after use and before restorage at deep freeze temperature, i.e. - 20°C.

8.3.6.4. Summary. Although there are reported HPLC procedures for the determination of folates, there are many problems to be solved before this approach can be used reliably and routinely. Sensitivity and selectivity are problems and improvements in semi-preparative sample extract 'clean-up' and pre-concentration of folates, chromatography and detection are required. Selective absorption and concentration using immobilised folate binding protein is a promising area for further study. HPLC determination of folates is a fertile area for study.

8.3.6.5 MBA of folates. Folates are usually extracted prior to MBA using a buffer solution containing antioxidant under reflux or autoclave conditions. Samples containing starch can form gels during this extraction stage and so inclusion of an incubation step with amylase is advantageous. Incubation with a deconjugase preparation (as previously discussed) is necessary prior to dilution and setting up of the assay. The test organisms *Streptococcus faecalis* (ATCC 8043, NCIB 6459) or *Lactobacillus casei* (ATCC 7469, NCIB 6375) can be used; the assay ranges are 0–5 ng and 0–1 ng folic acid per assay tube, respectively. *L. casei* is most commonly used because of the increased sensitivity obtainable. Incubation is at 37°C for 22–24 h if using *L. casei* (Bell, 1984; AOAC, 1990g). Because of the sensitivity of folate assays they can be very difficult to perform. All reagents, glassware, equipment and the laboratory must be scrupulously clean and staff must be thoroughly trained. Folate assays are susceptible to negative and positive drift caused by compounds in the sample extract which inhibit or stimulate the growth of the test organism in comparison to the standard curve. The cause of these effects is often difficult to identify and isolate.

As stated earlier, *L casei* shows reduced response to folates containing more than three glutamate residues and therefore MBAs include treatment with a deconjugase enzyme to yield mono- and di-glutamate. Tamura *et al.* (1972) reported *L. casei* response to mono-, di- and tri-glutamates to be equal but recently Goli and Vanderslice (1992) reported that the growth response to the di- and tri-glutamate were 88% and 59%, respectively of the response to the monoglutamate. If this is the case use of a conjugase which produces monoglutamates is to be recommended.

Wright and Phillips (1985) reported studies in three papers on the growth response of *L. casei* to different folate monoglutamates and reported that at commonly used assay concentrations 5-Me-THF may elicit only half the growth response as folic acid. This obviously has significant consequences for the determination of folate in foods where the major source of the vitamin is 5-Me-THF. The authors reported that changing the pH of the assay medium from 6.8 to 6.2 resulted in an equal response from the folates. Goli and Vanderslice (1989) recently reported different responses for various forms of folate but they found that the effect of the medium pH on the relative growth rates of folic acid and 5-Me-THF was not as pronounced as reported by Wright and Phillips. Clearly there is a need for further investigation of these reports.

8.3.7 Vitamin B_{12}

Vitamin B_{12} is a member of a group of compounds known as cobalamins and exists in foodstuffs mainly in coenzyme forms which are wholly or partly bound to cellular protein constituents. Good sources of the vitamin are offal meats followed by muscle meats, dairy produce and fish. Vitamin B_{12} is present in foodstuffs at very low levels and MBAs are the only way to estimate the B_{12} content of food satisfactorily.

Numerous extraction procedures have been proposed but a common approach involves addition of an acetate buffer to the sample followed by a small amount of sodium cyanide and immersion of the mixture in a boiling water bath for thirty minutes. After cooling, the mixture is diluted to a suitable volume and filtered. The cyanide converts the unstable cobalamins into cyanocobalamins which are more stable and more closely match cyanocobalamin which is used to construct the assay calibration line.

The choice of test organism for the determination of vitamin B_{12} has been a topic of great debate over the years. The two most commonly used are *Ochromonas malhamensis* and *Lactobacillus leichmannii*. *O. malhamensis* has a greater specificity for cobalamins than *L. leichmannii* and is claimed to provide a truer estimate of B_{12} biological activity. Deoxyribosides and B_{12} analogues can act as a growth factor for *L. leichmannii* but these materials are not usually present at significant levels in most foodstuffs; they mainly occur in fermentation broths. The effects of these compounds on the results of an *L. leichmannii* assay can be estimated by taking an aliquot of the sample extract, adjusting the pH to 12, autoclaving for thirty minutes (121°C), and assaying alongside the untreated sample extract. The high pH and autoclave conditions will destroy vitamin B_{12} and so the difference in assay results indicates non-B_{12} stimulation of the assay organism. *O. malhamensis* is photosynthetic and assays must be performed under appropriate conditions of illumination for up to five days, whereas *L. leichmannii* assays are incubated for only 22–24 h at 37 °C; as a result *L. leichmannii* assays are most commonly employed. The assay range for *L. leichmannii* is typically 0–0.2 ng cyanocobalamin per assay tube (Bell, 1984; AOAC, 1990L). Because of this high sensitivity, reagents and apparatus must be even more carefully prepared than those described for folate assays; all precautions necessary cannot be detailed here.

8.3.8 Pantothenic acid

Pantothenic acid is present in a wide range of plant and animal tissue and its name is derived from the Greek word meaning 'everywhere'. Pantothenic acid is the vitamin moiety of coenzyme A and to obtain the total pantothenic acid activity of foodstuffs it is necessary to hydrolyse the coenzyme. Because of instability to acid and alkali, enzymatic hydrolysis must be used. Many enzymes have been investigated, but it was not until Novelli *et al.* (1949) reported on the use of a double enzyme treatment using alkaline phosphatase and an enzyme preparation from avian liver that pantothenic acid could be liberated quantitatively from coenzyme A. Both enzymes can be purchased as extract powders but it is advisable to purify the avian liver extract before use (Bell, 1984). Extraction of samples is usually performed using Tris buffer under autoclave conditions followed by incubation overnight with alkaline phosphatase and purified pigeon liver enzyme preparation. The test organism is *Lactobacillus plantarum* (ATCC 8014, NCIB 6376) and the assay range typically 0–100 ng pantothenic acid per assay tube; incubation is performed at 37°C for 22–24 h (Bell, 1984; AOAC, 1990i).

8.3.9 Biotin

As with many of the B-group vitamins, biotin occurs in foodstuffs mainly in a bound or conjugated form such as biocytin and enzyme complexes in which it is attached to proteins and peptides. Biotin must be liberated from these bound forms by hydrolysis and this is usually achieved using acid. Some procedures recommended autoclaving with 3 M H_2SO_4 for 1 h or longer but this can lead to some degradation of biotin. Bell (1984) recommended a 30 min extraction with 1.5 M H_2SO_4 at 121 °C, which produced consistent results on a range of samples.

Saccharomyces cerevisiae has been used to determine biotin but it was found that it responded to some compounds related to biotin in addition to biotin itself. The test organism now most commonly used is *Lactobacillus plantarum* (ATCC 8014, NCIB 6367). The assay range is typically 0–0.25 ng per assay tube, and incubation at 37°C for 16–20 h is generally satisfactory for turbidometric measurement of bacterial growth.

Fatty acids can stimulate growth of *L. plantarum* but they should be removed by the precipitation and filtration step after autoclaving. Solvent extraction of sample extracts prior to setting-up the assay can be used if necessary.

8.4 Vitamin C

8.4.1 Introduction

Vitamin C consists of two biologically active substances L-ascorbic acid and L-dehydroascorbic acid, both of which may be present in foodstuffs. The most abundant form is ascorbic acid, but dehydroascorbic acid will be present if storage or processing condition have allowed oxidation to take place; cooking and holding food hot can also increase the dehydroascorbic acid content. Many methods are available for the determination of ascorbic acid, but if the analyst wishes to determine 'total vitamin C' then ascorbic and dehydroascorbic acids should be determined.

8.4.2 Extraction of vitamin C

Ascorbic acid (AA) is easily oxidised to dehydroascorbic acid (DHAA) which further degrades to the non-biologically active diketogulonic acid, therefore extraction conditions must be selected to minimise this oxidation and degradation. Raw foodstuffs may contain oxidases which will degrade ascorbic acid during sample preparation, and trace elements present will catalyse oxidation. Extraction is usually best performed in a high speed blender with an appropriate extracting solution. Acids such as oxalic, trichloroacetic, acetic, citric, metaphosphoric and mixtures of metaphosphoric and acetic, and solutions of EDTA and homocysteine have been used for extraction. Metaphosphoric–acetic acid mixtures are the most commonly used and sulphuric acid may be included for samples which contain appreciable amounts of basic substances; it is important that low pH is maintained to ensure stability of AA.

It is essential that samples are blended with extracting solution immediately after any preliminary sample preparation to avoid oxidation of the vitamin, and if analysis is not to be performed immediately extracts must be stored in the dark at freezer temperatures overnight.

8.4.3 Determination of vitamin C

8.4.3.1 Chemical methods. Many chemical methods are available for the determination of AA, most of which utilise the reducing power of the vitamin. Of the many reagents employed, 2,6- dichlorophenolindophenol (2,6-DCP) is by far the most popular because it is easy to use and provides a rapid estimate of the AA in a sample using simple equipment. 2,6-DCP is deep blue in colour, but colourless when reduced, therefore fixed volumes of sample extract are titrated until a rose pink colour is obtained indicating a slight excess of 2,6-DCP. The 2,6-DCP is standardised by titration against solutions of AA of known concentration.

The AOAC (1990i) recommends extraction of samples with metaphosphoric–acetic acid; if the pH of the extract is greater than 1.2 appreciable amounts of basic substances are present and a metaphosphoric–acetic–sulphuric extracting solution must be used. Fixed volumes of these extracts are then titrated against standardised 2,6-DCP until a pink colour persists for more than five seconds.

There are many substances which can interfere with the 2,6-DCP titration such as iron, copper, sulphur dioxide, pigments and tannins which can mask the end-point of the titration, and reducing compounds other than AA, and so care has to be exercised when using this procedure for some foodstuffs. Acetone can be added to extracts to form an acetone bisulphite complex prior to titration if sulphur dioxide is known to be present in a sample. It must be appreciated that this simple procedure only determines AA.

One of the most specific methods for the determination of 'total vitamin C', i.e. AA and DHAA, is the fluorimetric procedure developed by Deutsch and Weeks (1965). The method is based on the coupling reaction of o-phenylenediamine across the *cis* hydroxyl groups of DHAA to form a fluorescent quinoxaline derivative. Samples are extracted with metaphosphoric–acetic acid solution and the AA oxidised to DHAA using activated charcoal; o-phenylenediamine is then added and fluorescence measured and quantitation achieved by comparison against standards taken through the same procedure at the same time. A blank is obtained by adding boric acid to sample extracts which couples to *cis* hydroxyls thus preventing reaction with o-phenylenediamine. If the oxidation stage is omitted this procedure can be used to determine DHAA; if this result is subtracted from the total vitamin C a value is obtained for AA. This procedure has been used routinely in the LGC for over 25 years in a manual and semi-automated format with excellent results. The method is an AOAC final action procedure (1990k) and the 2,6-DCP and the fluorimetric procedures are described by Kirk and Sawyer (1991a).

8.4.3.2 HPLC determination of vitamin C. Nelson reported the HPLC separation of AA from other food additives in 1973 and a vast number of LC procedures have been published since that date. Some published procedures determine only AA whilst others are designed to determine AA plus DHAA. The latter can be achieved by reducing any DHAA in the sample extract to AA and chromatographing total AA, or by chromatographically separating and determining AA and DHAA. AA has a strong UV absorbance and is easily detected with good sensitivity but DHAA has only weak UV absorbance, making it more convenient to reduce DHAA to AA prior to HPLC. This avoids the necessity for dual wavelength monitoring or post-column reactions of DHAA with o-phenylenediamine.

Ascorbic acid shows strong absorption with a maxima at 245 nm moving to 265 nm as pH increases; however AA has an isobestic point at 251 nm and monitoring at or near this wavelength is advisable, to avoid effects of pH on absorption.

Reverse-phase HPLC with ion-pairing has been used for the determination of ascorbic acid but it can be difficult to achieve sufficient retention and selectivity for food analysis, and chromatography can be irreproducible. Ion-exchange or bonded amino columns usually provide the best mode of chromatography for AA, and they are also capable of separating AA from erythorbic acid which may be present as an antioxidant in some foodstuffs. Erythorbic acid is thought to have only about 5 % of the biological activity of AA so that for nutritional studies it is important that the two are differentiated.

Table 8.16 shows some chromatographic conditions reported in the literature for ascorbic acid. Reverse-phase ion-pair methods have not been included, neither have methods using pre- or post-column derivatisation or electrochemical detection. The food analyst should be able to determine ascorbic acid in most sample

Table 8.16 Chromatographic conditions reported for ascorbic acid

Reference	Column	Mobile phase	Detection
Ashoor et al. (1984)	Aminex HPX-87 30 cm × 7.8 mm	0.005 M H_2SO_4	245 nm
Tuan et al. (1987)	LiChrosorb–NH_2 25 cm × 4 mm	CH_3CN – 0.05 M KH_2PO_4 pH 5.95 (75:25)	268 nm
Lloyd et al. (1988)	PLRP – 5.5 µM 25 cm × 4.6 mm	0.2 M sodium dihydrogenphosphate pH 2.14	244 nm
Albrecht and Shafer (1990)	µ Bondapak NH_2 30 cm × 3.9 mm	0.005 M KH_2PO_4 (pH 4.6)–CH_3CN (30:70)	254 nm
Morawski (1984)	Radial Pak µ Bondapak NH_2	Methanol – 0.01 M KH_2PO_4 (80:20)	254 nm
Dennison et al. (1981)	µ Bondapak NH_2 30 cm × 4.6 mm	Methanol – 0.25% KH_2PO_4 pH 3.5 (50:50)	244 nm
Rizzolo et al. (1984)	Partisil 10 SAX 25 cm × 4.6 mm	0.1 M sodium acetate pH 4.25	250 nm

extracts using the type of simple systems listed. In some circumstances it may be necessary to pass sample extracts through C_{18} cartridges prior to HPLC to remove potential interferences. Extracting solutions of metaphosphoric–acetic acids may be compatible with some HPLC procedures but alternatives have been reported. Ashoor et al. (1984) used 0.1 M sulphuric acid containing 0.05 % EDTA and reported that standards were stable in this medium for 48 h, while Morawski (1984) used 10 mM thiourea in acetonitrile. Lloyd et al. (1988) reported that 1 mM homocysteine was as effective as 5 or 10% metaphosphoric acid for stabilising AA over a period of 150 min, and stated that both were compatible with their HPLC procedure. Albrecht and Schafer (1990) used 3% metaphosphoric acid while Rizzolo et al. (1984) used 6%. Dennison et al. (1981) used DL-homocysteine and Tuan et al. (1987) used dithiothreitol to convert DHAA to AA prior to HPLC, whilst Sawamura et al. (1990) recently reported on the use of sodium hydrosulphide for the same purpose.

The reverse-phase ion-suppression procedure of Lloyd et al. (1988) was shown to be capable of baseline resolution of AA and erythorbic acid and this separation has been reported by several workers using amino-bonded columns (see, e.g. Tuan et al., 1987). Lloyd et al. were able to use ion-suppression at low pH because they used a polymeric reverse-phase packing which is stable to extremes of pH; a silica based column packing would soon degrade under such conditions.

8.4.4 Summary

1. Ascorbic acid can be rapidly determined by titration with 2,6-DCP but the analyst must be aware that only AA is being determined and components of the sample matrix can interfere with the oxidation–reduction titration and affect the result obtained.
2. The fluorescent procedure of Deutsch and Weeks provides a reliable and robust means for determining the total vitamin C content of samples and

individual AA and DHAA contents. This procedure is not generally prone to interferences as sample blanks are taken through the whole procedure. The method can easily be semi-automated.
3. HPLC can provide reliable results as long as analysts use appropriate extraction media, completely resolve AA from sample matrix interferences and ensure that all DHAA is reduced to AA prior to HPLC if total vitamin C content is to be determined.

References

Addo, C. and Augustin, J. (1988) Changes in the vitamin B6 content in potatoes during storage. *J. Food Sci.* **53(3)**, 749–752.

Ang, C.Y.W. and Moseley, F.A. (1980) Determination of thiamin and riboflavin in meat and meat products by high pressure liquid chromatography. *J. Agric. Food Chem.* **28(3)**, 483–486.

Ang, C.Y.W., Cenciarelli, M. and Eitenmiller, R.R., (1988) A simple liquid chromagraphic method for determination of B6 vitamers in raw and cooked chicken. *J. Food Sci.* **53(2)**, 371–375.

Albrecht, J.A. and Schafer, H.W. (1990) Comparison of two methods of ascorbic acid determination in vegetables. *J. Liq. Chromatogr.* **13(3)**, 2633–2641.

AOAC (1990a) Vitamin D in fortified milk and milk powder. *Official Methods of Analysis* 15th edition. Method 981.17, pp. 1068–1069.

AOAC (1990b) *Official Methods of Analysis*, 15th Edition. Volume 2.

AOAC (1990c) *Official Methods of Analysis*, 15th Edition, Section 943.23, pp. 1049–1051.

AOAC (1990d) *Official Methods of Analysis*, 15th Edition. Section 970.65, 981.15, 940.33.

AOAC (1990e) *Official Methods of Analysis*, 15th Edition. Section 944.13, pp. 1084–1085.

AOAC (1990f) *Official Methods of Analysis*, 15th Edition. Section 961.15, pp. 1089–1091.

AOAC (1990g) *Official Methods of Analysis*, 15th Edition. Section 944.12, pp. 1083–1084.

AOAC (1990h) *Official Methods of Analysis*, 15th Edition. Section 952.20, pp. 1082–1083.

AOAC (1990i) *Official Methods of Analysis*, 15th Edition. Section 945.74, pp. 1085–1086.

AOAC (1990j) *Official Methods of Analysis*, 15th Edition. Section 967.21, pp. 1058–1059.

AOAC (1990k) *Official Methods of Analysis*, 15th Edition. Section 967.22, 984.26, pp. 1059–1061.

Ashoor, S.N., Monte, W.C. and Welty, J. (1984) Liquid chromatographic determination of ascorbic acid in foods. *J. Assoc. Official Anal. Chem.* **67(1)**, 74–80.

Ashoor, S.H., Know, M.J., Olsen, J.R. and Deger, D.A. (1985) Improved liquid chromatographic determination of riboflavin in milk and dairy products. *J. Assoc. Official Anal. Chem.* **68(4)**, 693–696.

Bell, J.G. (1984) Microbiological assay of vitamins of the B-group in foodstuffs. *Laboratory Practice* **23**, 235–42.

Bell, J.G. and Christie, A.A. (1973) Gas-liquid chromatographic determination of vitamin D in cod-liver oil. *Analyst* **98**, 268–273.

Bell, J.G. and Christie, A.A. (1974) Gas-liquid chromatographic determination of vitamin D_2 in fortified full cream milk. *Analyst* **99**, 385–396.

Bertelsen, G, Finglas, P.M, Loughridge, J., Faulks, R.M., Morgan, M.R.A. (1988) Investigation into the effects of conventional cooking on levels of thiamin and pantothenic acid in chicken. *Food Sci. Nutr.* **42F**, 83–96.

Bitsch, R. and Moller, J. (1989) Analysis of B6 vitamers in foods using a modified high-performance liquid chromatographic method. *J. Chromatogr.* **463**, 207–211.

Botticher, B. and Botticher, D. (1986) Simple rapid determination of thiamin by HPLC method in foods, body fluids and faeces. *Internat. J. Vitamin Nutr. Res.* **56**, 155–159.

Brubacher, G., Muller-Mulst, W. and Southgate, D.A.T. (1985) (eds) *Methods for Determination of Vitamins in Food Recommended by COST 91*, Elsevier Applied Science, London, pp. 129–140.

Bui-Nguyen, M. and Blanc, B. (1980) Measurement of vitamin A in milk and cheese by high pressure liquid chromatography. *Experimentia* **36**, 374–375.

Bureau, J.L. and Bushway, R.J. (1986) HPLC determination of carotenoids in fruits and vegetables in the United States. *J. Food Sci.* **51(1)**, 128–130.

Christie, A.A., Dean, A.C. and Millburn, B.A. (1973) The determination of vitamin E in food by colorimetry and gas-liquid chromatography. *Analyst* **98**, 161–167.

Coburn, S.P. and Mahuren, J.D. (1983) A versatile cation-exchange procedure for measuring the seven major forms of vitamin B_6 in biological samples. *Anal. Biochem.* **129**, 310–317.

Dawson, K.R., Unklesbay, N.F. and Hedrick, H.B. (1988) HPLC determination of riboflavin, niacin and thiamin in beef, pork and lamb after alternate heat-processing methods. *J. Agric. Food Chem.* **36**, 1176–1179.

Day, B.P. and Gregory, J.F. (1981) Determination of folacin derivatives in selected foods by high performance liquid chromatography. *J. Agric. Food Chem.* **29**, 374–377.

Defibaugh, P.W. (1987) Evaluation of selected enzymes for thiamin determination. *J. Assoc. Official Anal. Chem.* **70**(3), 514–517.

Dennison, D.B., Brawley, T.G. and Hunter, G.L. (1981) Rapid high performance liquid chromatographic determination of ascorbic acid-dehydroascorbic acid in beverages. *J. Agric. Food Chem.* **29**, 927–929.

Deutsch, M.J. and Weeks, C.E. (1965) Microfluorimetric Assay for Vitamin C. *J. Assoc. Official Anal. Chem.* **48**(6), 1249–1256.

Egberg, D.C., Heroff, J.C. and Potter, R.H. (1977) Determination of all-trans and 13-cis vitamin A in food products by high pressure liquid chromatography. *J. Agric. Food Chem.* **25**, 1127–1132.

Engelhardt, R. and Gregory, J.F. (1990) Adequacy of enzymatic deconjugation in quantification of folate in foods. *J. Agric. Food Chem.* **38**, 154–158.

Fellman, J.K., Artz, W.E., Tassinari, P.D., Cole, C.I. and Augustin, J. (1982) Simultaneous determination of thiamin and riboflavin in selected foods by high performance liquid chromatography. *J. Food Sci.* **47**, 2048–2050.

Fernando, S.M. and Murphy, P.A. (1990) HPLC determination of thiamin and riboflavin in soybeans and Tofu. *J. Agric. Food Sci.* **38**, 163–167.

Finglas, P.M. and Faulks, R.M. (1984a) The HPLC analysis of thiamin and riboflavin in potatoes. *Food Chem* **15**, 37–44.

Finglas, P.M. and Faulks, R.M. (1984b) Nutritional composition of UK retail potatoes, both raw and cooked. *J. Sci. Food Agric.* **35**, 1347–1356.

Fisher, J.F. and Rouseff, R.L. (1986) Solid-phase extraction and HPLC determination of β-cryptoxanthin and α- and β-carotene in orange juice. *J. Agric. Food Chem.* **34**, 985–989.

Goli, D.M. and Vanderslice, J.T. (1989) Microbiological assays of folacin using a CO_2 analyser system. *J. Micronutrient Analysis* **6**, 19–33.

Goli, D.M. and Vanderslice, J.T. (1992) Investigation of the conjugase treatment procedure in the microbiological assay of folate. *Food Chem.* **43**, 57–64.

Gounelle, J-C, Ladjimi, H. and Prognon, P. (1989) A rapid and specific extraction procedure for folates. Determination in rat liver and analysis by high performance liquid chromatography with fluorometric detection. *Anal. Biochem.* **176**, 406–411.

Gregory, J.F. and Feldstein, D. (1985) Determination of vitamin B_6 in foods and other biological materials by paired ion high performance liquid chromatography. *J. Agric. Food Chem.* **33**, 359–363.

Gregory, J.F. and Ink, S.L. (1987) Identification and quantification of pyridoxine-β-glucoside as a major form of vitamin B_6 in plant derived foods. *J. Agric. Food Chem.* **35**, 76–82.

Gregory, J.F., Sartain, D.B. and Day, B.P.F. (1984) Fluorometric determination of folacin in biological materials using high performance liquid chromatography. *J. Nutr.* **114**, 341–353.

Gregory, J.F., Engelhardt, R., Bhandari, S.D., Sartain, D.B. and Gusatfson, S.K. (1990) Adequacy of extraction techniques for determination of folate in foods and other biological materials. *J. Food Composition and Analysis* **3**, 134–144.

Hakansson, B., Jagerstad, M. and Oste, R. (1987) Determination of vitamin E in wheat products by HPLC. *J. Micronutrient Analysis* **3**, 307–318.

Haimano, T., Mitsuhashi, Y., Aoki, N. and Yamamoto, S. (1988) Simultaneous determination of niacin and niacinamide in meats by high performance liquid chromatography. *J. Chromatogr.* **457**, 403–408.

Hasselmann, C., Franck, D., Grimm, P., Diop, P.A. and Soules, C. (1989) High performance liquid chromatographic analysis of thiamin and riboflavin in dietetic foods. *J. Micronutrient Analysis.* **5**, 269–279.

Hirayama, S. and Maruyama, M. (1991) Determination of a small amount of niacin in foodstuffs by high performance liquid chromatography. *J. Chromatogr.* **588**, 171–175.

Holt, D.L., Wehling, R.L. and Zeece, M.G. (1988) Determination of native folates in milk and other dairy products by high performance liquid chromatography. *J. Chromatogr.* **449**, 271–279.

Indyk, H. (1987) The rapid determination of carotenoids in bovine milk using HPLC. *J. Micronutrient Analysis* **3**, 169–183.

Indyk, H. and Woollard, D.C. (1984) The determination of vitamin D in milk powders by high performance liquid chromatography. *N.Z. J. Dairy Sci. Tech.* **19**, 19–30.

Khachik, F., Beecher, G.R. and Whittaker, N.F. (1986) Separation, identification and quantification of the major carotenoid and chlorophyll constituents in extracts of several green vegetables by liquid chromatography. *J. Agric. Food Chem.* **34**(4), 603–616.

Khachik, F., Beecher, G.R. and Goli, M.B., (1991) Separation, identification and quantification of carotenoids in fruits, vegetables and human plasma by high performance liquid chromatography. *Pure and Appl. Chem.* **63** (1), 71–80.

Kirk, R. and Sawyer, R. (1991a) *Pearson's Composition and Analysis of Foods*, 9th Edition. Longman Scientific and Technical pp. 243–244, 264.

Kirk, R. and Sawyer, R. (1991b) *Pearson's Composition and Analysis of Foods*, 9th Edition. Longman Scientific and Technical, pp. 647–648.

Kirsch, A.J. and Chen, T.S. (1984) Comparison of conjugase treatment procedures in the microbiological assay for food folacin. *J. Food Sci.* **49**, 94–98.

Kobayashi, T., Okano, T. and Takeuchi, A. (1982) The determination of vitamin D in foods and feeds using high performance liquid chromatography. *J. Micronutrient Analysis.* **2**, 1–4.

Lloyd, L.L., Warner, F.P., Kennedy, J.F. and White, C.A. (1988) Ion suppression reversed-phase high performance liquid chromatography method for the separation of L-ascorbic acid in fresh fruit juice. *J. Chromatogr.* **437**, 447–452.

Lumley, I.D. and Lawrance, P.R. (1990) Advances in the determination of vitamins in foodstuffs - methods used at the Laboratory of the Government Chemist. *J. Micronutrient Analysis* **7**, 301–313.

Lumley, I.D. and Wiggins, R.A. (1981) Determination of riboflavin and flavinmononucleotide in foodstuffs using high performance liquid chromatography and a column enrichment technique. *Analyst* **106**, 1103–1108.

Martin, J.I., Landen, W.D., Soliman, A-G.M. and Eitenmiller, R.R. (1990) Application of a tri-enzyme extraction for total folate determination in foods. *J. Assoc. Official Anal. Chem.* **73**(5), 805–808.

McRoberts, L.H. (1960) Determination of thiamin in enriched cereal and bakery products. *J. Assoc. Official Anal. Chem.* **43**, 47–57.

Monard, A.M., Berthels, J., Nedelkovitch, G., Draguet, M. and Bouche, R. (1986) Determination of the total vitamin D_3 content of cod-liver oil by high performance liquid chromatography. *Pharm. Acta. Helv.* **61**(7), 205–208.

Morawski, J. (1984) Analysis of dairy products by HPLC. *Food Technol.* **38**(4), 70–78.

Newman, E.M. and Tsai, J.F. (1986) Microbiological analysis of 5-formyltetrahydrofolic acid and other folates using an automated 96-well plate reader. *Anal. Biochem.* **154**, 509–151.

Nicolson, I.A., Macrae, R. and Richardson, D.P. (1984) Comparative assessment of high performance liquid chromatographic methods for the determination of ascorbic acid and thiamin in foods. *Analyst* **109**, 267–271.

Ollilainen, V., Mattila, P., Varo, P., Koivistoinen, P. and Huttunen, J. (1990) The HPLC determination of total riboflavin in foods. *J. Micronutrient Analysis* **8**, 199–207.

Pedersen, J.C. (1988) Comparison of gamma-glutamyl hydrolase and amylase treatment procedures in the microbiological assay of food folates. *Brit. J. Nutr.* **59**, 261–271.

Piironen, V., Syvaoja, E-L, Varo, P., Slaminen, K. and Koivistoinen, P. (1984) *Internat. J. Vitamin Nutr. Res.* **53**, 35–40.

Pocklington, W.D. and Dieffenbacher, A. (1988) Determination of tocopherols and tocotrienols in vegetable oils and fats by high performance liquid chromatography. *Pure Appl. Chem.* **60**(6), 877–92.

Quackenbush, F.W. and Smallidge, R.L. (1986) Non-aqueous reverse phase chromatographic system for separation and quantitation of provitamins A. *J. Assoc. Official Anal. Chem.* **69**(5), 767–772.

Rammell, C.G. and Hoogenboom, J.J.L. (1985) Separation of tocols by HPLC on an amino-cyano polar phase column. *J. Liq. Chromatogr.* **8**(4), 707–717.

Reyes, E.S., Norris, K.M., Taylor, C. and Potts, D. (1988) Comparison of paired-ion liquid chromatographic method with AOAC fluorimetric and microbiological methods for riboflavin determination in selected foods. *J. Assoc. Official Anal. Chem.* **71**(1), 16–19.

Reynolds, S.L. and Judd, H.J. (1984) Rapid procedure for the determination of vitamins A and D in fortified skimmed milk powder using high performance liquid chromatography. *Analyst* **109**, 489–492.

Rizzolo, A., Formi, E. and Polesello, A. (1984) HPLC assay of ascorbic acid in fresh and processed fruit and vegetables. *Food Chem.* **14**, 189–199.

Russell, L.F. and Vanderslice, J.T. (1992) Comments on the standard fluorimetric determination of riboflavin in foods and biological tissues. *Food Chem.* **43**, 79–82.

Sampson, D.A. and O'Connor, D.K. (1989) Analysis of B6 vitamers and pyridoxic acid in plasma, tissues and urine using high performance liquid chromatography. *Nutr. Res.* **9**, 259–272.

Sawamura, M., Ooishi, S. and Li, Z-F. (1990) Reduction of dehydroascorbic acid by sodium hydrosulphide and liquid chromatographic determination of vitamin C in citrus juices. *J. Sci. Food Agric.* **53**, 279–281.

Schultz, A., Weidemann, K. and Bitsch, I. (1985) Stabilisation of 5-Me-THF and subsequent analysis by reverse phase high phase liquid chromatography. *J. Chromatogr.* **328**, 417–421.

Sertl, D.C. and Molitor, B.E. (1985) Liquid chromatographic determination of vitamin D in milk and infant formula. *J. Assoc. Official Anal. Chem.* **68**(2), 177–182.

Speek, A.J., Temalilwa, C.R. and Schrijver, J. (1986) Determination of β-carotene content and vitamin A activity of vegetables by high performance liquid chromatography and spectrophotometry. *Food Chem.* **19**, 65–74.

Stancher, B. and Zonta, F. (1982) High performance liquid chromatographic determination of carotene and vitamin A and its geometric isomers in foods. *J. Chromatogr.* **238**, 217–225.

Stancher, B. and Zonta, F. (1984) High performance liquid chromatography of the unsaponifiable from samples of marine and freshwater fish; fractionation and identification of retinol and dehydroretinol isomers. *J. Chromatogr.* **287**, 353–364.

Stancher, B. and Zonta, F., (1986) High performance liquid chromatographic analysis of riboflavin (Vitamin B$_2$) with visible absorbance detection in Italian Cheeses. *J. Food Sci.* **51**(3), 857–858.

Takatsuki, K., Suzuki, S., Sato, M., Sakai, K. and Ushizawa, I. (1987) Liquid chromatographic determination of free and added niacin and niacinamide in beef and pork. *J. Assoc. Official Anal. Chem.* **70**(4), 698–702.

Tamura, T., Shin, Y.S., Williams, M.A. and Stokstad, E.L.R. (1972) Lactobacillus casei response to pteroylpolyglutamates. *Anal. Biochem.* **49**, 517–519.

Thompson, J.N. and Duval, S. (1989) Determination of vitamin A in milk and infant formula by HPLC. *J. Micronutrient Analysis* **6**, 147–159.

Thompson, J.N. and Hatina, G. (1979) Determination of tocopherols and tocotrienols in foods and tissues by high performance liquid chromatography. *J. Liq. Chromatogr.* **2**(3), 327–344.

Thompson, J.N., Hatina, G. and Maxwell, W.B. (1980) High performance liquid chromatographic determination of vitamin A in margarine, milk and skimmed milk. *JAOAC*, **63**(4) 894–898.

Thompson, J.N., Hatina, G., Maxwell, W.B. and Duval, S. (1982) High performance liquid chromatographic determination of vitamin D in fortified milks, margarine, infant formulas. *J. Assoc. Official Anal. Chem.* **65**(3), 624–631.

Toepfer, E.W. and Polansky, M.M. (1970) Microbiological assay of vitamin B6 and its components. *J. Assoc. Official Anal. Chem.* **53**, 546.

Toma, R.B. and Tabekhia, M.M. (1979) High performance liquid chromatographic analysis of B-vitamins in rice and rice products. *J. Food Sci.* **44**(1), 263–268.

Trugo, L.C., Macrae, R., Trugo, N.M.F. (1985) Determination of nicotinic acid in instant coffee using high performance liquid chromatography. *J. Micronutrient Analysis* **1**(1), 55–63.

Tuan, S., Wyatt, J. and Anglemier, A.F. (1987) The effect of erythorbic acid on the determination of ascorbic acid levels in selected foods by HPLC and spectrophotometry. *J. Micronutrient Analysis* **3**, 211–228.

Tyler, T.A. and Shrago, R.R. (1980) Determination of niacin in samples by HPLC. *J. Liq. Chromatogr.* **3**(2), 269–277.

Vanderslice, J.T. and Maire, C.E. (1980) Liquid Chromatographic separation and quantification of B$_6$ vitamers and plasma concentration levels. *J. Chromatogr.* **196**, 176–179.

van Niekerk, P.J., Smit, S.C.C., Strydom, E.S.P. and Armbruster, G. (1984) Comparison of high performance liquid chromatographic and microbiological method for the determination of niacin in foods. *J. Agric. Food Chem.* **32**, 304–307.

Vidal-Valverde, C. and Reche A. (1990a) Reliable system for the analysis of riboflavin in foods by high performance liquid chromatography and UV detection. *J. Liq. Chromatogr.* **13**(10), 2089–2101.

Vidal-Valverde, C. and Reche, A. (1990b) An improved high performance liquid chromatographic method for thiamin analysis in foods. *Lebensm Unters Forsch.* **191**, 313–318.

Vidal-Valverde, C. and Reche, A. (1991) Determination of available niacin in legumes and meat by high performance liquid chromatography. *J. Agric. Food Chem.* **39**, 116–121.

Villalobos, M.C., Gregory, N.R. and Bueno, M.P. (1990) Determination of vitamin D_2 and vitamin D_3. in Foods, feeds and pharmaceuticals, using high performance liquid chromatography: comparison of 3 different columns. *J. Micronutrient Analysis* **8**, 79–89.

de Vries, E.J. and Borsje, B. (1982) Analysis of fat soluble vitamins. High performance liquid chromatographic and gas liquid chromatographic determination of vitamin D in fortified milk and milk powder. *J. Assoc. Anal. Chem.* **65**, 1229–1234.

Watada, A.E. and Tran, T.T. (1985) A sensitive high performance liquid chromatographic method for analysing riboflavin in fresh fruit and vegetables. *J. Liq. Chromatogr.* **8**(9), 1651–1662.

Weber, E.J. (1984) High performance liquid chromatography of the tocols in corn grain. *JAOCS* **61**(7), 1231–1234.

White, D.R., Lee, H.S. and Kruger, R.E. (1991) Reversal phase HPLC / EC determination of folate in citrus juice by direct injection with column switching. *J. Agric. Food Chem.* **39**, 714–717.

Wiggins, R.A., Zai, E.S. and Lumley, I.D. (1982) LC determination of vitamins A and D in foods and animal feedingstuffs. *Chromatogr. Sci. Appl. Liq. Chromatog.* **20**, 327–341.

Wills, R.B.H., Wimalasiri, P. and Greenfield, H. (1985) Comparative determination of thiamin and riboflavin in foods by high performance chromatography and fluorimetric methods. *J. Micronutrient Analysis.* **1**, 23–29.

Wills, R.B.H., Nardin, H. and Wootton, M. (1988) Separation of carotenes and xanthophylls in fruit and vegetables by HPLC. *J. Micronutrient Analysis.* **4**, 87–98.

Wilson, S.D. and Horne, D.W. (1982) Use of glycerol-cryoprotected Lactobacillus casei for microbiological assay of folic acid. *Clin. Chem.* **28**, 1198–1200.

Wilson, S.D. and Horne, D.W. (1983) Evaluation of ascorbic acid in protecting labile folic acid derivatives. *Proc. Natl. Acad. Sci. USA* **80**, 6500–6504.

Wilson, S.D. and Horne, D.W. (1984) High performance liquid chromatographic determination of the distribution of naturally occuring folic acid derivatives in rat liver. *Anal. Biochem.* **142**, 529–535.

Woollard, D.C. and Blott, A.D. (1986) The routine determination of vitamin E acetate in milk powder formulations using high performance liquid chromatography. *J. Micronutrient Analysis* **2**, 97–115.

Woollard, D.C. and Indyk, H. (1989) The distribution of retinyl esters in milk and milk products. *J. Micronutrient Analysis* **5**, 35–52.

Woollard, D.C. and Woollard, G.A. (1981) Determination of vitamin A in fortified milk powders using high performance liquid chromatography. *N.Z. J. Dairy Sci. Technol.* **16**, 99–112.

Woollard, D.C., Blott, A.D. and Indyk, H. (1987) Fluorometric detection of tocopheryl acetate and its use in the analysis of infant formulae. *J. Micronutrient Analysis* **3**, 1–14.

Wright, A.J.A. and Phillips, D.R. (1985) The threshold growth response of Lactobacillus casei to 5-methyl-tetrahydrofolic acid: implications for folate assays. *Brit. J. Nutr.* **53**, 569–573.

9 Food fortification
D. P. RICHARDSON

Many nutrition surveys in developed and developing countries continue to indicate that an appreciable fraction of the population, particularly young children, adolescents, the elderly and women of child bearing age, can suffer from nutrient deficiencies at a borderline or pathological level. In the developed nations, nutrient needs and food choices of these specific population groups, the increasing use of dietetic and low energy products, the overall trend towards consuming fewer calories and the greater reliance on commercially prepared foods, are just some of the reasons why it is essential to re-examine policies and guidelines for the addition of nutrients to foods.

Advances in food technology and new manufacturing processes have also created the possibilities for increased and decreased exposure to some essential nutrients (Harris and Karmas, 1977; Bender, 1978; British Nutrition Foundation, 1987). Generally, the advantages of food processing and food preparation are to arrest spoilage, to make foods safe and edible and, hence, to minimise the deterioration in the nutritional quality of the food. Although the concerns about losses of specific nutrients in various processes have to be considered in the context of the diet as a whole, the addition of micronutrients has been used to maintain or enhance the nutritional value of many commonly consumed foods. Developments of 'novel' foods such as margarine, low fat spreads and vegetable protein meat analogues which simulate existing foods have also raised questions of 'nutritional equivalency' in terms of the amount and bioavailability of added nutrients (Ministry of Agriculture, Fisheries and Food, 1974).

The addition of nutrients is accomplished using the techniques of fortification or enrichment, restoration, standardisation and supplementation. Fortification or enrichment refers specifically to the addition of nutrients to a food above the level normally found in that food; restoration refers to replacement of nutrients whose losses cannot be avoided during handling, processing and storage of foods; standardisation refers to the addition of nutrients to a food to compensate for natural variations; and supplementation, a more general term, refers to the addition to a food of nutrients which are not normally contained in that food or only in minute quantities.

9.1 General policies for nutrient additions

Several nutrients have been added to food and drink products around the world, as public health measures and as cost effective ways of ensuring the nutritional quality of the food supply. Addition of some nutrients has also formed the basis of marketing strategies in product development. In general, the main criteria for selecting nutrients to be added to foods are that they are shown to be necessary, safe and effective (National Academy of Sciences, 1975). The addition of nutrients also requires careful attention to food regulations, labelling, nutritional rationale, cost, the acceptability of the product to consumers and a careful assessment of technical and analytical limitations for compliance with label declarations.

Nutrient enrichment of foods can help prevent nutritional inadequacies in populations where there is a risk of deficiencies and where intervention is needed to correct a proven deficiency in an identified segment of the population. In this case, the criteria of effectiveness are whether or not nutrition status and health have been improved. When nutrient additions are part of a marketing and promotional strategy, a fundamental question is whether their presence will sell more products. The effectiveness of a programme of nutrient additions is influenced by whether the food that is to carry the nutrient(s) is going to be acceptable, consumed by those who need or want it and be at a price they can afford. Also, the nutrients must be bioavailable and sufficiently stable under the normal, and perhaps unusual, conditions of storage and household use. Evidence and assurances must also be sought to ensure that consumption of foods containing added nutrients will not create a nutritional imbalance and that an excessive intake of the nutrients will not occur, bearing in mind the cumulative amounts from other sources in the diet. General principles for the addition of nutrients to foods have been established by the Codex Alimentarius Commission (1987) of the Food and Agriculture Organization/World Health Organization, and the United States Food and Drug Administration (1987b).

9.2 Legislation concerning addition of nutrients to foods

The regulations concerning the addition of nutrients to foods and the authority to make claims vary from one country to another, and many problems can arise for food manufacturers owing to the lack of harmonization in this field (du Bois, 1987). Differences in methods of analysis and the variations in the lists of recommended daily amounts (RDA) of the nutrients are also problems which are encountered by food manufacturers selling the same products on an international basis. In addition, the reference points for addition of nutrients can be per 100 g, per kg, per 100 kcal, per food serving (in grams) or the amounts of a food which can be reasonably consumed daily (in grams). The serving sizes of daily portions can, of course, vary considerably from one population to another depending on their eating habits.

Four categories of foodstuffs can be identified where the addition of vitamins and minerals is compulsory in some countries: (a) foods for special dietary uses, (b) foods having lost nutrients during manufacture, (c) foods resembling a common food (replacement products) and (d) staple foods representing ideal vehicles for nutrients.

9.2.1 Food for special dietary uses

As defined by the Council of the European Communities, foods for special dietary uses or foods for particular nutritional uses are foodstuffs which, owing to their special composition or manufacturing process, are clearly distinguishable from foodstuffs for normal consumption, which are suitable for their claimed nutritional purposes and which are marketed in such a way as to indicate such suitability (European Council Directive, 1987). A particular nutritional use must fulfill the particular nutritional requirements: (a) of certain categories of persons whose digestive processes or metabolism are disturbed; or (b) of certain categories of persons who are in a special physiological condition and who are, therefore, able to obtain a special benefit from a controlled consumption of certain substances in foodstuffs; or (c) of infants or young children in good health.

An excellent example is the development of nutritional standards for the composition of infant formulas which are complete foods intended for use, where necessary, as a substitute for human milk and should meet the normal nutritional requirements of infants up to 4–6 months of age. Nutrient standards for nutritionally complete formula foods for use in weight-control diets and standards for 'low energy' and 'reduced energy' foods are also under consideration in the European Economic Community, and the Codex Alimentarius Commission (1985), but the issues go beyond the scope of the present paper.

9.2.2 Foods having lost nutrients during manufacture

The best example of the compulsory restoration of the level of vitamins and minerals lost during manufacture in the UK is the addition of nutrients to all types of flour, other than wholemeal flour. The quality of flours is regulated by the Bread and Flour Regulations (1984) and low extraction flours have to be enriched by the addition of thiamin (0.24 mg), niacin (1.60 mg), iron (1.65 mg) and calcium (235 mg minimum–390 mg maximum)/100 g. The Department of Health and Social Security (DHSS) (1981) reviewed the nutritional aspects of bread and flour and recommended that these additions be no longer mandatory. However, taking into consideration, for example, the fact that flour based products provide about one fifth of the thiamin intake in the diet of the elderly (D. J. Thurnham, personal communication) it is not surprising that the DHSS (1981) recommendation was not accepted in drawing up the new law.

In Denmark, flour must contain vitamin B_1 (thiamin) (5 mg), vitamin B_2 (riboflavin) (5 mg), calcium (2 g) and iron (30 mg)/kg and rolled oats must contain calcium (2.3 g/kg) and phosphorus (1.8 g/kg). Similarly, in the USA, most low-extraction rate flours are enriched with thiamin, riboflavin, nicotinamide and iron up to 0.44, 0.26, 3.5 and 2.9 mg/100 g, respectively (Bender, 1978; du Bois, 1987).

The additions of vitamins A and D to skim milk powder and the addition of vitamin D to evaporated milk in the UK are also, in part, examples of voluntary forms of vitamin restoration.

9.2.3 Foods resembling a common food

The best example is the addition of vitamins A and D to margarine, which was first made statutory in the UK at the start of World War II. By law, margarine must contain 804 µg retinol and 8 µg vitamin D/100 g to give it a nutritive value similar to that of butter. This compulsory addition of nutrients (Margarine Regulations, 1967) was a major public health measure and formed an important part of the efforts to prevent rickets. The addition of vitamins A and D is also compulsory in Belgium, Denmark, The Netherlands, Norway and Sweden.

In the UK, the Committee on Medical Aspects of Food Policy (COMA) (DHSS, 1980) laid down the general principle that 'any substance promoted as a replacement or an alternative to a natural food should be nutritionally equivalent in all but unimportant aspects of the natural food which it would simulate'. Thus the Food Standards Committee (FSC) (Ministry of Agriculture, Fisheries and Food, 1974) and the COMA panel (DHSS, 1980) gave their recommendations and specifications for the nutritional quality and use of textured vegetable protein (TVP) foods which simulate meat. In the UK, the FSC recommended that TVP shall contain (per 100 g dry matter) not less than 2.0 mg thiamin, 0.8 mg riboflavin, 5.0 µg vitamin B_{12} and 10 mg iron. The nutritional aspects and acceptability of vegetable protein foods which simulate meat have been described in detail elsewhere (Richardson, 1982).

9.2.4 Staple foods

Vitamin A deficiency and xerophthalmia are among the most widespread nutritional disorders that result in blindness in man, particularly in tropical and subtropical countries. One of the remedies is the fortification of staple foods. In India, for example, since tea is consumed universally (even by small children), this drink is used as the vehicle for enrichment. Other staples used for enrichment with vitamin A include rice and table sugar in Guatemala and monosodium glutamate in the Philippines (Bender, 1978). As previously stated, vitamins A and D are customarily added to skim milk powders and evaporated milk.

Well established examples of the benefits of trace element additions include elimination of goitre through iodisation of salt, the decrease in dental caries through fluoridation of the water supply and the addition of iron to bread and cereals, table salt, monosodium glutamate, sugar, coffee, tea, oils and fats.

9.3 Claims for nutrients and labelling of fortified foods

In the UK, the addition of vitamins and minerals is permitted to all foodstuffs with the exception of alcoholic drinks. Claims may be made only with respect to the 'scheduled' nutrients as stated in the Food Labelling Regulations (1984). These nutrients are the vitamins: A, thiamin, riboflavin, niacin, folic acid, B_{12}, C and D, and the minerals: calcium, iodine and iron. If a claim is made on the label that a food is a 'rich' or 'excellent' source of vitamins or minerals, then the daily food portion (described as 'the quantity of food that can reasonably be expected to be consumed in a day') must contain at least half the RDA for that nutrient. For any other claim, e.g. 'a useful source' or 'with added ...', the daily portion must contain at least one sixth of the RDA. Table 9.1 summarises the labelling requirements for vitamin and mineral claims and lists the RDA for the scheduled nutrients.

In Switzerland and Germany, a reference on the label to a given vitamin may only be made if at least one third of the RDA is contained in the daily food portion,

Table 9.1 The types and conditions for vitamin and mineral claims on food labels and recommended daily amounts of nutrients as specified by the Food Labelling Regulations (1984). Conditions for mineral and vitamin claims are the same

(a) Nutrients in respect of which claims may be made

Nutrient	Recommended daily amount (RDA)
Vitamin A	750 µg
Thiamin (vitamin B_1)	1.2 mg
Riboflavin (vitamin B_2)	1.6 mg
Niacin	18 mg
Folic acid	300 µg
Vitamin B_{12}	2 µg
Vitamin C (ascorbic acid)	30 mg
Vitamin D	2.5 µg
Calcium	500 mg
Iodine	140 µg
Iron	12 mg

(b) Claims confined to named vitamins

Rich or excellent source, e.g. 'rich in vitamin C' $\frac{1}{2}$ RDA	Any other case, e.g. 'useful source of vitamin C' $\frac{1}{6}$ RDA

In a quantity of food that can be reasonably consumed in a day + % RDA in a quantified serving and number of such servings in package

(c) Claims not confined to named vitamins

Rich or excellent source, e.g. 'an excellent source of vitamins' $\frac{1}{2}$ RDA	Any other case, e.g. 'with added vitamins' $\frac{1}{6}$ RDA

Of TWO or more scheduled vitamins
In a quantity of food that can be reasonably consumed in a day + % RDA in a quantified serving of the vitamins enabling the claim to be made and number of servings in the package

and a specific claim, e.g. 'rich in', is permitted only if 100% of the RDA is contained in the daily portion. Priority in these countries is given to restoration, and the maximum addition of vitamins should not exceed three times the defined RDA (du Bois, 1987).

In the USA, when nutrition claims are made on the label or in advertising, and when nutrients are added to foods, full nutrition labelling becomes mandatory (Richardson, 1981). Nutrition information must be presented in a standard format. Briefly, the upper portion of the nutrition label shows the amount of energy (kJ or kcal) in a quantified serving of the food, and lists in grams the amount of protein, carbohydrate, fat and sodium. The lower portion of the label shows the percentage of the (US) RDA for protein, five vitamins (vitamins A, C, thiamin, riboflavin and niacin) and minerals (calcium and iron) provided in one serving. A manufacturer may elect to list further vitamins and minerals for which (US) RDA have been established (United States Food and Drug Administration (1987a) nutrition labelling regulations).

9.4 Restrictive regulations and policies on health claims

As a general policy, the indiscriminate additions of nutrients to foods should be discouraged and information on food labels should not overemphasise or distort the role of a single food or component in enhancing good health. In the UK, implicit claims for the presence of vitamins and minerals are already controlled by the Food Labelling Regulations (1984). Claims as to the suitability of a food for use in the prevention, alleviation, treatment or cure of a disease, disorder or particular physiological condition are prohibited unless they follow strict rules for such foods, as those for special dietary uses. More recently, however, there has been a trend towards more explicit health related and disease prevention claims, e.g. calcium and osteoporosis, vitamin C and the cure and prevention of colds, B-vitamins to release stress and tension, etc.

More recently, amid a storm of controversy, a report was published on the effect of vitamin and mineral supplementation on non-verbal intelligence of a group of Welsh school children (Benton and Roberts, 1988). Although, in the past, studies have linked dietary inadequacy with adverse psychological effects, the implication of these results is that the study must be repeated under more controlled conditions to confirm the findings and to understand any underlying mechanisms. Clearly, any long-term improvement to the nutritional status of the children should be through a change in diet rather than via nutrient supplementation (Taitz, 1988). Nevertheless, individuals and mothers with children are concerned to ensure adequate intakes of nutrients to promote good health, and this would account for the current market interest in the benefits of vitamin and mineral supplements.

While health and benefit claims and statements on labels and promotions can be an important way of conveying nutritional information to the public, there is concern that without sufficient control or guidance, such claims could furnish misleading information (Freckleton *et al.*, 1989). The basic problems are how to allow valid, appropriate health claims and statements without opening the door to misleading and fraudulent claims, and to ensure that disease prevention and health benefits are founded on, and are consistent with, widely accepted, well substantiated, peer-reviewed scientific publications.

Further complexities in implementing health claims and statements on food labels are related to the amount and kind of scientific information necessary to substantiate them, and the handling of conflicting claims (e.g. for a food containing components, some of which are perceived to be beneficial and others which are perceived to be harmful).

There is some concern about the difficulty in simplifying complex health messages to fit the limited space on labels, and the potential threat of nutritional 'power races' among food companies to gain a competitive edge. To maximise their effectiveness, healthier eating and nutrition claims and statements on labels need to be used in conjunction with other methods of nutrition education.

In France, to avoid possible indiscriminate fortification and possible imbalance of nutrients, there is a general reluctance to allow the widespread addition of nutrients to foods. According to French legislation, enrichment, standardisation and supplementation with vitamins are permitted only for foods for special dietary uses, e.g. infant foods or low energy foods. Nothing is added to margarine, and there are only two situations when normal foods for healthy adults may bear the label claim 'with a guaranteed vitamin content': (a) in the case of special processing which maintains the natural amount of vitamins unaltered, and (b) in the case of restoration with vitamins to compensate losses during manufacture (the aim of restoration is 80–200 % of the naturally occurring vitamin content in the raw materials).

Similarly, in the framework of a general anti-additive policy, The Netherlands, Norway and Finland have a very restrictive legislation where no optional addition of vitamins and minerals to food is permitted. The regulations governing the additions of nutrients in different countries have been addressed by Bender (1978) and du Bois (1987).

9.5 The stability of vitamins

During the manufacture, storage and preparation of foodstuffs, vitamins are exposed to a wide range of physical and chemical factors as shown in Figure 9.1 (Killeit, 1988). The stability of these nutrients is important to the manufacturer on at least four counts:

(a) the technologist needs to know the extent to which the food processes and the distribution systems could affect the retention of the micronutrients, and try to minimize losses;
(b) the quality and legislative specialist requires detailed information, particularly if claims for vitamin content are being made on the label;

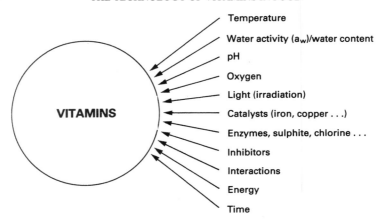

Figure 9.1 Factors influencing the stability of vitamins and their retention in foods (Killeit, 1988).

(c) the accountant will need to establish and justify expenditures on the potential for modification of processing techniques, the costs of nutrient premixes and the 'overages' of less stable vitamins;
(d) the nutritionist will have to assess the choice and ultimately the supply of the nutrient(s) to the consumer.

Obviously, with the varied nature of each vitamin, the possibility of each combination of added vitamins interacting in different ways with the food matrix, and the infinite number of ways of preparation, cooking and storage of foods, it is virtually impossible to generalise on the effects of individual factors on stability. Nevertheless, a huge amount of information has been collected on the behaviour of individual vitamins and, in the majority of cases it is possible to make good predictions as to the losses to be expected for new and modified products and processes. Table 9.2 summarises the sensitivity of vitamins to external factors (Killeit, 1988; Vitamin Information Service, 1988).

The addition of nutrients to a food product is usually achieved by the use of well defined premixes, nutricubes or pre-enriched ingredients. Some vitamins are also available in stabilised or encapsulated forms to improve retention in food processing. Guidelines for the forms of micronutrients, methods of adding, and engineering factors have been summarised by Bauernfeind and Brooke (1973). The vitamin fortification of foods has recently been reviewed by Counsell (1988).

9.6 Additions of iron sources to foods and drinks

One of the earliest recorded examples of fortification and therapeutic application of trace elements was in 4000 BC, when the Persian physician Melampus, medical adviser to Jason and the Argonauts, prescribed a diet including sweet wine laced with iron filings to strengthen the sailors' resistance to spears and arrows and

Table 9.2 Sensitivity of vitamins to external factors (Killeit, 1988)

Vitamin	Heat	Oxygen	Light	pH <7	pH 7	pH >7
A	-	-	-	-	+	+
D	-	-	-	+	+	-
E	-(+)	-(+)	-(+)	+(+)	+(+)	+(+)
K	+	+	-	-	+	-
C	-	-	-	+	-	-
Thiamin	-	-	+	+	-	-
Riboflavin	+	+	-	+	+	-
B$_6$	+	+	+	+	+	+
B$_{12}$	+	-	-	+	+	+
Biotin	+	+	+	+	+	+
Folic acid	+	+	-	-	-	+
Niacin	+	+	+	+	+	+
Pantothenic acid	-	+	+	-	+	-

+, Stable; -, unstable; (+), esterified as tocopheryl acetate.

enhance their sexual potency (Frazer, 1935). Efficacy of this particular tonic was never proven, but the story goes on to describe how, while in search of the golden fleece, Jason and his crew had many adventures and how they lingered too long with the women of Lemnos. As has been reported subsequently, the nutritional rationale for this prescription was amazingly far sighted and even today, none of the other trace elements attract more attention than Fe.

Benefits derived from added iron depend on the form of salts employed, uniformity of the food fortified, and composition of the meal and total diet in which the food article is eaten (Federation of American Societies for Experimental Biology, 1980). In general, iron compounds which exhibit the best bioavailability tend to cause adverse technological and nutritional effects in products. These effects may hinder consumer acceptability. Iron sources currently used in food fortification are shown in Table 9.3. The major chemical characteristics of iron sources which determine their behaviour in foods are:

(a) their solubility, ferrous salts are more soluble than ferric salts;
(b) their oxidative state, ferrous salts are more efficiently utilised than ferric salts (in man) and tend to be more reactive in food systems;
(c) their ability to form complexes which are not bioavailable; generally, ferric-iron has a greater tendency to form chelates than ferrous-iron.

The ferric form (Fe^{3+}) and its reduced form (Fe^{2+}) are the only states which occur naturally in foods. Elemental iron is found rarely in biological systems, but is used widely as an added nutrient. The presence of metal ions such as iron can:

(a) speed up vitamin degradation and loss of nutritional value of a product, particularly for vitamin C, thiamin and retinol;
(b) catalyse the oxidative rancidity of oils and fats;
(c) produce undesirable colours, colour fading, off-flavours and undesirable precipitates.

Table 9.3 Selected iron sources currently used in food fortification, together with their common names, formulas, iron contents and relative biological values (RBV) (Richardson, 1983)

Compound	Other common names	Formula	Fe content (g/kg)	Relative biological value* (RBV)
Ferric phosphate	Ferric orthophosphate	$FePO_4.XH_2O^\dagger$	280	3–46
Ferric pyrophosphate	Iron pyrophosphate	$Fe_4(P_2O_7)_3.9H_2O$	250	45
Ferric sodium pyrophosphate	Sodium Fe pyrophosphate	$FeNaP_2O_3.2H_2O$	150	14
Ferric ammonium citrate		$Fe_xNH_3(C_6H_8O_7)_x$	170	107
Ferrous fumarate		$Fe(C_4H_2O_4)$	330	95
Ferrous gluconate		$Fe(C_6H_{12}O_7)_x^{\ddagger}$	120	97
Ferrous lactate		$Fe(C_3H_5O_3)_2 3H_2O$	380	-
Ferrous sulphate		$FeSO_4.7H_2O$	320	$100^{\ddagger}$
Fe	Elemental Fe, ferrum reductum, metallic Fe	Fe	1000	-
Reduced Fe, H_2 or CO process		Fe	960	34
Reduced Fe, electrolytic		Fe	970	50
Reduced Fe, carbonyl process		Fe	980	67

*Iron-deficient rats are cured of iron deficiency by feeding them either a test iron sample or a reference dose of ferrous sulphate. The cure is measured by haemoglobin or packed cell volume repletion in the rats' blood and bioavailability of samples is reported against a value of 100 for ferrous sulphate. Thus, any iron sample which is less available than ferrous sulphate will have an RBV of less than 100 and one which is more available than ferrous sulphate will have an RBV greater than 100.
† Ferric orthophosphate contains from one to four molecules of hydration.
‡ The precise structures of the Fe salts are uncertain.

Many of these effects occur as a result of interactions between natural food components such as anthocyanins, flavanoids and tannins with metal ions. For example, all bioavailable iron compounds discolour dehydrated mashed potato; a black colour results when ferrous salts react with tannins present in cocoa products; and 'beer haze' is a term used to describe turbidity caused by a complex interaction between iron, copper, tin and calcium with tannins and protein in beer. The bioavailability and reactivity of iron sources may also change during processing and storage, as a result of changes in the chemistry of the iron. These aspects of the technology of iron fortification are further expanded by Richardson (1983).

At this stage in the discussion it may seem inconceivable to a food technologist that a nutritionist would wish to add trace elements to any food or drink products, because the majority of his time is likely to be spent trying to remove metal ions and prevent undesirable effects which cause product rejection. However, the food technologist can control and minimise adverse effects of metals in foods by using sequestrants which can hold the metal ion in solution in a non-ionisable, less reactive metal complex.

In attempting to fortify foods with iron and other minerals including calcium, it soon becomes evident that potential problems with colour, flavour, texture and quality control may be encountered. For any added nutrient, these problems may increase as a higher proportion of the RDA is included per serving, or as the serving size decreases. Close attention, therefore, must be paid to the effects of the addition of nutrients to foods in terms of their sensory attributes and shelf life stability.

9.7 Communicating nutrition

Results of consumer surveys in the UK (Jones Rhodes Associates, 1988) show that, although the majority of interest in nutrition stems from a perceived fear and apprehension about the food supply posing a danger to health in terms of too much fat, sugar, cholesterol and salt, there is still a considerable interest in the addition of nutrients to foods. In the US, examination of consumer responses on their willingness to pay more for nutrients suggested that addition was regarded more favourably in those products which were already perceived to be nutritious, such as milk products and breakfast cereals, but not in snack foods or soft drinks. These consumer perceptions are now reflected by fortification guidelines published by the Food and Drug Administration which discourage the indiscriminate addition of nutrients to foods (United States Food and Drug Administration, 1987b).

Although there is no doubt that the addition of nutrients can be a significant factor in establishing a market for a particular food product, there is, however, little published information in the UK which tells us whether the public understands what nutrient fortification means or whether consumers would be willing to pay more for a product with added nutrients.

Consumers whose interest in nutrition has been awakened will want to be able to assess the nutritional value of the foods they consume, and they will look increasingly to the food label to supply this information. The food manufacturing industries have responded by a marked revival of interest in the composition of foods, in improved methods of analysis, as well as in the establishment of nutrition databases for information about their products (Richardson, 1987).

9.8 Conclusion

Use of ready prepared foods continues to increase in this country; during preparation some nutrients may be removed or destroyed and others may be added. Food processing may result in changes in the bioavailability of some nutrients. In brief, the manufacturer must not only know the technology of combining ingredients to produce attractive, safe and nutritious foods, but in the current regulatory climate give serious attention to the formulation, labelling and cost implications of existing and proposed food regulations. Product developments, such as the fortification of foods, require the provision of sound advice on technical and legislative aspects from concept research right through to product launch (Freckleton *et al.*, 1989). If the addition of nutrients is desired or intended, questions that will need to be researched include: Do the ingredients support the claims? What labelling requirements may be triggered? How much processing and storage information is necessary to support on-the-shelf nutrition claims? What extra quality control will the new product need to support claims? Will the product contribute substantially to the diet? Is the product designed to be a replacement or alternative to a natural food, and should it be nutritionally equivalent to the natural food it would eliminate?

Changing lifestyles and decreased energy intake of the population during recent years have placed greater emphasis on the need to measure the amounts and the bioavailability of nutrients in the food supply and to promote research which defines forms of nutrients which can be absorbed and utilised by the body. Without this information, assessments of the degree of risk from nutrient excesses or deficiencies cannot be made. All these issues are particularly relevant to the addition of nutrients to foods, whether it is as a public health measure or as a selling point in product development.

Acknowledgement

This chapter is reproduced with permission from an article entitled *Food fortification* by David P. Richardson in *Proc. Nutr. Soc.* **49** (1990), Cambridge University Press.

References

Bauernfeind, J.C. and Brooke, C.L. (1973) Guidelines for nutrifying 41 processed foods. *Food Engineering* **45** (6), 91–97.
Bender, A.E. (1978) *Food Processing and Nutrition*. Academic Press, London.
Benton, D. and Roberts, G. (1988) Effect of vitamin and mineral supplementation on intelligence of a sample of schoolchildren. *Lancet* **i**, 140–143.
Bread and Flour Regulations (1984) *Statutory Instrument no. 1304*. H.M. Stationery Office, London.
British Nutrition Foundation (1987) *Food Processing—a Nutritional Perspective*. Briefing Paper no. 11. British Nutrition Foundation, London.
Codex Alimentarius Commission (1985) *Report of the 14th Session of the Codex Committee on Foods for Special Dietary Use*. Joint FAO/WHO Food Standard Programme, Rome.
Codex Alimentarius Commission (1987) *General Principles for the Addition of Essential Nutrients to Foods, Alinorm 87/26*, Appendix 5. FAO, Rome.
Counsell, J.N. (1988) Vitamin fortification of foods. In: *Food Technology International Europe* (ed. A. Turner) Sterling Publications, London, pp. 211–218.
Department of Health and Social Security (1980) *Foods which Simulate Meat. Report on Health and Social Subjects* no. 17. Report of the Panel on novel foods, Committee on Medical Aspects of Food Policy, H.M. Stationery Office, London, pp. 1–20.
Department of Health and Social Security (1981) *Nutritional Aspects of Bread and Flour. Report on Health and Social Subjects* no. 23. Report by the Committee on Medical Aspects of Food Policy, H.M. Stationery Office, London, pp. 1–64.
du Bois, I. (1987) Food enrichment: legislation in Europe. *Bibliotheca Nutrito et Dieta* **40**, 69–81.
European Council Directive (1987) Approximation of the laws of the Member States relating to foodstuffs intended for particular nutritional uses. *Official J. European Communities COM* **(87)**, 241.

Federation of American Societies for Experimental Biology (1980) *Evaluation of the Health Aspects of Iron and Iron Salts as Food Ingredients*. Life Sciences Research Office, Bethesda, Maryland, pp. 1–71.

Food Labelling Regulations (1984) *Statutory Instrument 1984, no. 1305, as amended*. H.M. Stationery Office, London.

Frazer, J.G. (1935) *The Golden Bough: the Magic Art and the Evolution of Kings*. Macmillan, New York.

Freckleton, A.M., Gurr, M.I., Richardson, D.P., Rolls, B.A. and Walker, A.F. (1989) Public perception and understanding. In: *The Human Food Chain* (ed. C.R.W. Spedding). Elsevier Applied Science, London, pp. 17–57.

Harris, R.S. and Karmas, E. (1977) *Nutritional Evaluation of Food Processing*, 2nd edition, Avi Publishing, Westport, Connecticut.

Jones Rhodes Associates (1988) *The National Health Survey*. Rhodes Associates, West Bridgeford, Nottinghamshire.

Killeit, U. (1988) *The Stability of Vitamins. A Selection of Current Literature*. Hoffman-La-Roche AG, Grenzach-Wyhlen, West Germany.

Margarine Regulations (1967) *Statutory Instrument no. 1867 as amended*. H.M. Stationary Office, London.

Ministry of Agriculture, Fisheries and Food (1974) *Food Standards Committee Report on Novel Protein Foods*, FSC/REP/62, H.M. Stationery Office, London. pp. 1–82.

National Academy of Sciences (1975) *Technology of Fortification of Foods, Proceedings of a Workshop*. National Academy of Sciences, Washington, DC.

Richardson, D.P. (1981) Nutrition labelling with special reference to the USA. *Proc. Inst. Food Sci. Tech.* **14**, 87–102.

Richardson, D.P. (1982) Consumer acceptability of novel protein products. In: *Developments in Food Proteins-1*, (ed. B.J.F. Hudson). Elsevier Applied Science, London, pp. 217–246.

Richardson, D.P. (1983) Iron fortification of foods and drinks. *Chem. Ind.* **13**, 498–501.

Richardson, D.P. (1987) Effects of nutritional guidelines on food marketing and new product development. In: *Food Acceptance and Nutrition* (eds J. Solms, D. A. Booth, R. M. Pangborn and O. Raunhardt). Academic Press, London. pp. 433–442.

Taitz, L. S. (1988) Which children need vitamins? *Brit. Med. J.* **296, 1753**.

United States Food and Drug Administration (1987a) *Code of Federal Regulations*. Title 21, *Nutrition Labelling of Food*, Part 101.9, *Sodium Labelling*, Part 101.13. Food and Drug Administration, Washington, DC.

United States Food and Drug Administration (1987b) *Code of Federal Regulations*. Title 21, *Nutritional Quality Guidelines for Foods*, Part 104.5. Food and Drug Administration, Washington, DC.

Vitamin Information Service (1988) *Press Releases*. Vitamin Information Service, London.

Appendix 1: Chemical and physical characteristics of vitamins

Vitamin A

Structural formula

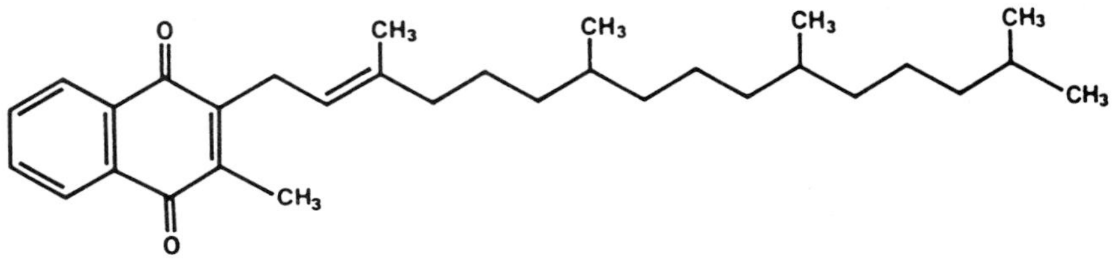

Principal commercial forms

Vitamin A acetate	R = COCH$_3$
Vitamin A palmitate	R = CO(CH$_2$)$_{14}$CH$_3$

	Empirical formula	Molecular weight
Retinol	C$_{20}$H$_{30}$O	286.45
Vitamin A acetate	C$_{22}$H$_{32}$O$_2$	328.50
Vitamin A palmitate	C$_{36}$H$_{60}$O$_2$	524.90

Standardisation

1 µg retinol equivalent corresponds to 3.33 international units (IU) of vitamin A activity.
1 international unit corresponds to the activity of 0.344 µg of pure crystalline vitamin A acetate.
The US Pharmacopoeia Unit (USP unit) is the same as the international unit.
The biological activity of pure vitamin A acetate is

2.904×10^6 IU/g

and of pure vitamin A palmitate is

1.817×10^6 IU/g

APPENDIX 1

Solubility

Retinol is soluble in fats and oils and practically insoluble in water and glycerol. Vitamin A esters are readily soluble in fats, oils, ether, acetone and chloroform. They are soluble in alcohol but insoluble in water.

Melting point

Vitamin A acetate	57–60°C
Vitamin A palmitate	28–29°C

Absorption spectrum

Vitamin A esters show a characteristic absorption spectrum, the position of the maxima depending on the solvent used: in cyclohexane the maxima is at 328 nm and in isopropanol it is at 326 nm.

β-Carotene

Structural formula

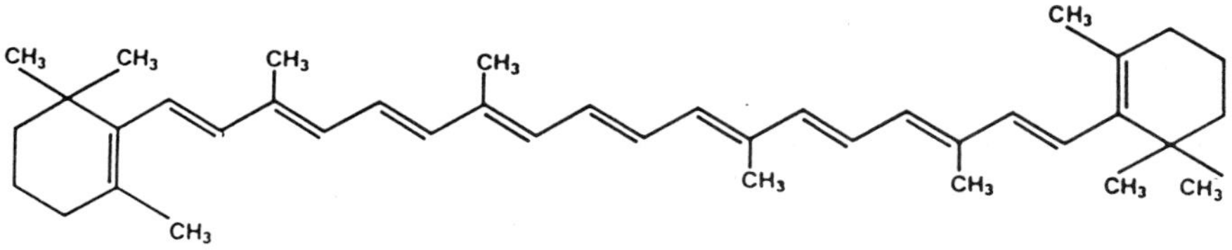

β-carotene

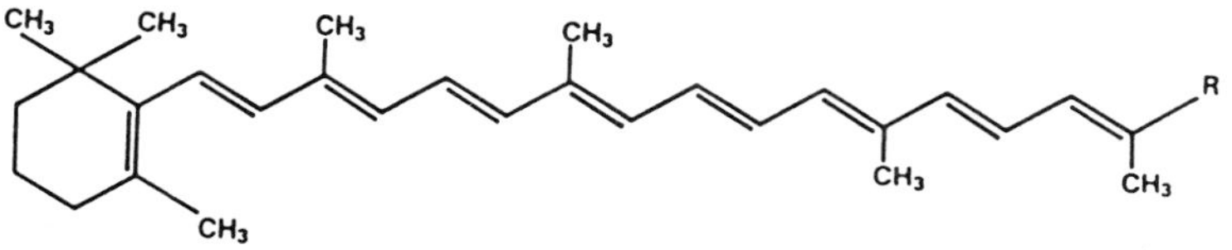

Apocarotenal: R=CHO
Apocarotenoic ester: R=COOC$_2$H$_5$

Principal commercial forms

β-Carotene
β- Apo - 8' - carotenal
Apocarotenoic ester

	Empirical formula	Molecular weight
β-Carotene	$C_{40}H_{56}$	536.9
Apocarotenal	$C_{30}H_{40}O$	416.6
Apocarotenoic ester	$C_{32}H_{44}O_2$	460.7

Standardisation

There have been a number of conventions to establish the relationship between the provitamin activity of β-carotene and vitamin A. The vitamin A equivalents based on retinol equivalents can be calculated as follows:

1 retinol equivalent
    = 1 μg retinol
    = 6 μg β-carotene
    = 12 μg other provitamin A carotenoids
    = 3.33 IU vitamin A activity from retinol
    = 10 IU vitamin A activity from β-carotene.

Solubility

β-Carotene is insoluble in water, sparingly soluble in alcohol, fats and oils. Apocarotenal and apocarotenoic ester are insoluble in water and sparingly soluble in fats, oils and alcohol.

Melting point

β-Carotene	176–182°C
Apocarotenal	136–140°C
Apocarotenoic ester	134–138°C

Absorption spectrum

Solutions in cyclohexane exhibit the following maxima:

β-Carotene	about 456 and 484 nm
Apocarotenal	about 461 and 488 nm
Apocarotenoic ester	about 134 and 138 nm

Vitamin D

Structural formulae

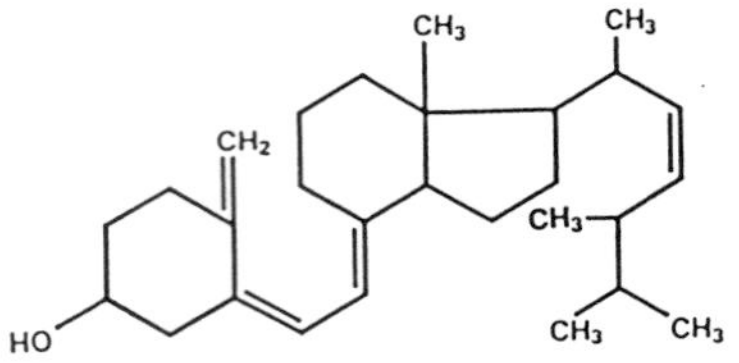

Vitamin D_2

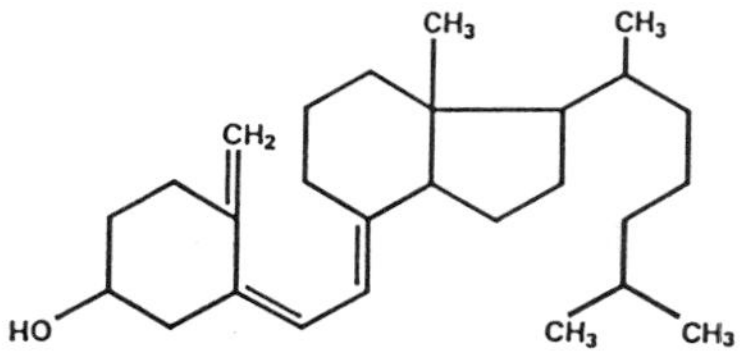

APPENDIX 1

Vitamin D₃

Principal commercial forms

Vitamin D_2: ergocalciferol
Vitamin D_3: cholecalciferol

	Empirical formula	Molecular weight
Ergocalciferol (D_2)	$C_{28}H_{44}O$	396.63
Cholecalciferol (D_3)	$C_{27}H_{44}O$	384.62

Standardisation

One international unit (IU) corresponds to the activity of 0.025 µg of either pure crystalline vitamin D_2 or D_3. The US Pharmacopoeia Unit of vitamin D corresponds to the international unit.

Solubility

Soluble in fats and oils, insoluble in water.

Melting point

Ergocalciferol (D_2)   113–118°C
Cholecalciferol (D_3)   82–88°C

Specific rotation

Ergocalciferol (D_2): $[a]_D^{20}$ = +102.5° to +107.5°
(c = 4 in absolute ethanol)
Cholecalciferol (D_3): $[a]_D^{20}$ = +105° to +112°
(c = 5 in absolute ethanol)

Absorption spectrum

Vitamins D_2 and D_3 exhibit an absorption maxima at 265 nm in alcoholic solution.

Vitamin E

Structural formula

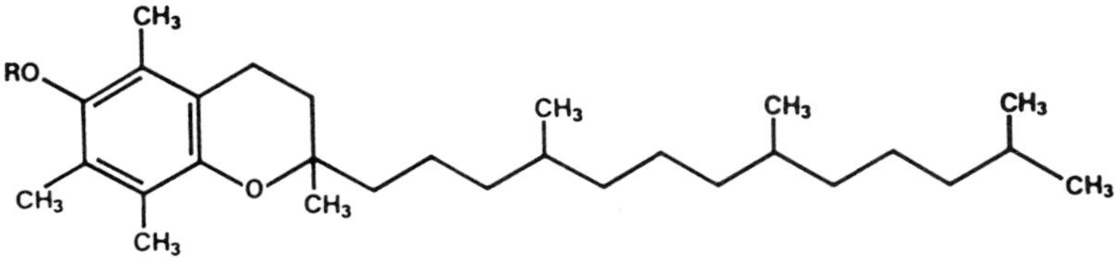

dl-α-tocopherol: R = H
dl-α-tocopheryl acetate: R = CH$_3$ CO

Principal commercial forms

	Empirical formula	Molecular weight
d-α-Tocopherol	C$_{29}$H$_{50}$O$_2$	430.7
dl-α-Tocopherol	C$_{29}$H$_{50}$O$_2$	430.7
d-α-Tocopheryl acetate	C$_{31}$H$_{52}$O$_3$	472.7
dl-α-Tocopheryl acetate	C$_{31}$H$_{52}$O$_3$	472.7
d-α-Tocopheryl succinate	C$_{33}$H$_{54}$O$_5$	530.8

Standardisation

The selected international unit (IU) of vitamin E is the biological activity of 1 mg of synthetic dl-α-tocopheryl acetate. The equivalents of the other forms of vitamin E are related to this standard. The relative activities of the most common forms of vitamin E are given below.

1 IU of vitamin E is equivalent to:

1 mg	dl-α-tocopheryl acetate
0.909 mg	dl-α-tocopherol (1.10 IU / mg)
1.12 mg	dl-α-tocopheryl succinate (0.89 IU / mg)
0.826 mg	d-α-tocopheryl acid succinate (1.21 IU / mg)
0.735 mg	d-α-tocopheryl acetate (1.36 IU / mg)
0.671 mg	d-α-tocopherol (1.49 IU / mg)
1.75 mg	d-β-tocopherol (0.57 IU / mg)
7.0 mg	d-γ-tocopherol (0.14 IU / mg)

Solubility

Tocopherols and their esters are insoluble in water but readily soluble in vegetable oils, alcohol and organic solvents.

Refractive index

α-Tocopherol   1.5030 – 1.5070 at 20°C
α-Tocopheryl acetates   1.4940 – 1.4985 at 20°C

Absorption spectrum

α-Tocopherol (in alcohol solution) maximum at 292 nm, minimum at 255 nm.
α-Tocopheryl acetate (in alcohol solution) maximum at 284–285 nm, minimum at 254 nm.

Vitamin K_1

Structural formula

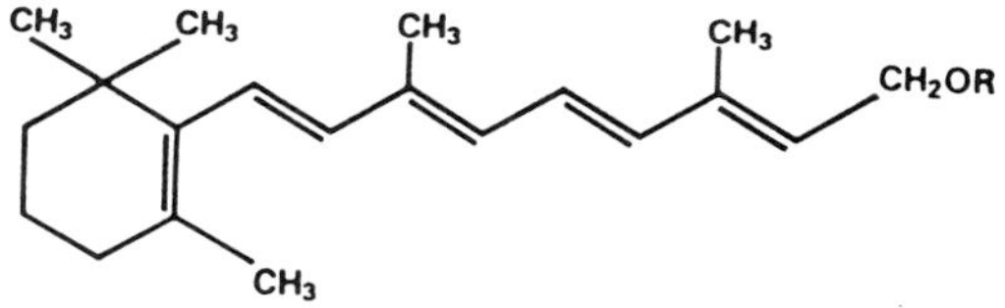

[Structural formula of Vitamin K_1 showing cyclohexene ring with CH_3 groups and polyene side chain ending in CH_2OR]

Principal commercial forms

	Empirical formula	Molecular weight
Vitamin K_1 (phytomenadione, phytonadione)	$C_{31}H_{46}O_2$	450.68

Standardisation

Analytical results are usually expressed as weight units of pure vitamin K_1 as no international standard for the biological activity of vitamin K has been defined.

Solubility

Vitamin K_1 is insoluble in water and sparingly soluble in alcohol. It is readily soluble in fats and oils.

Refractive index

$[n]_D^{20} = 1.525 - 1.528$

Absorption spectrum

Vitamin K_1 shows maxima at 243, 249, 261 and 270 nm and minima of 254 and 285 nm in cyclohexane.

Thiamin (Vitamin B$_1$)

Structural formula

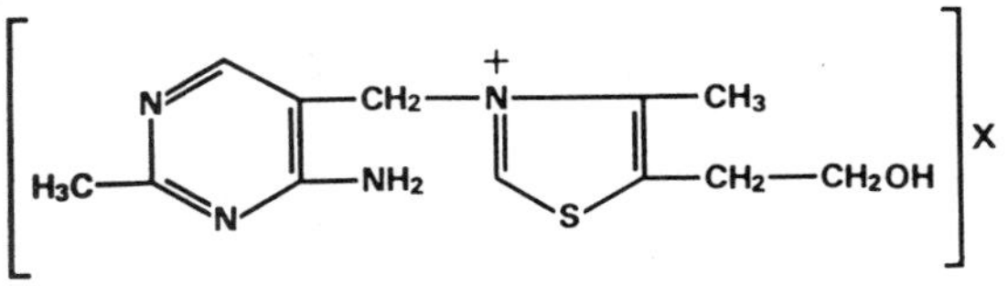

$$\left[H_3C-\underset{N}{\overset{N}{\diagup\diagdown}}-CH_2-\overset{+}{N}\underset{S}{\diagup\diagdown}\overset{CH_3}{\underset{CH_2-CH_2OH}{}} \right] X$$

Principal commercial forms

Thiamin chloridehydrochloride   X = Cl$^-$, HCl
(Thiamin hydrochloride)
Thiamin mononitrate   X = NO$_3^-$

	Empirical formula	Molecular weight
Thiamin hydrochloride	C$_{12}$H$_{17}$ClN$_4$OS.HCl	337.27
Thiamin mononitrate	C$_{12}$H$_{17}$O$_4$N$_5$S	327.36

Standardisation

To calculate the amount of thiamin cation from the salts (molecular weight of thiamin cation is 265.4):

From hydrochloride: divide the amount of hydrochloride by 1.271
From mononitrate: divide the amount of mononitrate by 1.234.

Solubility

Hydrochloride: readily soluble in water (about 1 g / ml), sparingly soluble in alcohol.
Mononitrate: slightly soluble in water (about 2.7 g / 100 ml) sparingly soluble in alcohol

Melting point

Hydrochloride: 250°C (decomposition)
Mononitrate: 190–200°C

Absorption spectrum

Thiamin shows a characteristic absorption spectrum in the region of 200–300 nm. In 0.1 N hydrochloric acid solution the absorption maxima of thiamin is around 245 nm. The positions of the maxima depend on the solvent and pH of the solutions.

Riboflavin (vitamin B$_2$)

Structural formula

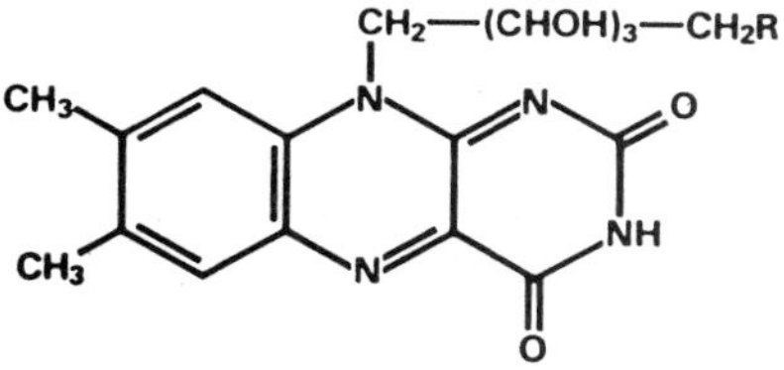

[Structural formula of riboflavin showing isoalloxazine ring system with CH$_3$ groups at two positions, N-CH$_2$-(CHOH)$_3$-CH$_2$R side chain, and NH, O groups]

Principal commercial forms

Riboflavin: R = OH
Sodium riboflavin-5′-phosphate:

$$R = -O - P \begin{array}{c} \diagup ONa \\ \diagdown OH \end{array}$$
$$\quad \quad \quad \; | \\ \quad \quad \quad \; O$$

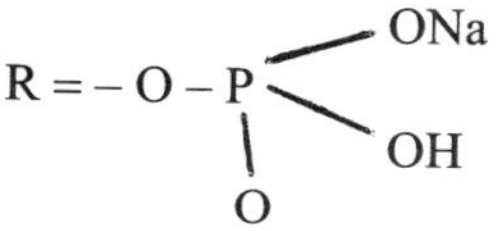

	Empirical formula	Molecular weight
Riboflavin	C$_{17}$H$_{20}$O$_6$N$_4$	376.36
Sodium riboflavin-5′-phosphate	C$_{17}$H$_{20}$O$_9$N$_4$PNa	478.34

Standardisation

1 g of sodium riboflavin-5′-phosphate = 0.730 g of riboflavin.

Solubility

Riboflavin: sparingly soluble in water (1 g dissolves in from 3000 to 15000 ml water depending on crystal structure). Readily soluble in dilute alkalis. Very sparingly soluble in alcohol.
Sodium riboflavin-5′-phosphate: soluble in water (112 mg / ml at pH 6.9, 68 mg / ml at pH 5.6 and 43 mg / ml at pH 3.8). Very sparingly soluble in alcohol.

Melting point

Riboflavin: decomposition at 280 – 290°C

Specific rotation

Riboflavin: $[a]_D^{20} = -122°$ to $-136°$
($c = 0.25$ in 0.05 N NaOH)
Sodium riboflavin-5′-phosphate: $[a]_D^{20} = +38°$ to $+42°$
($c = 1.5$ in 20% HCl)

Absorption spectrum

In 0.1 N HCl solutions riboflavin and riboflavin phosphate show absorption maxima at about 223, 267, 374 and 444 nm.

Vitamin B$_6$

Structural formula

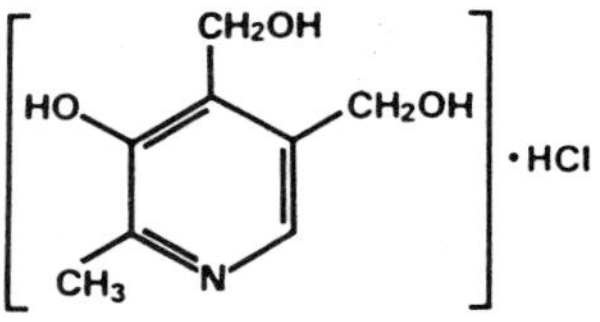

Pyridoxine hydrochloride

Principal commercial forms

	Empirical formula	Molecular weight
Pyridoxine hydrochloride	C$_8$H$_{12}$ClNO$_3$	205.64

Standardisation

1 mg of pyridoxine hydrochloride is equivalent to 0.82 mg pyridoxine or pyridoxamine and 0.81 mg pyridoxal.

Solubility

Readily soluble in water (about 1 g / 4.5 ml). Sparingly soluble in alcohol, soluble in propylene glycol.

Melting point

Decomposition with browning 205–212°C.

Absorption spectrum

In aqueous solution the absorption maxima are:

at acid pH:	291 nm
at neutral pH:	254 and 324 nm
at alkaline pH:	245 and 309 nm

Vitamin B_{12}

Structural formula

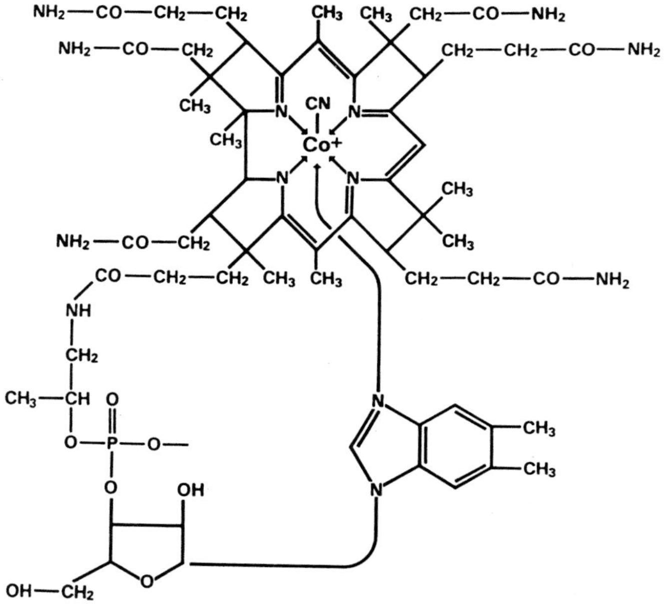
NH2—CO—CH2—CH2
CH3 CH3 CH2—CO—NH2
NH2—CO—CH2
CH2—CH2—CO—NH2
CH3
CN
N
N
CH3
Co+
N
N
CH3
CH3
NH2—CO—CH2
CO—CH2—CH2 CH3 CH3
CH2—CH2—CO—NH2
NH
CH2
CH3—CH O
O—P—O—
O
OH
N
N
CH3
CH3
OH—CH2
O

Principal commercial form

	Empirical formula	Molecular weight
Cyanocobalamin	$C_{63}H_{88}O_{14}N_{14}PCo$	1355.42

Standardisation

Analytical results are usually expressed as weight units of cyanocobalamin.

Solubility

Slightly soluble in water (about 1.25 g / 100 ml), soluble in alcohol.

Melting point

Cyanocobalamin chars at 210 –220°C without melting.

Absorption spectrum

The aqueous solution shows absorption maxima at 278, 361 and 550 nm.

Niacin

Structural formula

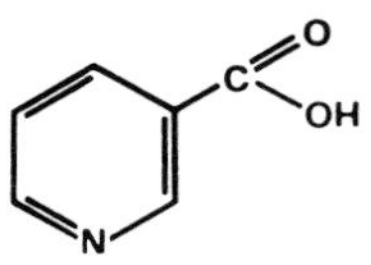

Niacin (nicotinic acid)	Niacinamide (nicotinamide)

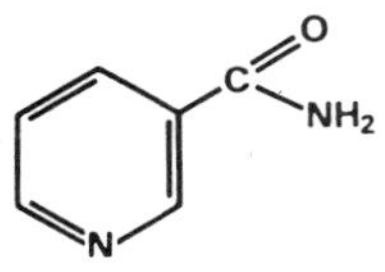

Principal commercial forms

	Empirical formula	Molecular weight
Niacin (nicotinic acid)	$C_6H_5NO_2$	123.11
Niacinamide (nicotinamide)	$C_6H_6N_2O$	122.13

Standardisation

Analytical results are normally expressed as weight units of niacinamide. Both forms possess the same vitamin activity. As the human body is capable of forming niacin from the amino acid tryptophan, niacin is often quoted in units of 'niacin equivalent' on the basis that 60 mg of tryptophan equals 1 mg of niacin equivalent.

Solubility

Niacin (nicotinic acid) is sparingly soluble in water (about 1.6 g / 100 ml) and alcohol (about 1 g / 100 ml). Readily soluble in alkali.
Niacinamide (nicotinamide) is very soluble in water (about 1 g / ml), slightly soluble in alcohol, soluble in glycerol.

Melting point

Niacin (nicotinic acid)	234–237°C (sublimation)
Niacinamide (nicotinamide)	128–131°C

Absorption spectrum

The acid and amide both show similar absorption spectra in aqueous solution with a maximum at about 261 nm and an extinction dependent on pH.

Pantothenic acid

Structural formula

$$CH_2OH-C(CH_3)_2-CHOH-CO-NH-CH_2-CH_2-R$$

Pantothenic acid: R = COOH
Panthenol: R = CH$_2$OH

Principal commercial forms

	Empirical formula	Molecular weight
Calcium pantothenate	$(C_9H_{16}O_5N)_2Ca$	476.53
Sodium pantothenate	$C_9H_{16}O_5NNa$	241.20
Panthenol	$C_9H_{19}O_4N$	205.25

Standardisation

Pantothenic acid is optically active with only the dextro-rotatory forms having vitamin activity. Although free pantothenic acid is extremely unstable, results are expressed in terms of weight units of pantothenic acid.

1 mg calcium pantothenate is equivalent to 0.92 mg pantothenic acid.
1 mg sodium pantothenate is equivalent to 0.91 mg pantothenic acid.
1 mg panthenol is equivalent to 1.16 mg calcium D-pantothenate.

Solubility

Calcium pantothenate is readily soluble in water (about 4 g / 10 ml), sparingly soluble in alcohol and soluble in glycerol.
Sodium pantothenate is very soluble in water and slightly soluble in alcohol.
Panthenol is very soluble in water, readily soluble in alcohol and slightly soluble in glycerol.

Melting point

Calcium pantothenate:	decomposition 195–196°C
Sodium pantothenate:	122–124°C
dl-Panthenol:	64.5–68.5°C

Specific rotation

Calcium pantothenate	$[a]_D^{20} = +26.0°$ to $+28.0°$ ($c = 4$ in water)
Sodium pantothenate	$[a]_D^{20} = +26.5°$ to $+28.5°$ ($c = 4$ in water)
Panthenol	$[a]_D^{20} = +29.5°$ to $+31.5°$ ($c = 5$ in water)

Folic acid

Structural formula

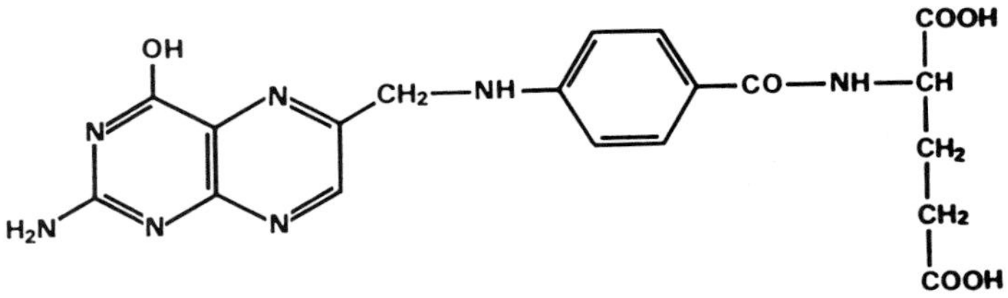

[Structure showing pteridine ring with OH, H₂N substituents connected via CH₂—NH to a para-aminobenzoyl group linked to glutamic acid: —CO—NH—CH(COOH)—CH₂—CH₂—COOH]

Principal commercial forms

	Empirical formula	Molecular weight
Folic acid	$C_{19}H_{19}N_7O_6$	441.40

Standardisation

Analytical results are generally expressed in weight units of pure folic acid as no international unit for the biological activity of this vitamin has been defined.

Solubility

Folic acid is sparingly soluble in water, readily soluble in dilute alkali, soluble in dilute acid and insoluble in alcohol.

Melting point

Darkens at 250°C followed by charring.

Specific rotation

$[a]_D^{20} = c.+ 20°$ ($c = 0.5$ in 0.1 N NaOH)

Absorption spectrum

Folic acid shows a characteristic absorption spectrum which is dependent on the pH of the solution. In 0.1 N NaOH the maxima are at 256, 283 and 365 nm.

Biotin

Structural formula

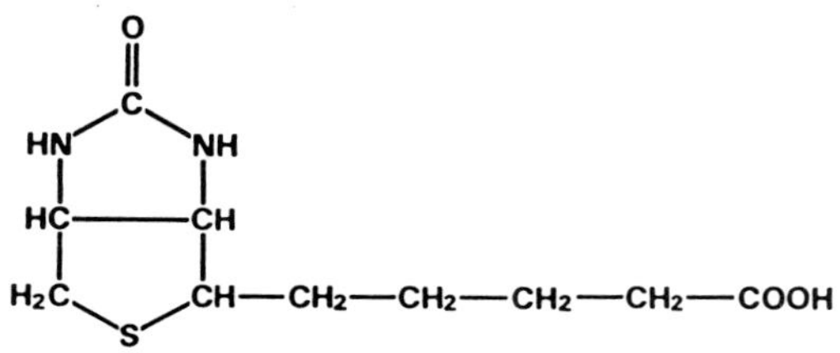

```
           O
           ‖
          C
       /     \
     HN       NH
      |       |
     HC ───── CH
     |        |
    H₂C      CH ─ CH₂ ─ CH₂ ─ CH₂ ─ CH₂ ─ COOH
       \    /
         S
```

Principal commercial form

	Empirical formula	Molecular weight
d-biotin	$C_{10}H_{16}O_3N_2S$	244.31

Standardisation

Analytical results are normally expressed as weight units of pure *d*-biotin.

Solubility

Very sparingly soluble in water (about 20 mg / 100 ml) and alcohol. Soluble in dilute alkali.

Melting point

228–232°C with decomposition.

Specific rotation

$[a]_D^{20} = +90°$ to $+94°$ ($c = 1.0$ in 0.1 N NaOH)

Vitamin C

Structural formula

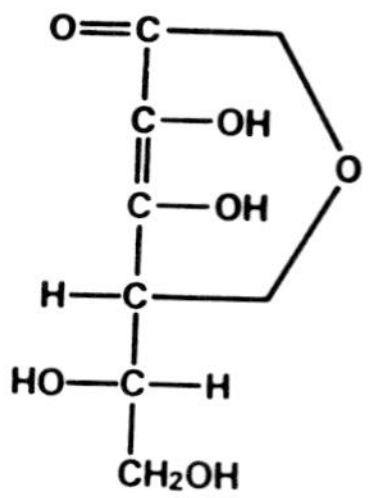

```
         O=C ─────┐
          |        │
          C ─ OH   │
          ‖        │ O
          C ─ OH   │
          |        │
         H ─ C ────┘
          |
      HO ─ C ─ H
          |
         CH₂OH
```

Ascorbic acid

Principal commercial forms

	Empirical formula	Molecular weight
Ascorbic acid	$C_6H_8O_6$	176.13
Sodium ascorbate	$C_6H_7O_6Na$	198.11
Calcium ascorbate	$C_{12}H_{14}CaO_{12}.2H_2O$	426.35

Standardisation

1 mg of sodium ascorbate is equivalent to 0.889 mg of ascorbic acid.
1 mg of calcium ascorbate is equivalent to 0.826 mg of ascorbic acid.

Solubility

Ascorbic acid is readily soluble in water (about 30 g / 100 ml), slightly soluble in alcohol, and sparingly soluble in glycerol.
Sodium ascorbate is very soluble in water (about 90 g / 100 ml) and almost insoluble in alcohol.
Calcium ascorbate is soluble in water and slightly soluble in alcohol.

Specific rotation

Ascorbic acid:	$[a]_D^{20} = -22°$ to $-23°$	($c = 2$ in water)
Sodium ascorbate:	$[a]_D^{20} = +103°$ to $+106°$	($c = 5$ in water)
Calcium ascorbate:	$[a]_D^{25} = +95°$ to $+97°$	($c = 2.4$ in water)

Absorption spectrum

Ascorbic acid in strongly acid solution shows an absorption maxima of about 245 nm which shifts at neutrality to 365 nm and at pH 14 to about 300 nm in UV light.

Appendix 2: Recommended nutrient reference values for food labelling purposes

Recommended dietary intakes in different countries. Adapted from FAO/WHO Joint Expert Consultation on Recommended Allowances of Nutrients for Food Labelling Purposes, Helsinki, 12–16 September, 1988.

	Thiamin (B$_1$)	Riboflavin (B$_2$)	B$_6$	Niacin	B$_{12}$	Folate	Pant.	Biotin	C	A	D	E	K
Units	mg	mg	mg	mg	μg	mg	mg	mg	mg	μg	μg	mg	mg
Argentina	1.3	1.9	—	21	2	0.2	—	—	30	750	2.5	—	—
Australia	1.0	1.2	1.3	16	2	0.2	—	—	30	750	—	—	—
Belgium	1.5	1.7	2.0	18	3	0.4	5.0	0.2	60	1000	7.5	10	0.10
Bolivia	1.1	1.5	2.2	18	3	0.4	4.0	0.1	60	750	5.0	10	0.70
Brazil	1.0	1.5	2.0	17	2	1.0	3.0	—	70	1500	10.0	5	—
Bulgaria	1.5	1.8	—	19	—	—	—	—	85	1200	—	—	—
Canada	0.4	0.5	0.2	7	2	0.2	5.0	—	60	1000	2.5	10	0.03
Caribbean	1.2	1.7	2.0	20	2	0.2	—	—	30	750	2.5	15	—
Chile	1.1	1.5	—	18	—	—	—	—	45	750	—	—	—
China (PR)	1.5	1.5	—	15	—	—	—	—	75	1300	10.0	—	—
Colombia	0.9	1.2	2.3	14	3	0.2	6.0	0.2	40	1600	5.0	10	0.10
Czechoslovakia	1.2	1.8	1.9	20	—	0.2	8.0	—	60	1000	—	12	—
Denmark	1.4	1.6	2.2	18	—	—	—	—	60	1000	5.0	—	—
Finland	0.5	0.7	1.1	7	—	—	—	—	33	440	—	—	—
France	1.5	1.8	2.2	18	3	0.4	7.0	0.1	80	1000	10.0	10	—
Germany (FR)	1.3	1.7	1.8	18	5	0.4	8.0	—	75	1000	5.0	12	—
Hungary	1.4	1.8	2.2	18	3	0.4	8.0	—	60	1000	5.0	12	—
India	1.4	1.5	—	19	1	0.1	—	—	50	750	5.0	—	—
Indonesia	1.0	1.4	—	17	—	—	—	—	30	1200	—	—	—
Ireland	1.2	1.6	2.2	18	3	0.3	—	—	60	750	7.5	10	—
Israel	1.0	1.6	1.8	17	—	—	—	—	70	1500	10.0	—	—
Italy	1.2	1.8	1.4	19	2	0.2	—	—	45	700	2.5	10	—
Japan	1.2	1.7	—	20	—	—	—	—	50	600	2.5	—	—
Korea (south)	1.3	1.5	—	17	—	—	—	—	55	750	10.0	—	—
Malaysia	1.0	1.4	—	16	2	0.2	—	—	30	750	2.5	—	—
Mexico	1.3	1.5	—	23	—	—	—	—	50	1000	—	—	—
Netherlands	1.4	1.6	2.0	18	3	0.4	—	—	50	1000	5.0	10	—
New Zealand	1.2	1.7	2.0	18	3	0.2	—	—	60	750	10.0	14	—
Norway	1.4	1.6	2.2	18	—	—	—	—	60	1000	5.0	—	—
Philippines	1.2	1.2	—	16	—	—	—	—	75	650	—	—	—
Poland	1.7	1.7	2.0	17	5	0.4	—	—	75	1500	10.0	30	—

Continued.

FAO/WHO Codex reference	B$_1$	B$_2$	B$_6$	Niacin	B$_{12}$	Folate	Pant.	Biotin	C	A	D	E	K
Portugal	1.5	1.8	2.2	18	3	—	—	—	75	1500	—	—	—
Singapore	0.5	0.8	1.0	6	—	—	3.5	0.1	—	360	5.0	10	—
South Africa	1.5	1.7	2.2	19	3	0.4	7.0	0.2	60	1000	10.0	10	—
Spain	1.2	1.8	—	20	2	0.2	—	—	45	750	2.5	—	—
Sweden	1.4	1.6	2.2	18	—	—	—	—	60	1000	5.0	—	—
Switzerland	1.2	1.8	1.6	15	1	0.1	10.0	—	75	1650	11.0	10	—
Taiwan	1.4	1.5	2.0	18	3	0.4	—	—	60	750	2.5	10	—
Thailand	1.0	1.4	—	16	—	—	—	—	30	750	10.0	—	—
Turkey	1.2	1.8	—	17	—	0.2	—	—	50	750	—	—	—
United Kingdom	1.2	1.6	2.0	18	—	0.3	—	—	30	750	10.0	—	—
Uruguay	1.2	1.6	2.2	19	2	0.2	—	—	30	750	2.5	10	—
USA	1.4	1.6	—	18	3	0.4	4.0	0.1	60	1000	5.0	10	0.07
Venezuela	1.2	1.6	—	20	3	0.2	—	—	30	750	2.5	—	—
Western Pacific	1.1	1.6	—	19	2	0.2	—	—	30	750	2.5	—	—
Range Minimum	0.4	0.5	0.2	6	1	0.1	3.0	0.1	30	360	2.5	5	0.03
Maximum	1.9	2.2	2.2	23	5	0.4	10.0	0.2	85	1650	11.0	30	0.70

European Community recommended daily allowances for food labelling purposes are expected to be published at the end of 1992. When accepted by the EC Parliament these will supersede the individual national requirements.

Index

alcohol, metabolism of 5
amino acids
  balances of 2
  metabolism of 6, 11
anaemia
  folic acid deficiency 48
  haemolytic vitamin E deficiency 4, 44
  megaloblastic vitamin B_{12} deficiency 10, 48
  pernicious vitamin B_{12} deficiency 48
  pyridoxin deficiency 8
antioxidant
  effect on lipids 4
  vitamin E 3, 4, 114
ascorbic acid *see* vitamin C
Ashbya gossypu, production of riboflavin 73
ATP, generation of 5

Bacillus, fermentation of vitamin C 82
β-carotene (provitamin A)
  absorption spectrum of 248
  analysis of 183–186
  as a food additive 136
  commercial isolation of 68, 69
  extraction of 183, 184
  in natural sources 20, 68
  location of 74
  melting point of 248
  production of 68, 69, 70
  solubility of 248
  stability of 94, 117, 127
  structural formula of 247
beans, thiaminases 96
beri-beri, 5, 46
beverages
  addition of vitamins 115–118
  carbonation of 116
biological assays 172
biotin (vitamin H) 12, 28, 224
  analysis of 224
  content in common foods 37
  deficiency in 12, 48
  fermentation of 87
  manufacture of 35, 85, 86
  MBA of 224
  melting point of 259
  solubility of 259
  specific rotation of 259
  stability of 99
  standardisation of 259
  structural formula of 259
blood
  coagulation of 4
  synthesis of 45
bone, calcification of 3
brain, metabolism of 7
bread
  loss of thiamin in 95
  removal of vitamins 124
  stability of riboflavin in 96, 127
  stability of vitamins in 126
  vitamin fortification of 125
butter, stability of vitamins in 92

calcification, vitamin D 3
calcitriol, vitamin D 3
calcium
  absorption 3
  ions 3
  release from bone 3
  renal absorption 3
calcium pantothenate, pantothenic acid 74, 75
Candida flaveri, production of riboflavin 73
carbohydrate
  metabolism of 1, 5, 7, 11, 45
  oxidative metabolism of 6
carnitine, deficiency of 13
carotenoids 1, 19, 143, 183
  analysis of 183–186
  as a food additive 154–168
  natural sources of 21, 23
cereals
  addition of vitamins 118–125
  vitamin B_6 in 212
  hot products 118
  RTE products 118
  stability of vitamins 119
chewing gum 141
  fortification of 141
  overage in 141
  vitamin C in 141
chocolate, fortification of 139

cholecalciferol, vitamin D_3 3
choline 12
coating
  of cereals 121
  of vitamins 110, 112
coenzyme A 11, 35
  biotin 12
  flavin adenine dinucleotide (FAD) 6
  flavin mononucleotide (FMN) 6
  nicotinamide adenine dinucleotide (NAD) 8
  nicotinamide adenine dinucleotide phosphate (NADP) 8
  pantothenic acid 97
  pyridoxal-5-phospate 7
  pyruvate conversion 5
  riboflavin 6
cofactors
  NAD, NADP 8
  thiamin phosphates 5
confectionery, fortification of 138, 139
conjunctiva, dryness of 2
convulsions, pyridoxin deficiency 8
copper ions, stability of thiamin 96
cornea
  dryness of 2
  vascularisation of 7
cyanocobalamin, vitamin B_{12} activity 9

dairy products *see* specific examples
decarboxylation, by thiamin phosphates 5
development, retinol in 1
differentiation, retinol in 1
DNA, synthesis of 8

eggs
  biotin binding by 12
  pantothenic acid in 97
  stability of thiamin 95
  vitamin A 20
encapsulation of vitamins 110
energy, metabolism 47
ergocalciferol *see* vitamin D_2
Erwinia, fermentation of vitamin C 82
extrusions cooking 119
eyelids, riboflavin deficiency 7

fat, metabolism 1, 5, 6, 11
fat soluble vitamins 19, 20, 43, 91, 137
fatty acid
  extraction of PUFAs from 85
  transport of 13
fish
  liver oils 20
  roes 20
folic acid (pteroylglutamic acid) PGA, also folates 10, 28, 48, 218
  absorption spectrum of 258
  analysis of 218-223

content in common foods 38
  deficiency in 11, 48
  extraction of 218, 219, 220
  manufacture of 78, 79
  MBA of 222, 223
  melting point of 258
  occurrence of 35-38
  solubility of 258
  specific rotation of 258
  stability of 97, 116, 123, 124, 218
  standardisation of 258
  structural formula of 258
fondant, fortification of 139
free radicals, damage by 1, 14, 15, 16, 17, 56, 57
fruit
  loss of vitamins in processing 103
  vitamin B_6 in 212
fruit drinks, stability of vitamins in 95, 117

gas liquid chromatography (GLC) 172
Gluconobacter, fermentation of vitamin C 82
glucose, transketolation of 5
glycogen, NAD 9
glycoproteins, RNA synthesis of 2
growth
  retinol in 1
  folic acid in 10

hard boiled candies 138
  addition of vitamins in 139
  loss of vitamins in 139
heart
  deficiency in vitamin K 5
  riboflavin storage in 4

high performance liquid chromatography (HPLC) 172
high-temperature short-time (HTST) 119
  stability of vitamins during 119

ice cream
  addition of vitamins to 136
  stability of vitamins in 136
immune system, vitamin E 4
immunoassay 172
immuno-defence systems 56
intestinal bacteria, vitamin K production 4
intestinal mucosa, riboflavin phosphorylation 4
intestine
  absorption of colabamins 9
  absorption of folic acid 10
  calcium absorption 3
iron
  absorption 14, 241
  addition to food 240, 241

INDEX

irradiation, deterioration of vitamins 105, 106

jaundice, vitamin K deficiency 4

kidney
 pantothenic acid in 97
 provitamins in 94
 riboflavin storage in 5
 vitamin D conversion in 3
 vitamin K deficiency in 5

Laboratory of the Government Chemist of UK (LGC) 172
Lactobacilli, determination of B-group vitamins 191, 192, 193
Lactobacilli casei
 MBA of riboflavin 206
 MBA of folates 222, 223
Lactobacillus fermenti, MBA of thiamin 200
Lactobacillus leichmanni, MBA of vitamin B_{12} 223
Lactobacillus plantarum
 MBA of biotin 224
 MBA of niacin 212
 MBA of pantothenic acid 224
light
 effect of exposure to 2
 laboratory environment 172
 stability of vitamins 90
lipid
 antioxidant effect on 3
 metabolism of 7
lipid membranes, stabilisation of 4
lips, riboflavin deficiency 7
liver
 carnitine deficiency 13
 niacin in 97
 pantothenic acid 97
 provitamins in 94
 riboflavin storage in 5
 vitamin B_6 in 98
 vitamin D conversion in 3

margarine, enrichment of 136, 137
meat
 loss of vitamins in processing 104
 niacin in 97
 vitamin B_6 in 98, 212
metabolism, retinol 1
microbiological assays (MBAs) 172, 191
milk
 exposure to light 96
 fortification of 128–131
 loss of vitamins in 130
 pantothenic acid in 97
 processing of 129
 provitamins in 94

stability of vitamins in 92, 96, 97, 131
storage of 131
mitochondria
 oxido-reduction in 8
 stabilisation of lipid membranes 4
moisture, stability of vitamins 90
Mortierella fungus, production of PUFAs 85
muscle weakness
 ascorbic acid deficiency 14
 pantothenic acid deficiency 12
 vitamin D deficiency 3

nerves, deficiency in vitamin K 5
niacin (nicotinic acid, nicotinamide) 8, 28, 47, 143, 206, 207
 absorption spectrum 256
 analysis of 207–212
 as a food additive 169
 content in common foods 31, 32
 deficiency in 9, 47
 extraction of 207, 208, 209
 manufacture of 73
 MBA of 212
 melting point of 256
 metabolism of 133
 solubility of 256
 stability of 96, 117
 standardisation of 256
 structural formula of 256
 synthesis of 6, 8, 47
nicotinamide *see* niacin
nicotinic acid *see* niacin
nitrogen, metabolism of 2

Ochromonas malhamensis, MBA of vitamin B_{12} 223
oil soluble vitamins, determination of by HPLC 173
overages 91, 107, 108, 109, 110
oxidative metabolism, flavoproteins 6
oxidising agents, stability of vitamins 90, 95, 97
oxido-reductive systems 6, 56
 niacin 8
 riboflavin 5, 6
oxygen, stability of vitamins 90

pantothenic acid 11, 28, 224
 analysis of 224
 content of in foods 36
 deficiency of 12, 48
 manufacture of 74
 MBA of 224
 melting point of 257
 solubility of 257
 sources of 35
 specific rotation of 257
 stability of 97

pantothenic acid *(continued)*
  standardisation of   257
  structural formula of   257
parathyroid glands, vitamin D synthesis   3
pasta, effect of production on vitamins   127
pasteurisation
  concentrates, nectars, fruit juices   116
  ice cream   136
  milk   98, 129
  yoghurt   133
pellagra, niacin deficiency   9, 47
peripheral tissues   1, 2
pH, stability of vitamins   90
phosphate
  absorption from intestine   3
  release from bone   3
photopsin   2
potatoes, deterioration of   90, 91
processing
  fruits   103
  loss of vitamins   103
  meat   104
  milk   104
  vegetables   103
prohormone, vitamins as   1
protection of vitamins   110
protein
  metabolism 1, 7
  synthesis of   3, 45
prothrombin, synthesis of   45
provitamin A   *see* β-carotene
pyridoxin   *see* vitamin B_6

ready-to-eat (RTE), cereal products   118
recommended daily allowance (RDA)   49–57, 58
recommended daily intake (RDI)   49
reducing agents
  stability of riboflavin   96
  stability of vitamins   90
reduction, cGMP   2
retinal, conversion of vitamin A to   70
retinaldehyde, vitamin A activity   43
retinoids, natural sources of   21, 22
retinol, (retinoic acid vitamin A)   1, 2, 43
  RNA synthesis   2, 3
  stability of   91
retinyl esters   1, 91
rhodopsin   2
riboflavin (vitamin B_2)   5, 28, 46, 143, 200
  absorption spectrum of   254
  analysis of   200–206
  as a food additive   136, 168
  chemical synthesis of   72, 73
  commercial forms of   253
  content in common foods   30
  deficiency in   7, 46
  extraction of   200, 201, 202, 203

  in metabolism   6, 7
  manufacture by fermentation   72
  MBA of   191, 206
  melting point of   253
  microbial production of   73
  oxido-reductive processes   5, 6
  phosphorylation of   5
  solubility of   253
  sources of   31
  specific rotation of   253
  stability of   96, 117, 127
  standardisation of   252
  storage of   5
  structural formula of   253
RNA
  retinol carrier   2
  synthesis of glycoproteins   2
roller drying, of liquid cereals   123

Saccharomyces carlsbergensis, MBA of vitamin B_6 217, 218
Saccharomyces cerevisiae, MBA of biotin   224
safety
  manufacture of vitamins   64
  RDA of vitamins   53–58
  use of vitamins   53, 54, 57, 58
scurvy, vitamin C deficiency   14, 42, 48, 49
sea food, thiaminases   96
shelf-life, deterioration of vitamins   106–110
spleen
  provitamins in   94
  riboflavin storage in   5
spray drying, of milk   129
starvation, vitamin deficiency   42
steroid hormone
  formation of   11
  vitamin D synthesis   3
Streptococcus faecalis, MBA of folates   222, 223
sunlight, vitamin D intake   3

taurine, deficiency in   12, 13
temperature, stability of vitamins   91
thiamin (vitamin B_1)   5, 28, 193
  absorption spectrum of   252
  analysis of 193–200
  as a food additive   94, 121
  commercial forms of   252
  content in common foods   29
  deficiency in   5, 45, 46
  determination of   196, 197, 198, 199
  extraction of   193, 194, 195, 196
  manufacture of   70, 71, 72
  MBA of   200
  melting point of   252
  phosphorylation of   5
  retention of in cereal products   121

INDEX

thiamin (vitamin B_1) *continued*
  solubility of 252
  sources of 28
  stability of 94, 116, 117, 123, 124, 126, 127, 133
  standardisation of 252
  structural formula of 252
  toxicity, vitamin D_2 3, 58
tryptophan, niacin synthesis 6, 8

ulceration, of eyes 2
ultra high temperature (UHT), of milk 129
ultraviolet, formation of vitamin D 3

vegetables, loss of vitamins in processing 103
  vitamin B_6 in 212
vegetable oils, tocopherols in 92
vision, deficiency of retinol 2, 43
vitamins
  addition to foods 233, 234, 243, 244
  coated forms of 64
  determination in foodstuffs 172–228
  fortification 114
  health claims 238, 239
  manufacture of 63, 64, 65, 66
  natural sources of 63
  overages 114, 133, 136, 138
  premix 64, 115
  protection of 110
  restoration of 114
  stability of 91, 103, 239, 240
  standardisation of 114
  synthesis in body 1
  variations in 19
  vitamin interactions 101, 102
  vitaminisation 114
vitamin A (retinol) 1, 179
  absorption spectrum of 247
  activity in foods 20
  addition to food 121, 137, 236
  analysis of 179–183
  commercial forms of 114, 246
  conversion to 70
  deficiency of 2, 43
  determination of in foods 20
  extraction of 179
  fortification of 115, 179
  in development 1
  in differentiation 1, 43
  in pregnancy 59
  manufacture of 66, 67, 68
  melting point of 247
  natural sources of 20, 179, 180
  solubility of 246
  stability of 91, 92, 116, 122, 123, 124, 126, 131
  standardisation of 246
  structural formula of 246

vitamin B-group (vitamin B-complex) 1, 45
  analysis of 190–193
vitamin B_1 *see* thiamin
vitamin B_2 *see* riboflavin
vitamin B_6 (pyridoxine) 7, 28, 47, 98, 212
  absorption spectrum of 254
  analysis of 212–218
  commercial forms of 254
  content in food 33, 34
  deficiency in 8, 47
  extraction of 213, 214
  in protein metabolism 7
  manufacture of 75, 76, 77, 78
  MBA of 217, 218
  melting point of 254
  natural sources of 212
  solubility of 254
  stability of 98, 117, 124, 212
  standardisation of 254
  structural formula of 254
vitamin B_{12} (cyanocobalamin) 9, 28, 47, 223
  absorption spectrum of 255
  analysis of 223
  commercial forms of 255
  content in food 34, 35
  deficiency in 10, 47
  fermentation of 79
  MBA of 223
  melting point of 255
  natural sources of 223
  retention of in cereal production 121
  solubility of 255
  stability of 98
  standardisation of 255
  structural formula of 255
vitamin C (ascorbic acid) 1, 13, 14, 28, 48, 115, 143, 224
  absorption spectrum of 260
  analysis of 225–228
  as a food additive 115, 121, 143, 144–154
  commercial forms of 260
  content in foods 39
  deficiency in 14, 48, 49
  extraction of 225
  fermentation of 82
  manufacture of 63, 79, 80, 81
  natural occurrence of 39, 224
  solubility of 260
  specific rotation of 260
  stability of 91, 99, 100, 116, 117, 123, 124, 133
  standardisation of 115, 260
  structural formula of 259
  variations in 40
vitamin D (D_2 ergocalciferol, D_3 cholecalciferol) 3, 23, 24, 44, 93, 173
  absorption spectrum of 249

vitamin D *(continued)*
  activity in foods 24
  addition to food 128, 137, 236
  analysis of 173–179
  commercial forms of 249
  content in foods 24
  deficiency in 3, 44
  extraction of 173, 174
  manufacture of 83, 173
  melting point of 249
  natural sources of 23, 24, 83, 173
  solubility of 249
  specific rotation of 249
  stability of 93, 131
  standardisation of 249
  structural formulae of 248
  toxicity 3
  variations in 23
vitamin E (tocopherol) 3, 143, 186
  absorption spectrum of 251
  activity of 24, 25
  analysis of 186–190
  as a food additive 137, 169
  as an antioxidant 114
  commercial forms of 250
  content of refined oils 25
  content of some foods 26
  deficiency in 4, 44
  extraction from vegetable oils 83, 84, 186
  natural sources of 186
  refractive index of 250
  solubility of 250
  stability of (tocopherol) 92, 123, 124
  standardisation of 250
  structural formulae of 250
  synthesis of 83, 84
  variations in 24, 26
vitamin F group, polyunsaturated fatty acids (PUFAs) extraction from natural sources 85
vitamin K
  absorption spectrum of 251
  activity 4, 45
  commercial forms of 251
  deficiency in 4, 45
  manufacture of 87, 88
  natural sources of 93
  refractive index of 251
  requirements 4
  solubility of 251
  sources of 27
  stability of 93, 94
  standardisation of 251
  structural formula of 251

water soluble vitamins 28, 45
  stability of 94

yoghurt
  addition of vitamins 132
  stability of vitamins in 133, 134, 135

xerophthalmia, vitamin A deficiency 2